The Common Sense Universe

Vivian NE Robinson*

* Dr Robinson was the first person to publish a paper explaining the physics upon which Einstein's gravity theory was based. It baffled some of the world's best mathematicians and physicists for over 100 years.

ORCID: 0000-0002-4117-6534

ISBN: 978-0-6454125-5-0

Edition: 1.0

Edited by Arnie Benn

Quicycle Books

Published by ETP Semra Pty Ltd
ABN: 87 001 578 883

Email: info@etpsemra.com.au

Throughout this book, omissions are due to brevity. Errors are unintentional. All dates used refer to the common era (CE). Neither the author nor the publisher will be held responsible for misinterpretations by the reader.

Cover image: Hubble Ultra Deep Field (Wikipedia)

There are not many books that try to present the whole universe, from its smallest sub atomic particles to its large scale structure. Most concentrate on either the small, sub-atomic, end or the large scale structure of galaxies and the universe as a whole. In those publications, most use such complex maths that mere mortals like us have no chance of understanding the topics presented.

In this book, I present the relatively simple physics that controls all aspects of the universe. Where necessary, the simplest of maths shows that the magnitude of the effect matches observation. That is the common sense approach. It shows that what are called complex topics like quantum mechanics, elementary particle and nuclear physics, Einstein's gravity and cosmology are quite simple to understand in this common sense universe. The maths and physics used are about the mid-level of most undergraduate maths and physics courses. That level is used without compromising technical accuracy.

Despite its simplification, it still requires a reasonable knowledge of physics and maths, to about mid undergraduate university level. Other career professionals, engineers, physical technologists and similar, should be able to understand a lot of it.

Those who want to understand it, but don't have the necessary skills are referred to a shorter version of this book, **The Simple Universe** tells the "how" everything works, without the "why" or the equations.

The Common Sense Universe gives a better match with observation than do the complex standard models.

Table of Contents

Acknowledgements

This study was made possible because I was self-employed and working in a scientific environment. That allowed me to form my own ideas on how things could work. It was made possible by the thousands of customers who purchased my Robinson Detectors and other equipment. A hearty thanks goes to former business associates, staff members and customers who made it all possible. Life in the "real world" does wonders to stimulate knowledge of realism and what is possible and what is not.

I would like to acknowledge the support received from family members over the decades it took to research and compile this manuscript.

Special mention must be made of Dr John Williamson for helpful discussions. These included some mathematics behind the "twist" to photons, which "twist" forms the basis of the electrons, protons, neutrons, and neutrinos that make up the building blocks of the universe.

Foreword

During the 19th century, classical physics based upon Newtonian mechanics, electromagnetism, Maxwell's equations and their extensions, made great progress. Everything was thought to be made of atoms. They were considered indivisible, the smallest entities possible. They were known to follow classical rules.

The end of the 19th century and the early 20th century saw improvements in measuring skills that revealed atoms were composed of smaller particles. Initially it was negative electrons surrounding a positive nucleus. In turn the nucleus was found to be made of positive protons held together by neutral neutrons. Maxwell's electromagnetic waves were also found to have a smallest quantity, now known as photons.

Those particles were discrete entities that did not obey the laws associated with a continuum of properties. Classical principles seemed no longer to apply.

Consider water as an example. Each molecule is made of two hydrogen atoms bound to one oxygen atom. As long as the size of the molecules was sufficiently small, they presented a continuum to any measurement. Classical physics applied. At smaller dimensions, the properties of individual molecules were different from those of bulk water because each molecule was a separate particle. The hydrogen and oxygen atoms in individual water molecules do not behave at all like water. Going to smaller dimensions, the electrons, protons, and neutrons don'tt behave anything like water.

Each of those particles is a separate entity with specific properties. They were quantized. There were only a limited number of them. They could only be changed in quantized steps. Quantum mechanics was developed to explain their behaviour.

Those particles had such strange properties that their structure and properties appeared to defy all logic. It was known they behaved like waves as well as particles. Theoreticians took to describing their actions in terms of mathematical wave functions. Waves have interesting properties. They can extend over great distances and can be manipulated in many ways. It was "easy" to get a match with some observations. Quantum mechanics became wave mechanics with complex mathematics that most people couldn't follow. Common sense disappeared as counter intuitive results were obtained.

The beginning of the 20th century saw Einstein's theories of relativity. Derived from purely mathematical considerations, the mathematics was exceedingly difficult for most people to follow. The validity and accuracy of his calculations are still debated over 100 years later, even though they have been verified in experimental situations against which they have been tested. With no knowledge of the physics that underpins his calculations, manipulating his mathematics to extend his work became common practice. There seemed no physical logic in his mathematics.

The leading fields of physics were quantum mechanics and relativity. Those fields developed a mystique about them. Theoretical studies could only be done with advanced mathematics. That allowed their proponents to point out how accurate was their work. Everyone else had to accept their results.

In this presentation I point out that there is nothing magical, mystical, or difficult about those fields. The apparently complex behaviour of sub-atomic particles is due to their unique structure. Understand that structure and classical concepts still apply. That includes quantized electron orbits in atoms, tunnelling, and other apparently peculiar quantum mechanical behaviour.

The same holds true with Einstein's relativity theories. The structure of every individual sub-atomic particle means its motion with respect to another observer automatically subjects it to Einstein's special relativity correction. It clearly shows what is energy, what is mass and why $E = mc^2$. It derives a fourth correction Einstein missed, which aids greatly in understanding the behaviour of individual particles.

His general relativity theory and gravitational field equations are of great interest. His maths was so complex that, over 100 years later, none of the "experts" knew the physics of how mass distorted space-time to generate gravity.

In 2021 I published a paper in an Institute of Physics journal, https://doi.org/10.1088/2399-6528/abee2f , pointing out that physics was photon redshift. Distances are measured by photon wavelengths and time by photon frequencies. Photons are redshifted as they move away from a massive object. That is how mass distorts space-time to produce gravity.

In simplifying Einstein's maths, I showed the "experts" made three mathematical errors when they derived their solutions that predict black holes. Einstein didn't believe they could be predicted from his equations. He was correct. An exact mathematical solution to that physics predicts the shapes astronomers have observed and indicated are black holes.

This work also shows that gravity is weaker than inverse square. If gravity was inverse square, an infinite steady state universe would collapse in on itself. This common sense presentation, supported by appropriate maths, shows that gravity is weaker than inverse square. When gravity is weaker than inverse square, an infinite steady state universe won't collapse under its own mass. That totally changes all aspects of cosmology. The observed redshift of light from distant galaxies is due to an effect that slows down cosmic rays. The SNe1a magnitudes at different redshifts are exactly what is expected from an infinite steady state universe. The universe is not even expanding, let alone expanding at an accelerating rate. There is no dark energy.

In the same manner, galaxies are rotating as expected, without the need for a additional mass. Their rotations are explained by gravity and other known effects. There is no dark matter.

Chapter 12 sums it up the best. The current descriptions of the universe, the standard models, are based upon mathematics, which most people don't understand. Things like quarks and gluons, black holes, the Big Bang, dark energy, and dark matter are mathematicians' descriptions of the universe. They do not relate well to physical reality. For those interested, an expanded version of chapter 12 is available as a separate book "The Simple Universe". Its only mathematical equation is $E = mc^2$. It is used to explain the relationship between mass and energy.

Mathematics doesn't make things happen. Physical principles make things happen. This presentation outlines the physical principles involved at different levels of the universe. It uses enough simplified yet accurate mathematics to show the magnitude of the physical effect matches observation. That is a physicist's universe.

In order to explain the how, what, and why of the universe. It is easiest to start at the beginning and build it up, step by step. That is what **The Common Sense Universe** is all about. It starts from universe zero, from absolutely nothing — empty space, no particles, no fields. Nothing. A totally clean slate, as some would say. From that we build up a picture of the universe, particle by particle, putting them all together as appropriate.

It shows how free space supports photons, the fundamental particles that are the building blocks of everything in the universe. It moves on to the structure of electrons, protons, neutrons, and neutrinos, the only long-term stable particles. It shows how they have properties such as mass, spin, charge, magnetic moment, duality, and more. From their structure it shows why many of their properties are quantized, why they are subject to the special relativity corrections when they move with respect to observers, and of course, why $E = mc^2$. It gives a brief introduction to the origins, structure, properties, and role of the myriad sub nucleon particles discovered from particle accelerators, with good descriptions of a few.

It progresses through the mechanisms by which the various particles interact to form atomic nuclei, followed by why the electrons have such apparently peculiar behaviour in atoms. From atoms to the solar system, gravity excluded, modern science has done a particularly good job in describing and beneficially using many of its intricacies.

The Common Sense Universes resumes again. Chapter 8 explains the origins of Einstein's gravity and what mass distorting space–time means. Chapter 9 shows the properties of matter that produce gravity. Chapter 10 answers some riddles associated with galaxies.

That gives us all the matter and all the forces that are needed to add the hundreds of billions of galaxies seen in the "visible by telescope" universe. It shows why an infinite steady state universe will not collapse. The universe is not even expanding, let alone expanding at a more rapid rate. Both the physics and mathematics that are used to describe it hold up against logical scrutiny.

1. Space and Photons

Abstract:

Empty space has properties that enable it to transmit electromagnetic radiation. This presentation describes how these properties support photons and give them their properties.

1.1. Introduction

That the universe exists is obvious. We are part of it. We look around and experience our little corner of it in so many ways. Why it exists is another story. One of the original explanations was "*In the beginning God created the heavens and the earth*" (Gen **1**, 1). The book of Genesis went on to explain how it was completed by the seventh day. The first serious variation to that theory came from Muhammad. He pointed out that *al-'ilāh* ("the god" in Arabic) completed the task in seven periods of time. He did not define the length of those periods. Early humans saw the universe as the Earth, the moon, the Milky Way and other fixed stars, a few wandering stars, and a couple of fuzzy patches (nebulae) that could be seen in the night sky with the unaided eye.

The emergence of modern science saw other explanations for the universe's existence. Telescopes showed that the nebulae were other "island universes" like our Milky Way galaxy. More powerful telescopes showed more distant galaxies. The universe was much larger than previously imagined. By the early 20th century CE, the generally held view was that the universe was infinite and static. Further studies suggested an infinite universe would collapse under its mass.

Observations showed that the light from distant galaxies was redshifted. The further away were the galaxies, the greater was the redshift. (Hubble, 1929).

That led many scientists to believe the universe was expanding and that it started from a Big Bang (Lemaitre, 1931, 1936). By the 1970's, the Big Bang theory for the formation of the universe was widely accepted. It goes something like: "*In the beginning there was nothing (or something – we do not know yet). 13.98 billion years ago it exploded in a Big Bang, which created space and time. The whole universe expanded, stretching space, time, and light. What astronomer can see represents only 4% of what is required to explain its observed properties. The remainder is dark energy and dark matter. Scientists have been searching for signs of either and have not found any trace of them after many decades of searching.*" And on it goes.

This is all rather confusing. If you really want to find out what is going on, the best way is to describe the universe step by step, from scratch. To do that, it is best to start out with what we all know exists. Everybody knows about the three space dimensions, x, y, and z. Nobody has detected any other dimensions. We will not use them. People are familiar with time. It ticks by at the rate of 60 seconds per minute, 60 minutes per hour, 24 hours per earth day, etc. It ages us all. We will not worry about hypothetical particles like quarks and gluons that have not been independently isolated and detected. We will use measured and known physical constants, physical principles, and the most common component in the universe, empty space. From then on, reason and logic, which should be known as common sense, take over.

1.2. Empty Space

Empty space refers to an absence of what is called baryonic matter, i.e., protons, neutrons, and their associated electrons. It is occupied by electric, magnetic, and gravitational fields. In turn, those fields are generated by charged and neutral particles. No particles, no fields. So, to build a universe we must first build some particles. The only thing we have with which to do that is empty space — nothing.

At this point it helps to understand the universe by knowing that it is made from only two (2) entities, mediums, items or "stuffs", by whatever title you like to call them. They can't be called substances or matters because one of them, empty space, is nothing, the absence of anything.

Empty space is not entirely empty or free from all matter. Intergalactic space contains a background of cosmological microwave radiation and neutrinos. Whether they are the nothingness of empty space, or the presence of those backgrounds, that "nothing" has some interesting properties. It can support and transmit electric, magnetic, and gravitational fields. The electric fields are facilitated by a property of free space called its electric permittivity, denoted ε_0. Magnetic fields are facilitated by the magnetic permeability of free space, denoted μ_0. Their importance was well presented by Maxwell (1865).

Einstein (1915, 1916, 1952 b) showed that gravity is facilitated by the fabric of space-time. We need first to consider the electric and magnetic properties of free space to derive matter with mass. Then we can consider the effect of mass on the structure of space and time and its facilitation of gravity. It is shown later that the relationship between space, time and electromagnetism is quite straight forward.

There are important features of electric permittivity, ε_0, and magnetic permeability, μ_0. ε_0 can transmit and resist the passage of an electric field through free space. It can store electric energy in free space. Similarly, μ_0 can transmit and resist the passage of magnetic fields through free space and store magnetic energy. The other important feature is that the two are totally interconnected. Whenever there is a change in the electric field in one direction, there will be a change in the magnetic field in a direction perpendicular to it, and vice versa, a changing magnetic field will produce a changing electric field perpendicular to it.

Maxwell went on to point out that the combination of a changing electric field in one direction in free space, say "x", would generate an associated magnetic field change in a second perpendicular direction, say "y". The assembly would travel in the third direction, say 'z', at the speed of light c. That is determined from the relation

$$\varepsilon_0\mu_0 = \frac{1}{c^2} \qquad (1.1$$

Magnetic permeability is defined as $\mu_0 = 4\pi.10^{-7}$ newtons/ampere^2 (N/A^2). The speed of light in matter free space is defined as c = 299,792,458 meter/second (m/s). From those values and equation 1.1, the electric permittivity of matter free space was calculated to be ε_0 = 8.8541878128 (±13) x 10^{-12} Farad/meter (F/m).

Many experiments established that there were indeed waves of electromagnetic energy (eg, Hertz, 1887; Marconi, 1896). They were experimentally verified by diffraction measurements, the generation of radio waves, and other observations. They had a frequency, υ, and wavelength, λ, related through

$$\upsilon\lambda = c = (\varepsilon_0\mu_0)^{\frac{-1}{2}} \qquad (1.2$$

The work of Planck (1900 a, b) and Einstein (1905 a, b, 1952 a) established that those electromagnetic waves were discrete electromagnetic packets. Each packet, now called a photon, has energy

$$E = h\upsilon \qquad (1.3$$

where h is Planck's constant, 6.62607... x 10^{-34} m^{2}kg.sec^{-1}. Equation 1.3 was experimentally verified by predictions of the shape of black body radiation curves and the photoelectric effect. Photons still retain the properties of wavelength and frequency. They are the smallest quantity of electromagnetic radiation that can exist at a particular frequency.

Having established some of their properties, it is necessary to describe a structure of photons that can give them those and other properties. The starting point is ε_0 and μ_0. Although photons have been extensively studied, no proposed model is universally accepted. The two major photon models used are based around the Copenhagen Convention interpretation, and that forwarded by Einstein and de Broglie (1957).

The Copenhagen Convention interpretation treats photons as massless point particle to which known properties are attached mathematically. That model is useful in mathematical calculations. Maths don't make things happen. It does not give any indication about their physical reality.

The Einstein-de Broglie model treats photons as a wave function, Ψ. No indication of the form of Ψ was given. Einstein (1905b, 1952a) determined that photons had mass and transmitted inertia. That was verified experimentally (Pound and Rebka,1959; Pound and Snider, 1961; Pound 2000; van der Mark and t'Hooft, 2000). Photons with those properties are used in this study.

This presentation rejects the Born model and its associated Copenhagen Convention that the photon is a point particle that has its various properties mathematically attached in a Hamiltonian (Chandrasekar, 2012). That enables the correct answer to be obtained mathematically. It does not give any insight into the photon's structure or how it interacts within the universe. There is no common sense associated with its use.

The universe does not operate on mathematical principles. It operates on physical principles. These can be worked out by a reasoned, logical approach to those principles. Mathematics is required to check the magnitude of the physical principle. If the mathematics associated with the physical principle matches observation, the physical theory has some validity. Mathematics by itself does not provide a physical reason for an event to occur. This representation starts with the Einstein (1905 a, b, 1952 a) and de Broglie (1957) model (Vigier, 1992; Evans and Vigier, 1994) that a photon can be described as a wave function, Ψ, by giving a mathematical and physical description of Ψ using classical concepts.

1.3. How ε_0 and μ_0 Support Photons

That free space transmits electromagnetic energy is easy to prove. You see objects through the light transmitted from them. For photons to travel, the energy in one part of space must be temporarily stored in space and transmitted to another part of space as the photon travels. Photons could not exist if free space could not store electromagnetic energy. The energy is stored in space while the photon is passing through it and returned to the photon when it has passed.

Figure 1.1 A gives a schematic outline of part of such a plane-polarised photon. It starts out from the right-hand side with the positive electric field, E, pointing down the page and the north magnetic field pointing into the page. As the photon travels in the direction indicated

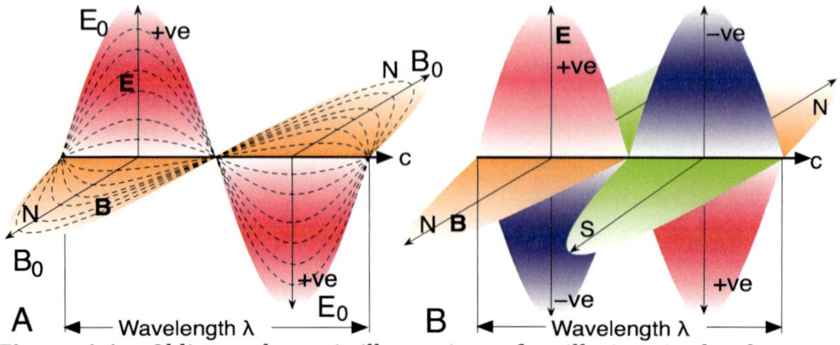

***Figure 1.1:** Oblique schematic illustrations of oscillations in the electric, E, and magnetic, B, fields associated with plane-polarised photons.*

by the arrow, their field strengths vary in a sinusoidal manner. Both reach their maximum of E_0 and B_0 respectively at $\frac{1}{4}\lambda$, before decaying back to zero at $\frac{1}{2}\lambda$.

The fields then reverse their direction. The positive field points up the page and the north magnetic field points out of the page. They again reach their maximum strengths of E_0 and B_0 at $\frac{3}{4}\lambda$, before decaying back to zero at the end of the wavelength. The whole assembly moves at the speed of light, c, in the direction perpendicular to both fields, as indicated by the arrow. The dashed lines represent the lines of equal field intensity.

A positive electric field in one direction equates to a negative field in the opposite direction. Similarly, a north magnetic field in one direction equates to a south magnetic field in the opposite direction, as illustrated in figure 1.1 B. The arrowed line forms the axis about which the electric and magnetic fields oscillate as they travel at c.

Such a structure obeys Maxwell's equations. A change in the electric field in one direction generates a related change in the magnetic field perpendicular to it. Similarly, a changing magnetic field generates a corresponding changing electric field perpendicular to it. When generated, the structure moves perpendicular to both fields according to equation 1.1.

As derived by Einstein, photons have mass. How electro-magnetic oscillations of the type shown in figure 1.1 possess mass is found in a definition of mass as its ability to resist inertia. The electric permittivity, ε_0, and magnetic permeability, μ_0, of free space store and release energy in a finite time, not instantaneously. That occurs because it takes time for the electric and magnetic fields to spread out from their axis of travel.

In that manner, free space provides an inertial resistance to the passage of photons. When photons are absorbed, the energy of the electric and magnetic fields held by the electric permittivity and magnetic permeability respectively is transferred to the absorbing particle. That gives photons mass and allows them to impart inertia between an emitter and absorber. If energy storage and release were instantaneous, photons would have no mass and would travel at infinite speed. The universe would then always be Euclidean.

A moving mass has momentum, mass multiplied by speed. Momentum can impart inertia to another body. The Einstein-de Broglie model implies that photons have momentum, given by

$$p = \frac{E}{c} = \frac{hv}{c} \qquad (1.4$$

Photons have momentum. Momentum is a property of mass multiplied by velocity. From known properties of free space and observation, Equation 1.4 gives photons a mass

$$m_p = \frac{E}{c^2} = \frac{hv}{c^2} \qquad (1.5$$

Equations 1.1 to 1.5 establish some fundamental properties of photons:

i. Photons are transmitted through space by its properties of electric permittivity and magnetic permeability.
ii. A photon consists of an electric field, *E*, oscillating in one direction with an inseparable magnetic field, *B*, oscillating in unison perpendicular to it, and the combination moving at c through space in the third direction perpendicular to both *E* and *B*. The oscillation is characterised by its frequency, v, and wavelength, λ, as shown in equation 1.2.
iii. ε_0 and μ_0 both transmit and resist the passage of electromagnetic energy, as well as store it. This energy comes in

discrete electromagnetic packets called photons. Each photon has energy given by $E = hv$.

iv. Their energy gives them momentum, indicated in equation 1.4.

v. In turn, their momentum gives them mass, given in equation 1.5.

One of the properties of mass is its ability to resist the application of a force, and hence, energy. Energy is applied to ε_0 and μ_0. Combined, they resist the passage of that energy. (If they did not, the energy would travel with $c = \infty$. Although c is very rapid, it is not infinitely fast.)

Definitions of ε_0 and μ_0 include their ability to resist the application of energy. This ability gives the property of mass to the photon. This gives it inertial mass and velocity, i.e., momentum. That momentum can be imparted to an object upon which the photon impinges, increasing its mass.

This presentation shows the structure of a photon's wave function, Ψ, and how that structure gives photons many of their properties. These properties of photons originate from the properties of free space, the properties of "nothing."

They support electromagnetic radiation because nothing does not have an ability to prevent its passage. They store energy because the electric and magnetic fields spread out from the axis of the photon's direction of travel. As photons move through space, the fields they generate collapse at one point and reappear within a further wavelength of the photon's travel. That occurs with the electric field always perpendicular to the magnetic field. The time it takes for the electric and magnetic fields to move away from the axis and collapse back to it gives space its properties of electric permittivity and magnetic permeability.

Support for that idea comes from the refractive index of optically transparent media, such as glass. The presence of the matter with the appropriate values increases the refractive index of that material, taking it from 1 in free space to $n > 1$. The speed of light in a material with refractive index n is given by c/n. From equation 1.1 this means that $n = \sqrt{\frac{\varepsilon_0\mu_0}{\varepsilon_n\mu_n}}$, where ε_n and μ_n are the respective electric permittivity and magnetic permeability of the substance that has a refractive index of n. The increased refractive index means it takes longer for the electric and magnetic fields to spread out from the photon's axis of travel. The photons will therefore move slower, at c/n.

1.4. Physical and Mathematical Models of Ψ

Having established those fundamental properties of photons, it is necessary to establish the form of their wave function, Ψ. Figure 1 illustrates the simplest form of a single wavelength oscillating plane-polarised electromagnetic wave. In this form, a photon is generated when energy, *E*, is imparted to ε_0 and μ_0, generating an oscillation of frequency υ and wavelength λ. The oscillation starts out from zero, building the electric field, *E*, to a maximum E_0 and magnetic field, *B*, to a maximum B_0, each of one polarity, before decaying back to zero. The oscillation then repeats in the opposite polarity before again going back to zero. In its simplest form, that describes a single wavelength plane-polarised photon, as illustrated in figure 1.1. An oblique presentation is used to give a three-dimensional effect. The whole oscillation travels on the central axis perpendicular to both fields at the speed of light, c.

Those fields are strongest close to the axis and diminish with distance from it. In the absence of other information, it is suggested the electric and magnetic field intensities E_d and B_d at distance d from the central travel axis are given by

$$E_d = E_0 e^{-\frac{d}{\lambda}} \text{ and } B_d = B_0 e^{-\frac{d}{\lambda}} \qquad (1.6$$

The wave function $\Psi_{1,p}$ for a single wavelength plane-polarised photon

$$\Psi_{1,p} = \begin{matrix} E_{X,d(z,t)} = E_0 e^{-\frac{d}{\lambda}}.sin(kz - \omega t) \\ B_{Y,d(z,t)} = B_0 e^{-\frac{d}{\lambda}}.sin(kz - \omega t) \\ E_Y = B_x = 0 \end{matrix} \qquad (1.7$$

where k is the wave number and ω is the phase frequency. Other expressions can be used. As figure 1.1 B shows, a positive field in one direction means a negative field in the opposite direction. Electromagnetic energy is imparted to start the oscillation and stored in

ε_0 and μ_0. As it travels, it rises to a maximum, goes back to zero and restarts in the opposite direction.

As mentioned in equations 1.4 and 1.5, this gives photons their properties of momentum and mass. When the oscillation is absorbed, its momentum is imparted to the absorbing particle, also altering its mass. Integrating over one cycle gives $E_X = E_Y = B_x = B_Y = 0$.

Other interpretations of single wavelength plane-polarised photons are possible. For the fields to extend beyond the photon's travel axis, they must impart some of their energy into that surrounding space. One possible way in which this can occur is if the fields are spread out by secondary photons. That appears to place limits on the distance that the fields can travel away from the photon's travel axis. As the energy of the secondary photons diminish, their wavelengths increase. That could extend the distance of the field's influence. Alternatives include the fields spreading laterally faster than c.

Photons remain enigmatic particles. They are not point particles with properties attached mathematically. They are particles composed of electromagnetic field oscillations over a limited extent. That is one of the reasons why the exact position of a photon can never be determined experimentally. Each photon has a length and breadth within which it is located. Not being a point particle, it cannot be defined more accurately than its length and breadth dimensions. That is the origins of Heisenberg's uncertainty principle, $\Delta p \Delta x \geq h$, where Δp and Δx are the uncertainties in the photon's momentum and position respectively, and h is Planck's constant.

An electric charge placed near the photon's axis and in the electric field plane would experience an attraction towards and repulsion from the photon as it passes. Similarly, a magnetised object, monopole or otherwise, would experience the photon's magnetic field if it were similarly positioned near the photon's magnetic field plane. Under quantum electrodynamics (QED), the force moving the charge or monopole is due to photon exchange. As the first half of a photon passes a charged (or magnetised) particle, a photon is exchanged with the particle. That causes the attraction (or repulsion). As the second half of the photon passes, another photon is exchanged, returning the particle to its original position.

A feature of this interpretation of the Einstein/de Broglie model of photons is that they have mass. The mass is spread throughout the fields shown in figure 1.1.

There are many variations of photons, of which figure 1.1 is only one. It indicates the basic properties of wave-particle duality. Their particle nature is due to their limited extent. Their wave nature is due to their oscillating electric and magnetic fields.

Figure 1.1 shows a plane-polarised photon in which the electric and magnetic fields each remain in their own plane. The most important variation of that structure is photons that are circularly-polarised. They are photons in which the electric and magnetic fields remain perpendicular to each other, yet both make one complete revolution around the axis per wavelength.

Figure 1.2 gives examples of circularly-polarised photons. Figure 1.2 A illustrates an oblique view of the positive field making three complete 360° revolutions over three wavelengths. In this case, the arrows represent the strength and direction of the positive electric field. The curve joining their tops represents a line of equal electric potential near the extremities of the field.

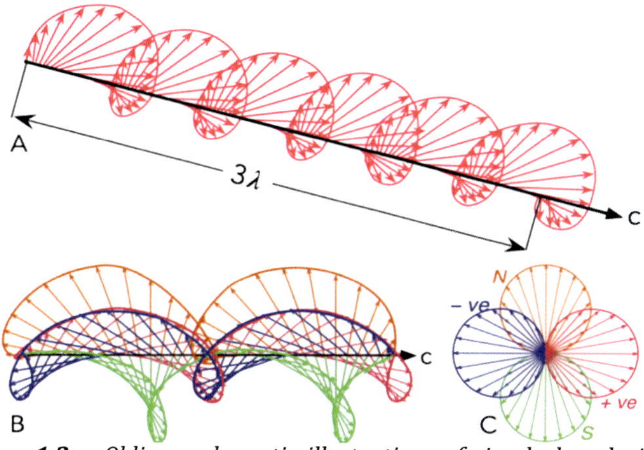

Figure 1.2: *Oblique schematic illustrations of circularly-polarised photon oscillations in the electric, E, and magnetic, B, fields, with those fields rotating 360° in one wavelength.*

The field starts from zero and rises to a maximum at ¼λ. It goes back to zero at ½λ after having rotated through 180°. This process is repeated once more to complete one wavelength. Because the polarity reverses after a half wavelength, the polarities always remain in the same quadrant, as seen in figure 1.2 C.

Figure 1.2 A shows the positive field repeated as the photon travels through three wavelengths. Figure 1.2 B shows an oblique view of all four fields, each making one 360° revolution in one wavelength. This gives the wave function, $\Psi_{1,c}$, for a single wavelength circularly-polarised photon as

$$\Psi_{1,c} = \begin{matrix} E_{X,d}(z,t) = E_o e^{\frac{-d}{\lambda}} sin^2(kz - \omega t) \\ E_{Y,d}(z,t) = E_o e^{\frac{-d}{\lambda}} sin(kz - \omega t) cos(kz - \omega t) \\ B_{X,d}(z,t) = B_o e^{\frac{-d}{\lambda}} sin(kz - \omega t) cos(kz - \omega t) \\ B_{Y,d}(z,t) = B_o e^{\frac{-d}{\lambda}} sin^2(kz - \omega t) \end{matrix} \quad (1.8$$

Equations 1.6 to 1.8 inclusive are approximations intended to indicate the format of the photons. Among the approximations are the diminution of the lateral field strengths with distance, d, from the axis of travel. The term $e^{\frac{-d}{\lambda}}$ implies the field strength diminishes as the inverse square of the distance. That applies to both the electric and magnetic fields. Both are open ended and a change in one induces an equivalent change in the other.

It should be noted that photons cannot travel sideways. Equations 1.7 and 1.8 only allow for the axis of the photon to move in the direction perpendicular to its E and B fields. The arrows illustrated in figure 1.2 are the E and B fields. They could be considered as photons carrying the fields away from the travel axis. Photons can and only travel in that linear direction between being emitted and absorbed. That is also a property of any other photon wave function. While a photon's trajectory can be altered, e.g., by a change in refractive index, it cannot do that by moving sideways.

All photons emitted by either electrons or nuclei are circularly-polarised. The reason for that will become apparent in future chapters. Their polarization state can be changed by filters, reflection, and possibly other means. Antenna emissions such as radio and television signals form plane-polarised photons.

Figure 1.3 shows an end-on image of a single photon (Romero, 2019). Its four lobed structure has a similar appearance to figure 1.2 C. It shows the electric fields, left and right, and the magnetic fields, top and bottom.

The magnetic fields are joined because a north pole pointing in one direction is the same as a south pole pointing in the other direction. The two link up and reinforce each other. The close proximity of a positive electric field to a negative electric field causes a cancellation of the fields when they are close to each other. That shows up in figure 1.3 as the dark regions between the two charges. It is displayed in figure 1.1 B by the diminishing colour intensity of the charges close to the photon's axis of travel.

1.5. Properties of Photons with Wave Function Ψ

The first feature to note about equations 1.7 and 1.8 is that they have a term $\omega = 2\pi v$. That defines a wavelength given by $\lambda = \frac{c}{v}$. The range of wavelengths varies from greater than 1 kilometre (km) to less than 1 femtometre (fm), as illustrated in figure 1.4. The visible range covers just a small section of the total.

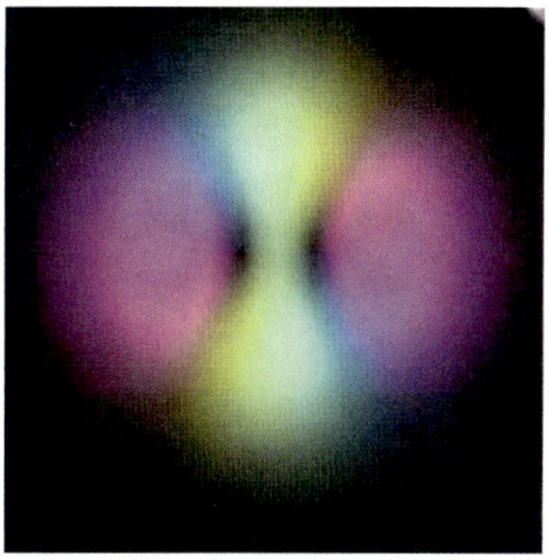

Figure 1.3: *End view of a circularly-polarised photon (Romero, 2019)*

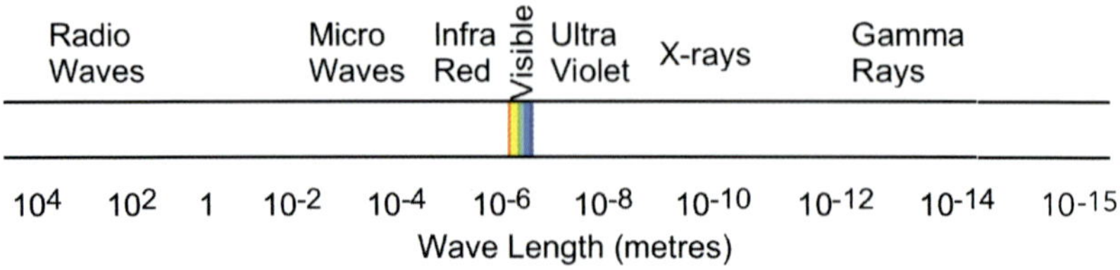

Figure 1.4: *Wavelength range of photons*

1.5.i. Photon mass

The above has described possible structures of photons. They exist because space has the property that it can transmit, store, and resist the passage of electric and magnetic fields through it. They are inherent properties of electric permeability, ε, and magnetic permeability, μ. For matter free space these are designated ε_0 and μ_0 respectively. The ability of ε_0 and μ_0 to facilitate the passage of an electromagnetic field enables a photon to travel through space. Their ability to store energy means they can hold energy for the time when that energy is passing through them, returning the energy to the photon once it has passed. The ability of ε_0 and μ_0 to resist the passage of an electromagnetic field means that it requires energy to commence an oscillation. Once started the energy will continue through free space. If it were not resisted, the photon would travel infinitely fast.

Photons having mass was referenced earlier and is used in this model. Through $p = E/c$, this energy, $E = h\upsilon$, gives the photon its property of momentum, p. A photon travelling between an emitter and absorber takes energy and momentum from the emitter and imparts them to the absorber (Einstein 1905 b, 1952 b). Dividing its momentum by its velocity gives the photon its mass $m_p = p/c = h\upsilon/c^2$.

That mass only exists when the photon is travelling at the speed of light through its travel medium. If that medium is free space it will travel at c. As mentioned above, if it is a medium such as glass that is transparent to visible photons, the electric permittivity and magnetic permeability are larger than ε_0 and μ_0. Those photons will travel at a slower speed given by

$$c_n = \sqrt{\frac{\varepsilon_0 \mu_0}{\varepsilon_n \mu_n}} \quad (1.9$$

where ε_n and μ_n are respectively the electric permittivity and magnetic permeability of the medium of refractive index n given by

$$n = c/c_n \quad (1.9a$$

As the photon moves from ε_0 to ε_n and μ_0 to μ_n, it slows down from c to c/n. It still retains its same energy, E. Because Planck's constant h has not changed, its frequency, υ, is increased and its wavelength, λ, is shortened. The energy is stored in ε_n and μ_n. When the photon travels through the medium and back into the ε_0 and μ_0 of free space, the photon resumes its travel through free space at c. The higher values of ε_n and μ_n over ε_0 and μ_0 imparts the energy the photon requires to increase its speed to c.

A photon cannot exist at rest and has no rest mass. All experimental attempts to determine a rest mass for photons have produced results experimentally indistinguishable from zero (Olive et al., 2015).

1.5.ii. Photon spin

As illustrated in the derivation of equation 1.5, a photon has mass $m_p = \frac{h\upsilon}{c^2}$. That mass is distributed in the electromagnetic fields of the photon. Whether the photon is plane or circularly-polarised, the mass is distributed around the central axis, see figures 1.1 B and 1.2 C. In linear travel, the photon's momentum can be regarded as centred on its axis of travel.

A circularly polarized photon has its whole mass rotate, or twist, through 360° every wavelength. A rotating mass has angular momentum.

The measured spin of photons is $\hbar = h/2\pi$. Attributing spin as the angular momentum of the rotating electromagnetic field gives a reason for the measured $\hbar$ property. The photon travels linearly at c. Its centre of mass, a mathematical point, travels faster than c because it has to travel further as it twists off axis through 360° per wavelength.

That off axis twist is equivalent to the photon's mass rotating about its travel axis. Photon spin is not a peculiar quantum mechanical property. It is the measured angular momentum of a circularly polarized photon's mass twisting through 360° every wavelength.

The moment of inertia, *I*, of a rotating hoop of mass m is given by $I = mr^2$, where *r* is the averaged radius of the rotating mass, i..e., its centre of mass. Equation 1.5 gives the mass of a photon as $m_p = \frac{hv}{c^2}$. Angular momentum, *L*, is given by $L = I\omega$, where ω is its angular velocity given by $\omega = \frac{v}{r}$. The circumference travelled by the centre of mass with radius $\lambda/2\pi$, is λ. To travel that distance the centre of mass needs to travel at $v = c$, with respect to the photon's linear travel. That gives $I\omega = \frac{hv}{c^2}.\frac{\lambda^2}{4\pi^2}.\frac{2\pi c}{\lambda} = \frac{hv\lambda}{2\pi c}$ which, because $\lambda = \frac{c}{v}$, simplifies to

$$I\omega = \frac{h}{2\pi} \qquad (1.10$$

A photon's centre of mass rotating at an average distance of $\frac{\lambda}{2\pi}$ from its axis, will have the observed angular momentum of $\hbar$. The important feature here is that a circularly-polarised photon has angular momentum, $\hbar$. The photon itself travels at c, taking its rotating electromagnetic field with it.

The spin of particles was originally measured as angular momentum. With the introduction of quantum mechanics that treated particles as points, angular momentum was changed to spin because a point particle cannot have angular momentum. Spin became a quantum mechanical property, which its proponents said had no classical analogy. It is easy to check if the angular momentum is the origin of spin. Circularly polarised photons have spin. Plane polarised photons do not have spin.

Figures 1.1 to 1.3 illustrate that photon's electric and magnetic fields are distributed around their central axis of travel. Their mass is not located on their axis of travel. From that it follows that plane-polarised photons have no angular momentum, and hence, no spin. Circularly polarised photons will have angular momentum. The extent of the field, and hence, the mass distribution of a photon to give the observed angular momentum of $\hbar$ i is at radius $r = \frac{\lambda}{2\pi}$.

Circularly-polarised photons are the preferred form of photons emitted from atoms and nuclei. Plane polarised photons are emitted from linear sources such as radio transmitters. Filters are available that can change the polarization state of photons.

1.5.iii. Wave-particle duality

This structure makes it easy to resolve the photon's wave-particle duality. Figures 1.1 and 1.2 show that photons are electric and magnetic field oscillations supported in free space by ε_0 and μ_0. These oscillations give photons wave properties in the same way that oscillations in other media give wave properties in those media. A photon's particle property comes from the limited extent of the electromagnetic field oscillations in ε_0 and μ_0. Photons are packets of electromagnetic energy, or oscillations, that carry momentum, p, and energy $E = pc = h\upsilon$ through space.

Photons have a structure. They are not single points. They have extended dimensions. Their lateral extent exceeds their axis of travel, although with reduced intensity. It is impossible to define the position of a photon within those dimensions. Each position on a photon can interact through its field.

A photon is an electromagnetic structure spread out over a distance, λ, in its direction of travel. Its lateral extent appears to be greater than λ. A wave pattern is imposed upon it, with a length periodicity of λ and a time periodicity of υ^{-1}. This places limitations on a photon's location. It is not a single point. Its position can never be accurately determined. This spread of a photon's electromagnetic field enables it to "experience" the world around it and react to different circumstances, just like any other wave.

In that manner, the one structure gives photons both their wave and particle properties. Their limited extent causes them to behave like particles. That is seen in the emission of light and other electromagnetic radiation, and in its absorption and the photo-electric effect. The oscillatory nature of its electromagnetic field is a wave and generates diffraction and interference patterns.

Both waves and particles reflect with the angle of incidence equal to the angle of reflection. If a photon encounters a surface with discontinuities over its wavelength, its wave properties override its particle properties. If a photon has sufficient energy that, when it is absorbed by an electron, it imparts its energy to that electron, its particle property of energy $E = h\upsilon$ change the energy of the electron. The photoelectric effect is the best-known physical example of this. The change of colour of some materials when exposed to sunlight is one of the most frequently observed chemical effects of this phenomenon. A photon imparts sufficient energy to an electron to either break a chemical bond responsible for one colour or form another bond responsible for a different colour.

Photons are entirely electromagnetic. Their electromagnetic field gives them mass. The angular rotation or twist of their electromagnetic field through 360° in one wavelength gives circularly-polarised photons a twist angular momentum, an intrinsic twist, of $\hbar$. In quantum mechanics, this angular momentum is referred to as spin. Plane polarised photons do not have this spin.

The images and wave equations presented show that photons are not point particles. Photons are spread throughout their structure. Their position can't be accurately defined to a point. That is why there will always be an uncertainty associated with either the position or momentum of a photon.

A photon is a particle and a wave in one structure. When it encounters an obstruction, its wave properties cause its trajectory to be diffracted. The closer is the photon's trajectory axis to the obstruction, the greater will be its deflection. With several photons passing a sharp edge at different distances, the different diffracting angles will result in the sharp edge appearing diffuse. When a photon's trajectory axis significantly overlaps the obstruction, it will be absorbed.

Photons emerge from the obstruction with a change of direction. This is the origins of the diffraction and interference effects associated with waves. In diffraction, the photon encounters a single obstruction. Its direction is altered depending upon how it encounters that obstruction. Individual photons will emerge from that obstruction in a slightly different direction that depends upon the proximity of their axis to the obstruction's edge. Multiple photons will emerge from it with a high probability that each will emerge in a slightly different direction. This also gives the edge of a sharp object imaged from a single source a slightly diffuse appearance on a distant screen.

When a photon impinges upon a single slit it will pass through it and diffract the same as any other wave. Not all photons impinging upon the slit will be transmitted. The smaller the slit's width, the greater will be the scattering because more photons will be closer to the edge, giving a greater average change to the direction of individual photons. The smaller slit also means a lower probability that a photon will be transmitted, giving a lower intensity.

The situation with photons passing through two slits can be more complicated. If the slits are too far apart for an individual photon's electromagnetic oscillations to pass through both, a beam of light will only show two diffuse slit images on a screen on the other side. There are two possibilities when photons impinge upon slits that are close enough for each photon to pass through both slits. In the absence of other effects, it retains the same phase and recombines to produce the typical interference pattern.

It is not a question of which slit the photon passes through. Portions of each photon's lateral extent pass through both slits. When the photons emerge on the other side, their wave patterns interact in the normal manner, producing the interference pattern. Photon interference is a single particle event.

1.6. Variations

Although photons have been widely studied, they are not well understood. The above has described photons as a single wavelength oscillation. Many other variations have been forwarded (Van der Mark, 2017). Among those is the possibility that photons are composed of more than one wavelength. Figure 1.5 shows proposed examples of a linearly-polarised photon, A, and circularly-polarised photon, B, both consisting of seven wavelengths. In both cases, the fields commence small, build to a maximum, and decay to zero as the photon passes a single point. Figure 1.5 B shows only the arrowed representation of the

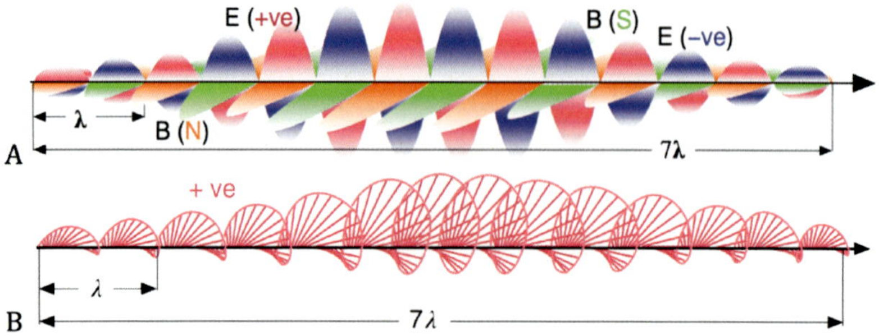

***Figure 1.5:** Schematic representations of a linear, A, and circular, B, polarised photons containing seven wavelengths per photon.*

positive charge. Both are shown in slightly different oblique presentations.

Multiple wavelength photons exist because the time taken for an electron to change from one orbit to another can be longer than the emitted photon's frequency. Other variations can be introduced by filters. They include elliptically and circularly-polarised photons in which the fields rotate through 360° in two wavelengths.

Photons are enigmatic particles that appear to have a property of changing to suit different conditions. The features they all have in common are their oscillating electromagnetic fields, which give them their wave properties, and their limited extent, which gives them their particle properties. They can also be described by a wave function, Ψ.

This marks the first major difference between the standard model and the present work. The standard mathematical model treats the photon as a point particle, to which its properties are attached. In contrast, this physicists' model treats photons as particles whose physical form and structure gives them their properties.

1.7. Summary

Despite over a century of study since their existence was determined by Planck and Einstein, the properties of photons are still not well understood. There is still widespread treatment of photons as point particles to which their properties are mathematically appended.

The above has presented a physical model and associated mathematical equations for the wave functions, Ψ, for both plane and circularly-polarised photons. Photons propagate because of the electric permittivity and magnetic permeability of the space through which they travel. In matter free space these are denoted by ε_0 and μ_0 respectively. In the presence of matter, they change to ε_n and μ_n respectively, where n is the refractive index of the space through which they are travelling.

Under this model, photons are entirely electromagnetic. Their wave and particle properties co-exist simultaneously in the photon's wave function, Ψ. Their wave properties are due to the undulations supported by ε_0 and μ_0. Their particle nature is due to their limited extent. They have energy $E = hv$. As determined by Einstein, their energy gives them mass $m_p = \frac{hv}{c^2}$. Their mass is contained within their electric and magnetic fields. These are distributed off their propagation axis.

In circularly-polarised photons, these fields rotate through 360° in every wavelength. That gives circularly-polarised photons a twist angular momentum, $\hbar$. In quantum mechanics it is referred to as spin. Plane polarised photons do not have that intrinsic twist. The model, as physically illustrated in figures 1.1, 1.2 and 1.5, depends upon the electric permittivity and magnetic permeability of free space. Having shown how these properties of free space support photons and their energy, we can now set about converting these photons into particles. Particles are the fundamental building blocks of the universe. Once we have built them, we can start to build the universe.

The huge difference between this presentation and the standard model is the treatment of photons. The standard model calls them a point particle and attaches their properties mathematically. It does not give them mass. It is a non sensical maths approach. This presentation describes their structure and shows how that structure gives them their known properties. That is the common sense approach. The remainder of this book shows just how much that difference makes in our understanding of all aspects of the universe.

An important feature in the above is that photons come in an assortment of different states and that space supports them and the fields they generate. They are the foundations of everything in the universe.

1.8. List of References

Chandrasekar N, (2012); "Quantum Mechanics of Photons", Advanced Studies of Theoretical Physics, **6**, pp 391 – 397

de Broglie L, (1957); "Mechanique Ondulatoire du Photon", et Theorie Quantique des Champs 2nd Edition, (Gauthier – Villars) Paris

Einstein A, (1905 a); "Uber der lichtelektrische Wirkung; von P Lenard", Annalen der Physik, 17, pp 132 – 139

Einstein A, (1905 b); "Ist die Trägheit eines Körpers von seinem Energiegrehalt Abhängig?", Annalen der Physik, **17**, pp 639 – 641

Einstein A, (1915); "Zur Allgemeinen Relativätstheorie," Sitzungsberichte der Königlich Preussischen Academie der Wissenschaften, Part 2, pp 778 – 786.

Einstein A, (1916); "Die Gründlage der Allgemeinen Relativätstheorie," Annalen der Physik, **42**, pp 769 – 822

Einstein A, (1952 a); "Does the Inertia of a Body depend upon its Energy-Content?" in "The Principle of Relativity", Einstein A, Lorentz HA, Weyl H and Minkowski H, Translated by Perrett W and Jeffery GB, Dover Publications, New York, pp 69-71

Einstein A, (1952 b); "The Foundation of the General Theory of Relativity," in Lorentz H.A., Einstein A., Weyl H. and Minkowski H., The Principle of Relativity, translated by W. Perret and G. B. Jeffery, Dover Publications, New York, NY, pp 109-164.

Evans M and Vigier J-P (1994); "The Enigmatic Photon, Volume 1: The Field $\mathbf{B}^{(3)}$", Fundamental Theories of Physics, Kluwer Academic Publishers, Dordrecht, Ch 3

Hertz HR, (1887); "Uber sehr schnelle electrische Schwingungen", Annalen der Physik, **267**, pp 421 – 448

Lemaître G, (1931) "L'Hypothèse de l'atome primitive" in , Lemaître, The Primeval Atom – an Essay on Cosmogony, D. Van Nostrand Co, 1946.

Lemaître G, (1936); "l'Universe en Expansion", Annales de la Societié Scientifique de Bruxelles, **A53**, pp 51 – 85

Marconi G, (1896); "Improvements in transmitting electrical impulses and signals in an apparatus thereof", British patent no. 12,039

Maxwell JC, (1865); "A Dynamical Theory of the Electromagnetic Field", Philosophical Transactions of the Royal Society London, **155**, pp 459 – 512

Olive KA et al., (2015): http://pdg.lbl.gov/2015/listings/rpp2015-list-photon.pdf

Planck M, (1900 a); "Auf das Gesetz der Energieverteilung im Normalspektrum" Annalen der Physik, **1**, pp 719 – 736

Planck M, (1900 b); "Über eine Verbesesrung der Weinschen Spektralgleichung", Verhandlungen der Deutschen Physikalischen Gesellschaft, **2**, pp 202 – 204

Pound RV and Rebka GA, (1959); "Gravitational Red-shift in Nuclear Resonance", Physical Review Letters, **3**, pp 439 – 441

Pound RV and Snider JL, (1961); "Effect of Gravity on Nuclear Resonance", Physical Review Letters, **13**, pp 539 – 540

Pound RV (2000); "Weighing Photons", Classical and Quantum Gravity, **17**, 2303 – 2311

Romero J, (2019); "Shaping up in high-dimensional quantum information, Australian Physics, **56** pp 84 – 86

van der Mark MB and 't Hooft GW, (2000); "Light is Heavy", van Atom tot Quantum, Nederlandese Natuurkundige Vereninging, pp 1 – 6

van der Mark MB (2017); Private communication

Vigier J-P (1992); "Present Experimental Status of the Einstein - de Broglie Theory of Light", Proc. I.S.Q.M. 1992 Workshop on Quantum Mechanics, Tokyo

2. From Photons to Particles

Abstract:

The hypothesis is forwarded that every individual subatomic particle is composed of a photon confined such that it makes two revolutions within its wavelength. It is shown how this gives rise to the known properties of particles, to some of their behaviours, as well as to the special relativity corrections that relate to them.

2.1. Introduction

Most of the visible universe is made up of five major components. They are protons, neutrons, electrons, neutrinos, and photons. Other particles are known to exist, such as the anti-particles of protons, neutrons, electrons, and neutrinos. There are also short lived "elementary" particles. Other forms of mass, such as dark matter, have been postulated in order to explain galaxy rotation. They have never been detected. Similarly other forms of energy, such as dark energy have been postulated in order to explain what are considered to be anomalous type 1a supernovae intensities. Again, this type of energy has never been detected. For this presentation, we will build our universe using only particles that have been detected. We will also use only the three detected space dimensions and time. String theory is definitely excluded.

The previous chapter showed how photons exist because of the properties of empty space. Photons are energy. It is well known in the science community that mass and energy can be interchanged. When a particle and its anti-particle meet, they annihilate each other and produce photons. When a high-energy photon impinges upon a heavy nucleus, it can generate a particle/anti-particle pair. It would thus be a reasonable assumption that particles and photons had the same constituents. The description of photons in chapter 1 suggests that they are fundamental particles with no sub-constituents. They can only be separated into other photons. They were explained in terms of three space dimensions, time, and the properties of matter-free space. No additional dimensions are required in order to explain any of their properties.

The best proof of this proposal is the experimental results from all particle accelerators like CERN's Large Hadron Collider or Brookhaven's particle accelerators. Their experiments start with charged particles, protons or electrons. They are accelerated to high speeds by adding photons to them. Energies of many TeV have been achieved. They are then smashed into other particles. The result is a shower of particles and photons. When everything has died down, all that is left are the original particles, plus possibly a few neutrino/anti-neutrino pairs that aren't easily detected. All other particles decay, directly or eventually, into photons. They, along with the photons emitted at the same time, have long since been absorbed as heat. Nothing else has ever been detected. When a particle and its anti-particle are brought together, they annihilate each other by producing photons. Common sense suggests that two way interchange between photons and particles is the best experimental evidence that everything is made of photons.

Apart from photons, the only stable particles ever detected are electrons, protons, neutrons, and neutrinos. This chapter shows the proposed relationship between photons and particles. The success of the model is determined by how well it agrees with observation. This presentation shows this match extends from the smallest subatomic particles to the universe's large-scale structure.

2.2. The Model of Matter

With that as the background, how can photons travelling linearly at *c* form particles that can stay in one place? It is suggested they do this by making two revolutions per wavelength. Williamson and van der Mark (1997) were the first to publish that this is the structure of electrons. They were followed by Qui-Hong (2007) and Robinson (2011). Bourgoin (2008) suggested the possibility of photons forming toroids.

Photons of the forms illustrated in figures 1.1 and 1.2, making two complete revolutions per wavelength, are illustrated in figure 2.1.

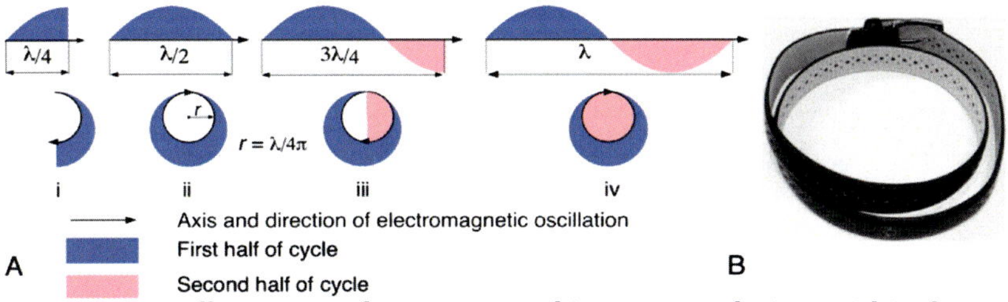

Figure 2.1: *Illustration of structures making two revolutions within their wavelength.*

The different colours simply represent the two halves of a wavelength. The top illustrations show a linear photon. The bottom illustrations show the rotating photon in phase with the linear photon above it. By the time a linear photon has made a quarter wavelength, the rotating photon has turned through 180°. When the linear photon has made a half wavelength, ii, the rotating photon has turned through 360°.

This continues, completing $\frac{3}{4}\lambda$ at iii, and two complete revolutions at iv. In both linear and rotating situations, the photon's axis travels at c. The linear photon travels through its next wavelength at a different position in space. The rotating photon still travels at c, but remains in the same position when not moving with respect to an observer.

Under the Einstein-de Broglie model, as predicted by Einstein and observed, photons have mass. It is contained within their oscillating electromagnetic field. A photon of mass m_p will rotate in a circle of radius:

$$r = \frac{\lambda}{4\pi} \qquad (2.1$$

It is a feature of this model that a rotating photon is only one wavelength long.

There is a special feature associated with a linear structure that makes a 360° twist per wavelength. As shown by Williamson (2019a), when the ends connect, a double loop is formed with one side of the structure always facing toward the outside. This is depicted in figure 2.1 B. It shows a flat, two-shaded belt where one end is twisted through 360° before being inserted into the buckle and fastened in place. The belt thus forms a double-loop structure. Note that only the dark surface of the belt faces the outside.

Figure 1.3 shows the magnetic field is continuous, while the electric field is broken in the middle. The photon's magnetic field is much stronger than its electric field. The photon bends around the magnetic field. In making exactly two revolutions per wavelength, the magnetic field is always in one direction. The photon's north and south poles are always synchronized. That leaves the electric field on the outside. Figures 2.1 A and B show there will only be one electric field on the outside.

Note that a linear photon has its magnetic fields open ended. Rotating photons have their magnetic field closed. This gives rotating photons a lower energy state than linear photons. Where possible, photons prefer the rotating form to the linear form. In the rotating form, they present one electric field to the outside world and have a strong magnetic field through their centre.

Because it has a rotating mass, it will have angular momentum, $L = I\omega$, where I is its moment of inertia and $\omega = \frac{c}{r} = \frac{4\pi c}{\lambda}$, its angular velocity. It is known that each individual subatomic particle has a property called spin, which has a value ½$\hbar$, where $\hbar$ is Planck's constant, h, divided by 2π. It is suggested that the property called spin is angular momentum. Uhlenbeck & Goudsmit (1925; 1926) indicated that electrons had an intrinsic angular momentum of ½$\hbar$, to explain the Zeeman effect. It is the particle's intrinsic spin. Every single sub-atomic particle has that internal angular momentum of ½$\hbar$.

The moment of inertia of a rotating hoop of mass m and radius r is given by $I = mr^2$. In these calculations, it requires that the electric and magnetic fields are distributed around the axis. That allows this calculation to be made with the centre of mass located on its rotation axis. Starting with $I\omega$ = ½$\hbar$ and multiplying both sides by ω gives:

$$mr^2.\frac{c^2}{r^2} = \frac{h}{4\pi}.\frac{4\pi c}{\lambda} = \frac{hc}{\lambda} = hv = E = mc^2 \qquad (2.2$$

Equation 2.2 shows the physical meaning of $E = mc^2$. Energy E is a photon travelling at c in a straight line. Mass m is the same photon making two revolutions per wavelength, while still travelling at c. The mass, m_p, of the rotating photon that is the particle is related to its frequency υ through

$$m_p = \frac{h\upsilon}{c^2} \tag{2.2a}$$

It has a radius

$$r = \frac{\hbar}{2mc} \tag{2.2b}$$

2.3. Photons Travel in a Circle

In order to give a physical reason for a photon to make two revolutions per wavelength, we need to look at figure 1.2 and the discussion on circularly-polarised photons. Circularly polarised photons have "spin" $\hbar$. Yang, et al. (2020) referred to that spin as angular momentum. To avoid confusion, that spin shall be referred to as the photon's twist spin, or intrinsic twist. It is generated by its electric and magnetic fields twisting through 360° in one wavelength. That angular momentum of $I\omega = \hbar$ is the origin of a circularly polarized photon's spin, herein called its twist. Figures 1.1 and 1.2 show that these fields are off the axis. When the photon twists as shown in figure 2.1, the mass contained within the energy fields also twists. Section 1.5 ii showed the off axis centre of mass has a radius $r = \frac{\lambda}{2\pi}$.

Figure 2.1 B shows that when a flat belt was twisted through 360° and its ends joined, there was only one surface on the outside. When a photon with intrinsic twist of 360° in one wavelength makes two revolutions per wavelength, the magnetic field causes the ends to automatically meet. They do not have to join. Figures 1.3 and 2.1 mean there will only be one electric field on the outside, with the magnetic field perpendicular to its rotation plane.

A photon with intrinsic twist $\hbar$, has a preferred configuration. If possible, it will seek to make two revolutions within its wavelength. Its intrinsic twist automatically brings the ends of the wave together. The alignment of the magnetic fields and the natural state of two revolutions per twist cause the photon to form the more stable structure when it has a single charge field on the outside and its magnetic field through the middle. The touching ends give maximum magnetic field strength. Once formed, conservation of both forms of angular momentum, supported by the interlocking magnetic fields and the unit electric field on the outside keeps the structure together. Some of the mathematics of how this is achieved has been described by Williamson (2014, 2019b) and van der Mark and Williamson (2021).

Equation 2.2 showed that this structure gives the physical relationship between energy and mass through the equation $E = mc^2$. The following shows how this model explains the origins of the physical corrections associated with Einstein's special relativity theory.

2.4. The Special Relativity Corrections

Consider what happens to a particle under this model as it moves. Figure 2.2 A illustrates a hoop moving along its plane of rotation, like a wheel rolling along a surface. Figure 2.2 B illustrates the same hoop moving along its axis of rotation perpendicular to that rotation plane. All other possible motions should be considered as a combination of these two. In this study, only the path shown in figure 2.2 B will be considered. The rotating photon moves through space in the direction of its rotational axis. That requires each particle of this type to instantaneously orient itself such that the direction of rotation of the photon would always be perpendicular to the particle's travel direction.

As mentioned in chapter 1, photons cannot move sideways. The only way such a structure can move is to spiral along a helical path. Figure 2.3 shows images of the rotating photon (particle) at rest, A, and moving,

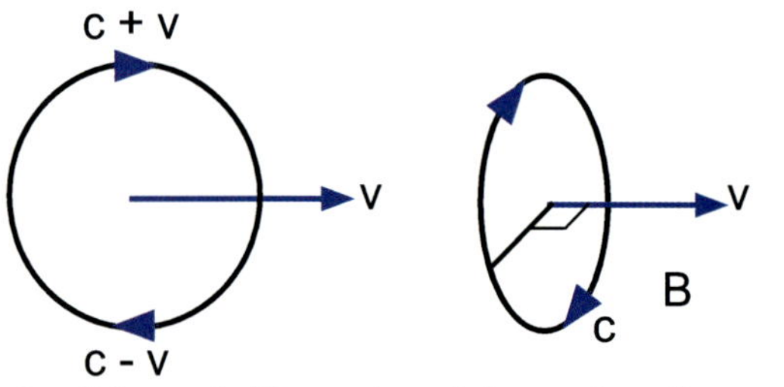

Figure 2.2: *Schematic illustration of the motion of an oscillation moving in the directions parallel, A, and perpendicular, B, to the plane of the rotation of the electromagnetic oscillation.*

B. In all cases, images i, ii, and iii represent the rotating photon viewed end on, obliquely, and parallel to the plane of rotation respectively.

The photon's travel axis is seen as a circle when viewed end-on, as shown in figure 2.3 B i. The oblique view in figure 2.3 B ii shows the helical travel path of photon's axis. When viewed side-on, as in figure 2.3 B iii, the photon's travel axis looks like a "sawtooth" wave, but it is still a helix.

The front still travels distance ct in time t, given by

$$(2\pi r_v)^2 + (vt)^2 = ct^2 \qquad (2.3$$

where r_v is the radius at velocity v. Equation 2.3 can be rewritten as $\frac{(2\pi r_v)^2+(vt)^2}{(2\pi r_0)^2} = \frac{ct^2}{(2\pi r_0)^2}$. Since $2\pi r_0 = ct$, that simplifies to

$\frac{(2\pi r_v)^2}{(2\pi r_0)^2} + \frac{(vt)^2}{(ct)^2} = 1$. It can be rewritten as $\frac{r_v^2}{r_0^2} + \frac{v^2}{c^2} = 1$, for which one solution is

$$r_v = r_0\sqrt{1 - \frac{v^2}{c^2}} \qquad (2.4$$

In this rotating photon model, a particle will contract laterally as it moves. The degree of lateral contraction is given by equation 2.4.

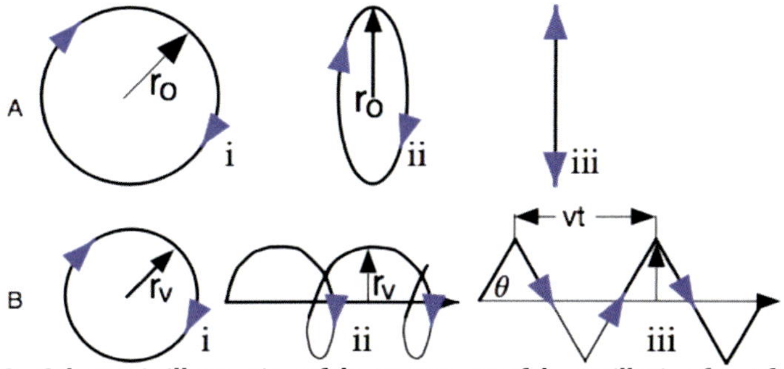

***Figure 2.3:** Schematic illustration of the movement of the oscillation front for a particle at rest, A, and moving at velocity, v, B, illustrated from end on, i, at an angle, ii, and side on, iii. Note that B ii and B iii depict a helix.*

When this type of particle moves at increasing speed, its radius contracts while its angular momentum, $I\omega$, remains constant since none has been added or subtracted. Thus, angular momentum, $I\omega$, remains constant at ½$\hbar = mr^2.\frac{csin\theta}{r}$ for all velocities. θ represents the angle between the direction of travel of the particle and the angle traversed by the spiralling photon, as shown in figure 2.3 B iii. The $sin\theta$ term comes about because the motion of the photon is now no longer completely circular.

Some of the velocity, c, is in the direction of the propagation of the particle and some in the direction of rotation. It is only the component, $sin\theta$, in the direction of rotation, which contributes to the angular momentum. That gives $m_0r_0 = m_vr_vsin\theta$, where m_0 and r_0 are the rest mass and rest radius of the electron, and m_v and r_v are its mass and radius at velocity, v. In this rotating photon model of a particle, the photon cannot spin faster. To match angular momentum, its mass must increase. As such, $m_v = \frac{m_0r_0}{r_vsin\theta}$. Since $sin\theta$ is given by $\frac{r_v}{r_0}$, this gives $m_v = m_0.\frac{r_0^2}{r_v^2}$, which simplifies to

$$m_v = \frac{m_0}{\left(1-\frac{v^2}{c^2}\right)} \qquad (2.5$$

Equation 2.5 means that, as the velocity of a particle of this type increases, its mass increases at a rate faster than is allowed under the previously derived relationship, $E = mc^2$. That is at odds with the principle of the conservation of mass-energy. It can be avoided if the rotation rate of the particle's photon slows down. Then equation 2.5 can be re-written as

$$m_v = \frac{m_0}{\sqrt{1-\frac{v^2}{c^2}}}.\frac{1}{\sqrt{1-\frac{v^2}{c^2}}} \qquad (2.5a$$

indicating that the time correction required is

$$\frac{1}{\sqrt{1-\frac{v^2}{c^2}}} \qquad (2.6$$

In the above description of the rotating photon, classical concepts of length and time have been used. If time on the moving particle were to slow down according to equation 2.6, the mass would increase at the slower rate given by

$$m_v = \frac{m_0}{\sqrt{1-\frac{v^2}{c^2}}} \qquad (2.7$$

That agrees with predictions from Einstein's special theory of relativity. This rotating photon model requires the time interval, t_v, at velocity, v, to slow down according to the relationship

$$t_v = \frac{t_0}{\sqrt{1-\frac{v^2}{c^2}}} \qquad (2.8$$

This also agrees with the special theory of relativity, as well as experimental observation. The photon, however, is still required to travel at c. With its rotation rate slowed down, it must now have a linear forward travel component in its motion. This is schematically illustrated in figure 2.4.

The thicker lines in figure 2.4 are meant to indicate the portion of the spiral that is out of the plane of the page; the thinner lines indicate the portion of the spiral that is into the plane of the page. All show the photon making three spirals, travelling 1½ wavelengths. Figure 2.4

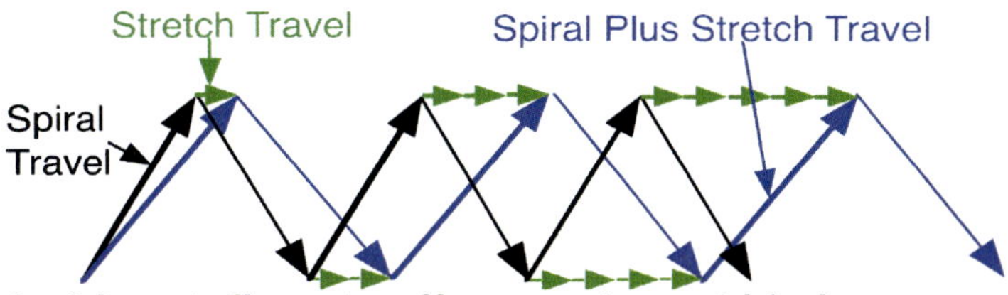

Figure 2.4: *Schematic illustration of how a moving particle's photon must both spiral, as well as stretch its way through space.*

shows a side-on view of the photon. The "spiral travel" lines (in black) show the spiralling photon without the "forward stretch."

The small "stretch travel" arrows (in green) show the introduced forward motion. The spiral plus stretch travel line (in blue) shows the trajectory the photon takes as it travels.

It is not that the spiralling photons move sideways as indicated in figure 2.4. The rotating photon is still spiralling forward. Each stretched spiral travels a little further than the rotating spiral would, were it not for the necessity that its rotational speed should slow down.

The radius reduction in equation 2.4 occurs because of the constant speed of light. Figure 2.5A shows the close packed spirals of a slow moving electron with an energy of a few keV. Figure 2.5B shows the same number of spirals of an energetic electron, many MeV. The distance travelled by the photon for the same six spirals is constant for an external observer. The stretched out photon in B must have a lower radius than that of A. The thicker lines in B indicate the photon is of higher mass because of its relativistic mass increase.

Figures 2.4 and 2.5 show that the special relativity corrections are not something that is imposed on matter by some mathematical means. The constant speed of light means that they are an *automatic consequence* of this rotating photon model of the structure of matter. As the particle helically spirals its way through space, it travels the total distance seen by a stationary observer.

When it has a forward linear motion, it makes fewer revolutions to go from one position to another in the direction of travel. Making fewer revolutions means that the spiralling photon moves through a shorter distance to reach its destination. As can be seen from equation 2.6, the correction factor for this forward stretch motion is $\frac{1}{\sqrt{1-\frac{v^2}{c^2}}}$.

This results in the distance travelled by a rotating photon moving at velocity, v, being apparently shortened, according to the well-known relativistic correction for distance, namely:

$$l_v = l_0 . \sqrt{1 - \frac{v^2}{c^2}} \qquad (2.9$$

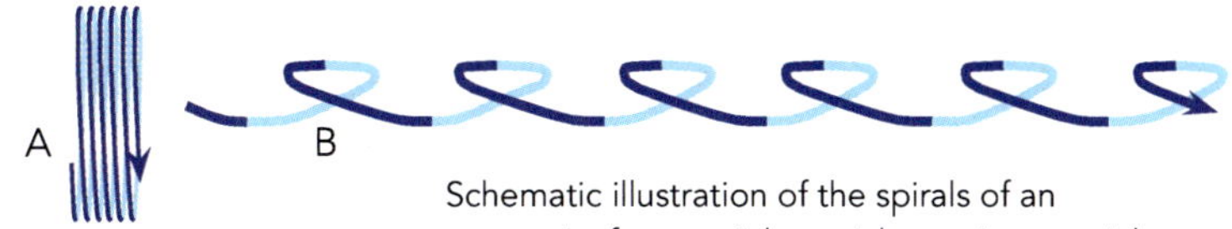

Figure 2.5: *Schematic illustration of the different spiral actions for a slow moving particle, A and a fast moving particle, B.*

To increase the velocity of a particle, it must be given energy, i.e., it must absorb photons. The rotating photon absorbs the incoming linear photon and its mass increases. But this is not a simple increase in mass that decreases radius. As the particle travels faster and its mass increases, it must further slowdown the rate at which it rotates. The slowing rotational rate makes it look, to an external observer, as if time is slowing down on the particle. It must still travel at the same speed, so its whole structure takes on an additional forward motion. The photon still spirals its way forward as it rotates, but it stretches forward at the same time.

The special relativity corrections are a simple consequence of the rotating photon model of the structure of matter. Length appears to be less, as if distance has been shortened, because a particle does not have to spiral the whole distance. Einstein (1905, 1952) calculated his equations based entirely upon mathematical reasoning emerging from the constancy of the speed of light. The present model does it from the rotating photon model of the structure of matter, the constancy of the speed of light, and Pythagoras' theorem.

It may help to regard the rotating photon's axis as the hypotenuse of a right angle triangle. It always has a fixed speed c. Wrap a fixed length of string exactly twice around a thick rod and its ends will not be far apart. Wrap the same piece of string exactly twice around a thin rod and the ends will be further apart. The distance between the ends of the string represents how far the photon travelled in one wavelength. That shows why the diameter of a particle decreases with its speed, as derived in equation 2.4.

This model automatically derives the features of the special theory of relativity, and yields the same relativistic corrections for mass, length, and time:

$$E = mc^2 \qquad (2.2$$

$$m_v = \frac{m_0}{\sqrt{1-\frac{v^2}{c^2}}} \qquad (2.7$$

$$t_v = \frac{t_0}{\sqrt{1-\frac{v^2}{c^2}}} \qquad (2.8$$

$$l_v = l_0\sqrt{1-\frac{v^2}{c^2}} \qquad (2.9$$

It provides the physical reason for these corrections, based purely upon the structure of individual particles. The circumferential speed of a rotating photon spiralling through space remains constant at the speed of light, c, and its radius must be altered to suit its motion. This automatically results in time slowing down, mass increasing, and the distance of spiral travel decreasing. Unlike the Einstein (1905) and Lorentz (1904) corrections, this rotating photon model also predicts a contraction in the width of an individual subatomic particle, such as an electron, according to:

$$r_v = r_0.\sqrt{1-\frac{v^2}{c^2}} \qquad (2.4$$

This has been detected, but not recognized as such. According to this model, the electron's rest mass gives it a rest radius of 1.93 x 10^{-13} m. This is quite small, but much larger than the dimensions that physicists can measure. Electrons have been accelerated to over 29 GeV (Bender et al., 1984). That increases an electron's mass by over 50,000 times, as it travels close to the speed of light. It was found that electrons behave as if they are point particles, having dimensions less than 10^{-17} m. Subsequent measurements with electrons accelerated to TeV energies have shown collision cross sections of $\approx 10^{-18}$ m, as predicted in equation 2.4.

From equation 2.6, the electron's angular momentum of ½$\hbar$ indicates that, at rest, it has dimensions consistent with a radius of the Compton wavelength divided by 4π, or 1.930796 x 10^{-13} m. The observed dimension of less than 10^{-17} m at GeV energies and $\approx 10^{-18}$ m at TeV energies are consistent with the radial contraction the electron experienced at large accelerating voltages.

This model describes the dimensions of an electron, having properties equivalent to a radius of some 1.93 x 10^{-13} m at rest, yet also having a dimension of less than 10^{-17} m at high energies. Its relativistic mass increase enables the electron to maintain its angular momentum of ½$\hbar$ and still be measured as a 'point particle.' This model clarifies that its diameter decreases with its energy, and hence, with its velocity. In doing this, it retains its angular momentum and charge.

It is only the width of individual subatomic particles that contract. This contraction does not apply to space as a whole. Lateral dimensions are maintained by electric charge. A hydrogen atom, which has a radius of 0.053 nm at rest with respect to an observer, will still have the same radius when traveling at relativistic velocities with respect to the same observer. Space is foreshortened in the direction of travel of the particle, but not in the other directions. As shall be shown later, electric charge is unaffected by velocity. However, the radius of an individual particle, such as an electron, decreases with increasing velocity. That is an important difference in modelling the structure of a particle such as the electron.

The above shows the special relativity corrections for mass, length, and time with velocity are caused by the rotating photon structure of each individual sub-atomic particle. Figures 2.3 to 2.5 show how those corrections are automatically induced into every particle, irrespective of the frame of reference.

2.5. Some Consequences

Figure 2.6 illustrates what happens when a circularly-polarised photon, A, and a plane-polarised photon, C, rotate twice within their wavelength. Figure 2.6 E shows a schematic plan view of the photon's central travel axis.

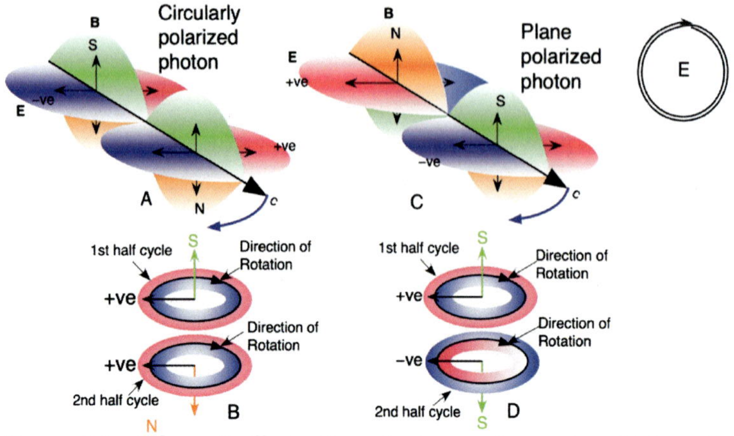

Figure 2.6: *Schematic illustration of primary photon, A and C travelling twice within their wavelength, generating the structures shown in B and D respectively.*

The circularly-polarised photon representation, A, has been simplified to show that the electric charge and magnetic fields are always in the one quadrant. As it rotates twice in the direction indicated by the arrow, one polarity, positive, is always on the outside and the other, negative, is always on the inside. One magnetic pole, south, always points up while the other pole, north, always points down.

Once formed, it is held there by the ½$\hbar$ angular momentum of the photon making two revolutions per wavelength, the twist spin $\hbar$ of the photon twisting through 360° as it rotates twice within its wavelength. The particle's magnetic field is always unidirectional. The external charge is always unipolar. It rotates exactly twice per wavelength because that gives the strongest magnetic field and only one external electric field.

When circularly polarized photons make two revolutions per wavelength, the magnetic fields are always in the one direction and form a closed loop. That helps to make rotating photons more stable than linear photons, in which their magnetic fields are open and hence less stable.

In comparison, the rotating plane-polarised photon in figure 2.6 D shows it has one polarity during the first half wavelength and the other polarity during the second half wavelength. Under this model, that would make it a neutral particle. The magnetic fields would also alternate with each half cycle. While it does not have any intrinsic photon twist, it still has the particle's intrinsic spin of ½$\hbar$ angular momentum as the photon makes two revolutions per wavelength. The differences between the structures in B and D suggest that charged particles will be more stable than neutral particles.

Under this model, neutral particles have equal positive and negative fields at different times in the rotating photon's cycle.

Figure 1.2 C, supported by figure 1.3 indicate that the fields originate from the photon's travel axis and spread out. The fields get weaker with distance from the travel axis. This feature will be retained when the photon's axis makes two revolutions per wavelength.

The photon that is the particle rotates at c in a very small diameter circle. For an electron or positron, that radius is 193 fm. It is much smaller for nucleons. The photon's rapid rotation subjects it to a significant centripetal force. As with all centripetal forces, loosely bound items will be "thrown off". Photons electric and magnetic fields have the same properties as photons. The centripetal force to which the fields are subjected causes them, i.e., photons, to be thrown off, or ejected, from the rotating photon's central axis. This is schematically illustrated in figure 2.7.

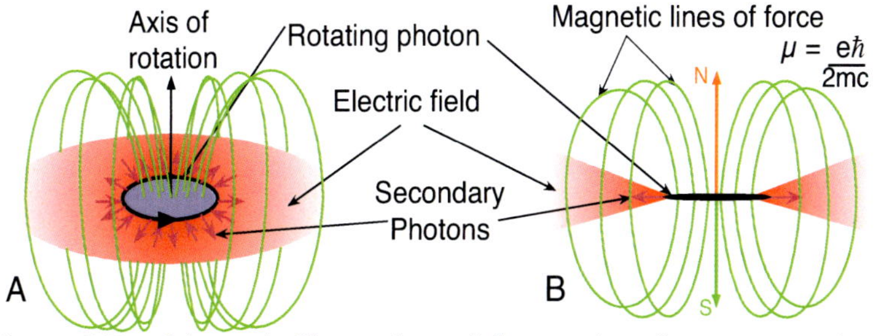

Figure 2.7: *Schematic illustrations of the rotating photon generating an electric field at its circumference of its plane of rotation, with the magnetic field perpendicular to it.*

Figure 2.7 A illustrates an oblique view of the important features of a structure produced by the rotating photon in figure 2.6 B. It shows the rotating photon travelling in the anti-clockwise direction. giving an external positive electric field and an internal negative electric field.

The ejected photons are the arrows labelled as secondary photons. For ease of interpretation, they are shown as originating perpendicularly from the rotating photon's axis.

If the secondary photons do not interact with another particle, they are re-absorbed to conserve mass. This re-absorption makes them virtual photons. Under quantum electrodynamics (QED) (Feynman, 1985), photons are the force-carrying particles for electric charge. The emission and absorption of those virtual photons at the rotating photon's circumference is the origins of electric charge.

The amount or quantity of virtual photons generated depends upon the angular momentum, $\frac{1}{2}\hbar$, of the rotating photon. Every individual sub-atomic particle is a photon making two revolutions per wavelength. That gives all such particles the same $\frac{1}{2}\hbar$ angular momentum. All individual charged particles will have the same electric charge, as is observed. All individual neutral particles will have half positive and half negative charge.

Electric charge is not uniformly distributed about a spherical particle. It originates at the rotating photon's axial circumference and in its plane of rotation. The side view, figure 2.7 B, shows the external electric field spreading out into the third dimension as the distance from the rotating photon's axis increases.

The ability of photon to both emit and absorb photons has long been known. Their ability to lose part of its energy has been established through Compton (1923) scattering. An X-ray can interact with an electron and lose part of its energy. The reverse has also been detected, where photon energy increases (Sunyaev and Zel'dovich, 1980). This effect requires the presence of matter particles. The detected increase in energy is also associated with photons near atoms. It is suggested that the rotating photon nature of particles makes it easier to understand this effect.

Electric charge is generated by the continual emission and absorption of secondary or virtual photons at the circumference of the rotating photon that is the particle. The polarity of that charge is determined by the polarity of the electric field that is on the outside of the rotating photon structure, as shown in figure 2.6 B. Electric charge is not uniform and three dimensional throughout the particle. It is planar, emitted into the two dimensions of the rotating photon's plane of rotation. Photons spread the charge field into the third dimension as they move away from the rotation plane.

That may seem a strange origin of electric charge. Figure 3.6 and other descriptions of the nucleons show it has been measured to an accuracy of $\approx 10^{-17}$ m, 10^{-2} fm.

The first aim of a theory must be to match observation. That match suggest it is a good description of the origin of electric charge in particles. A second aim is to make predictions that can be experimentally tested. Many examples are given in the following chapters.

Figure 2.6 A and B show that there will be a magnetic field perpendicular to the photon's rotation plane. Bohr (1913) determined that the magnetic field associated with an electron is $\frac{e\hbar}{2mc}$, the Bohr magneton. Equation 2.2b derived that the radius of a particle of mass, *m*, making two revolutions per wavelength is $r = \frac{\hbar}{2mc}$. Under this model, the magnetic moment of a particle is its electric charge multiplied by its radius.

When a photon makes two revolutions per wavelength about its magnetic field, the field becomes a closed loop. The inverse square law changes to an inverse cube law. The magnetic fields closed loop nature reduces its magnitude away from its immediate vicinity.

This gives a physical description of what happens when a linear photon is converted to a rotating photon. Williamson (2019b) gave a mathematical description of the process through his equation $D_\mu \Xi_G = 0$. In his work, the Ξ_G terms are a list of variables that include the electric and magnetic fields. The D_μ terms are a list of their derivatives with respect to each other. The equation indicates that the sum of all the changes is zero. Whenever there is one change somewhere, there is another change somewhere else to balance it out. Newton's third law of motion — *"For every action there is an equal and opposite reaction"* — is one expression of this in the mechanical world.

An indication of how this plays out in the present model can be seen by considering the annihilation of an electron with its anti-matter equivalent, a positron. Both are oppositely charged spin half particles. Their photons are rotating at c in a circle of radius $\frac{\hbar}{2mc}$, although in opposite directions with respect to their magnetic moments. In close proximity, their opposite charges and magnetic moments attract each other. On contact they unlock each other's angular momenta. The two oppositely rotating photons now become two linear photons, moving away in opposite directions in order to conserve linear momentum.

Before the interaction, there were two photons that had net zero momentum (angular and linear), charge and magnetic moment. They had mass = $2m_e$, where m_e is the electron's mass. After the interaction, they have net zero momentum (angular and linear), charge, and magnetic moment. Their mass is carried away with the photons, which have mass $m_e = \frac{h\upsilon_e}{c^2}$ where υ_e is the photon's frequency associated with the mass of an electron. The total change is zero. The equation also works in the reverse direction, for the conversion of photons into mass.

Another feature of interest is: "What keeps the particle intact?" The answer is found in its intrinsic angular momenta. The rotating photon's intrinsic spin of $\frac{1}{2}\hbar$ is the result of the whole mass of the particle rotating at the speed of light, twice within its wavelength. Once it is established, the conservation of angular momentum will keep it rotating.

That is only part of the story. The photon associated with a charged particle also twists once around its direction of travel every wavelength. This is what gives circularly-polarised photons an intrinsic twist of $\hbar$. The rotating photon is therefore twisting one revolution on its axis as it makes two revolutions, at light-speed, to complete the particle. The two momenta form the particle. It has a strong magnetic field formed by the combination of the two momenta acting on the photon. The conservation of those two angular momenta held in place by the magnetic field means that, once this circular motion is established, it is so strongly reinforced that it can only be unlocked by its anti-particle.

2.6. Electric Charge and Magnetic Moment

Quantum electrodynamics (QED) has demonstrated its ability to make some useful predictions. It is a complex theory that takes many years to learn and understand, and even more to make a meaningful calculation. Its prediction of the electron's anomalous magnetic moment is accurate when no mistakes are made. QED theory indicates photons are the force-carrying particles for electromagnetism. Their emission and absorption by a particle are the physical mechanisms that impart the property of electric charge to that particle.

Chapter 1 presented information on circularly and plane-polarised photons. Integrating the equations for plane-polarised photons gives no net electric or magnetic field. Integrating those for circularly-polarised photons gives residual electric and magnetic fields. If these photons were to rotate twice within their wavelength, as in figures 2.1 and 2.6, they will produce different results.

Figures 2.6 A and B show a circularly-polarised photon bending about its electric field axis. During the first half wavelength, the positive field will be on the outside, the negative on the inside, the south magnetic pole will point upwards, with the north magnetic pole downwards. The result would be a residual external charge and magnetic moment associated with the particle. In a rotating plane-polarised photon, seen in figure 2.6 C and D, there is a positive charge on the outside for the first half cycle and a negative charge on the outside for the second half cycle. The directions of the magnetic poles also alternate with each half cycle. A particle consisting of a rotating plane polarized photon will have mass and intrinsic spin of $\frac{1}{2}\hbar$. It will not have any electric charge or magnetic moment.

Electric charge is caused by the emission and absorption of secondary or virtual photons by a rotating photon, which is a particle. It occurs because of the centripetal action of the $\frac{1}{2}\hbar$ intrinsic spin's angular momentum to which the electric field is exposed. If these secondary or virtual photons are not captured by another particle, they will be reabsorbed by the original rotating photon particle. The red arrows in figure 2.7 A represent those secondary photons associated with the inner rotating photon. Whether by emission or absorption, their existence aids the rotating primary photon in its circular trajectory.

Rotating circularly-polarised photons reinforce their charge and magnetic moments. In rotating plane-polarised photons, these cancel each other out. It is therefore expected that the unit charge and magnetic moment reinforcement occurring in rotating circularly-polarised photons should make them more stable particles than neutrally charged rotating plane-polarised photons.

This introduces the magnitude of electric charge and the magnetic moment of charged particles. Each stable fundamental charged particle, i.e., proton and electron, has an angular momentum of $\frac{1}{2}\hbar$. As such, the number of photons emitted and absorbed is the same, irrespective of the particle's mass or radius. With this as the origin of charge, the charge on a particle would be expected to be the same under all conditions, including at relativistic speeds. It has been observed that the charge is the same, 1.60217648 x 10^{-19} coulomb on all charged particles of any mass and at all speeds.

That re-enforces the idea that charge is a property of a particle's structure, not of its mass. Nor is it something that is attached mathematically. It is the rotating circularly polarized photon's structure that generates it electric charge. Remove the rotation and the charge disappears.

This origin of charge precludes the possibility of fractional charge. The possible exception to this is a particle whose lifetime is too short to make a complete revolution. That would require a lifetime of the order of 10^{-26} sec. Particles such as the muon and tau lepton, which are stable for many millions of rotations, show the same electric charge value.

Under this rotating photon model, there are two possible contributions to the electric charge and magnetic fields of particles. Both are influenced by the residual electric, E, and magnetic, B, fields of the original photon. The above has suggested that the electric charge is due to the continual emission and absorption of secondary photons caused by the primary photon rotating twice within their wavelength. The residual E field is responsible for the polarity of the particle's charge. The particle's magnetic field also has two possible origins. One is the photon's residual B field. The other is the charge circulating the particle at radius $r = \frac{\hbar}{2mc}$, see equation 2.4. The Bohr (1913) magneton, given by

$$\mu_B = \frac{e\hbar}{2mc} \qquad (2.10$$

is equal to its electric charge e multiplied by its radius. Equation 2.10 will be used in all future calculations of the magnetic field associated with particles.

Electric charge and magnetic moment are an integral part of the structure of particles. Electric charge is generated by the photon's rotation causing it to emit and absorb secondary or virtual photons as it rotates in a circle. The electric charge originates at the axis circumference of the rotating photon that is the particle. That charge is generated at that axis and in the rotation plane. That is shown in figure 2.7 A as the arrows leaving the photon axis circumference. Figure 2.7 B shows the electric field spreads into the third dimension as the photon get further away from that axis.

The magnetic field is perpendicular to the electric field, as suggested by Maxwell's equations. Both are generated because the photon that is the particle rotates in the plane of its electric field. This leaves its magnetic field perpendicular to that rotational plane. These properties arise from the particle's structure, and not something of unknown origin that is added mathematically.

Equation 1.8 shows a photon's magnetic field varying the same as its electric field, namely as inverse square. The strengths of magnetic fields are known to vary as inverse cube. The magnetic fields associated with a photon are open ended. Those associated with magnets, including the Bohr magneton associated with particles, are closed loops. Closed loop magnetic fields are much weaker at greater distances because they curve around their central axis. That leaves a smaller residual to be detected further away.

The rotating photon's magnetic field closed loop structure means protons, neutrons and other single photon charged particles always have a magnetic field associated with them. That differs from many materials such as iron that exhibit magnetic properties when placed in a magnetic field.

No attempt has been made to calculate the strength of the magnetic field generated by the rotating electric charge. All that is said at this stage is that the combination of the two gives the Bohr magneton of electric charge multiplied by radius of rotation, namely $\frac{e\hbar}{2mc}$.

2.7. Positive and Negative Electric Charge

The requirement that a particle such as an electron must have an anti-particle of equal properties, but opposite charge, first came from the work of Dirac (1928). His prediction was confirmed a few years later when a particle with the same mass as an electron was discovered to have opposite charge. Any model of the structure of matter must therefore also explain the origins of the two different forms of charge, negative and positive.

From electrodynamics, it is known that the magnetic field generated by a current flowing in one direction around a circle will produce a magnetic field with the north magnetic pole pointing in the direction given by the "right hand rule." If the direction of rotation of the electric current is reversed, the north magnetic pole will point in the other direction. It follows that, if the photon that is the electron were to rotate in the opposite direction with respect to the north magnetic pole, it would have an opposite electric charge. This can be deduced from figure 2.6 B. Rotate the photon in the opposite direction and the negative charge will be on the outside.

The polarity of the charge emitted by the rotating photon is due to the direction of its rotation with respect to the magnetic field, as illustrated in figure 2.8. In this figure, a dot represents the north magnetic pole pointing out of the page. A cross represents the north magnetic pole pointing into the page.

Figure 2.8 shows clockwise and anti-clockwise rotations, the only possibilities. Likewise, there are only two ways a photon can twist its electromagnetic field. Those four images represent the only possibilities. An anti-clockwise rotation with the north magnetic pole pointing out of the page, figure 2.8 A, will generate a positive charge. The same rotation with the north magnetic pole pointing into the page, figure 2.8 C, will result in a negative charge.

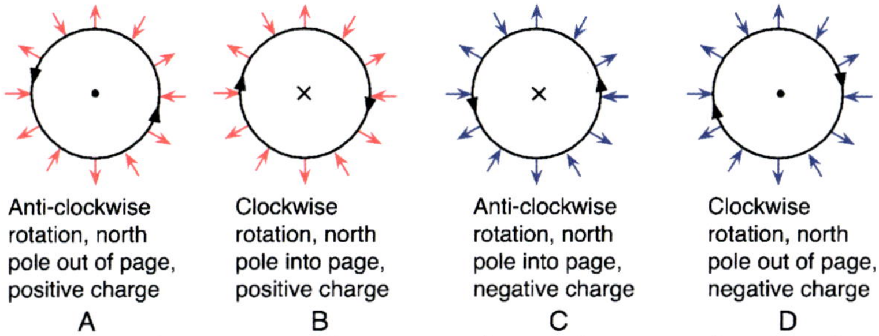

Figure 2.8: *Illustrations of the opposite charges that are produced by photons rotating in different directions with respect to their magnetic fields.*

A clockwise rotation with the north magnetic pole pointing into the page, figure 2.8 B, yields a positive charge. A clockwise rotation with the north pole pointing out of the page, figure 2.8 D, will yield a negative charge.

A positively charged particle is a mirror Image of its negatively charged anti-particle. The spin of a particle depends upon the direction from which it is observed. The direction of the photon's rotation with respect to its magnetic field determines its charge.

A property of a particle and its anti-particle is that, when they meet, they destroy each other, liberating vast amounts of energy in the process. In reality, a particle and its anti-particle are similar photons that are rotating in opposite directions with respect to their magnetic field. If a particle is rotating clockwise with respect to its north magnetic pole, its anti-particle is rotating anti-clockwise with respect to its north magnetic pole. The two will have opposite electric charge and will attract each other. As they come together, they align their magnetic fields to attract each other. When they meet, they unlock each other's angular momenta, and the resulting two linear photons travel away in opposite directions in order to conserve linear momentum.

Angular and linear momentum are both conserved because they start out rotating in opposite directions and finish up traveling linearly in opposite directions. A particle and its anti-particle meeting are simply the two unlocking each other's angular momenta, thus becoming two linear photons. As shown in equation 2.3, this yields Einstein's famous equation, $E = mc^2$, in which energy E represents the state of linear photons, and mass, m, represents the state of rotating photons.

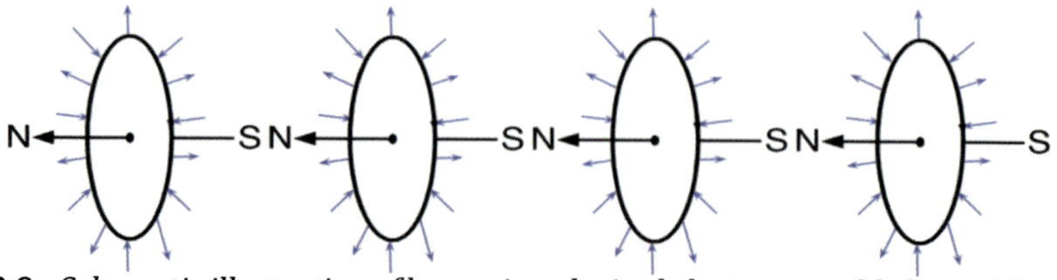

***2.9:** Schematic illustration of how spin polarised electrons could align without electric charge repulsion.*

When particles move, they do so with their magnetic field leading. If a beam of spin polarised particles were aligned, as shown in figure 2.9, they could travel in unison. Their magnetic fields would attract each other and maintain that line. They could not move too close together because their electric charge fields would repel each other. Figure 2.9 shows a situation where four electrons are entangled.

2.8. Spin, Chirality, and Parity

A particle, such as the electron, is a photon of the appropriate energy rotating in one direction with respect to its magnetic field. Its anti-particle, the positron, is the same photon rotating in the opposite direction with respect to its magnetic field. In this way, an anti-particle is a mirror image of a particle, as shown in figure 2.10.

Figure 2.10 A illustrates the trajectory of the rotating photon of a particle moving forward in time. Figure 2.10 B illustrates the same photon rotating in the opposite direction with respect to its magnetic field. Note that, apart from its charge, it is now indistinguishable from figure 2.10 B. Using a "thought experiment", imagine the photon in figure 2.10 A rotating in the opposite direction. This is equivalent to the particle traveling backward in time.

If the angular momentum of individual particles could be unlocked, vast amounts of energy could be released. This is not an easy task without having their anti-particles.

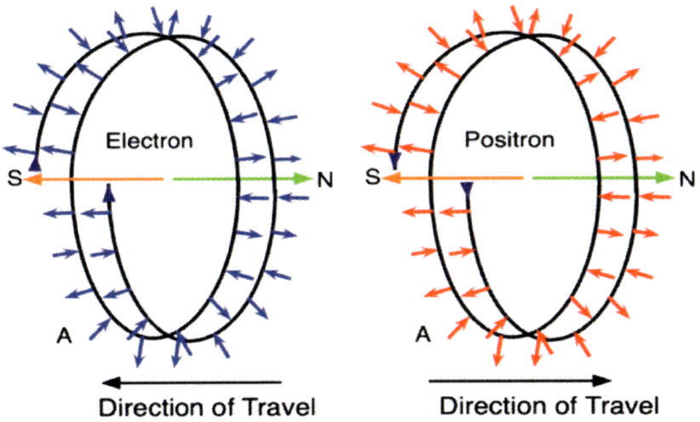

Figure 2.10: *Illustrations of the equivalence of an electron moving forward in time and a positron moving backward in time.*

The reasons for the masses of the electron and proton are not known to the author. Some mathematical reasons for their masses have been forwarded under standard models. They appear more as exercises in numerology than as physical explanations.

One certainty is that, if an appropriate linear photon is crumpled or compressed on collision with a nucleus, its magnetic fields will seek to form a closed loop by making the photon rotate twice within its wavelength.

The linear momentum of the photon is converted into angular momentum, generating two photons. Conservation of angular momentum requires them to rotate in opposite directions. When a closed-loop particle and anti-particle have been established, their resulting structures automatically give rise to opposite charges. This fulfills the requirements for the conservation of spin, parity, and chirality. Spin is conserved because the photons are rotating in opposite directions.

Chirality is illustrated in figure 2.11. A and C represent the left hand and right hand respectively. The thumb points in the direction of the north magnetic pole. The fingers represent the direction of the photon's rotation. Figures 2.11 A and C both show the north pole pointing up. Figure 2.11 B shows that a clockwise rotation generates a negative charge and has left-handed chirality.

Figure 2.11 D shows that an anti-clockwise rotation generates a positive charge and has right-handed chirality.

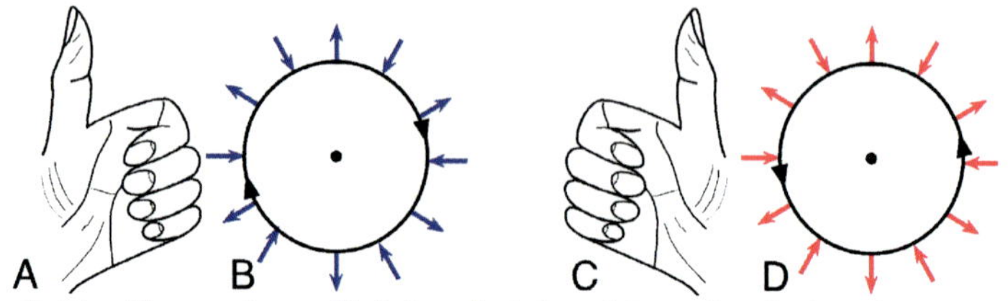

Figure 2.11: *Illustrations of left-handed, A and B, and right-handed, C and D, spin, and chirality, relative to the north magnetic pole pointing out of the page.*

In this model, the spin of a particle is the angular momentum of the rotating primary photon travelling in a circle and making two revolutions to complete one cycle, giving it an angular momentum of $\frac{1}{2}\hbar$. If the photon were traveling such that it made one revolution per cycle, it would have twice the angular momentum, namely $\hbar$. It could not be stable. Angular momentum, and hence spin, can only have one of two values, either +1 (x $\frac{1}{2}\hbar$) or –1 (x $\frac{1}{2}\hbar$). These can be designated as 'up' and 'down,' depending upon the nomenclature used. Spin cannot have another value because it can only spin one way or the other.

If a particle has a spin that is different from $\frac{1}{2}\hbar$, it can only be in units that are multiples of $\frac{1}{2}\hbar$. Spins are either added or subtracted when a combination of one or more individual particles' spins are added. Fractions of $\frac{1}{2}\hbar$ spins are not possible. There is no need to postulate the quantization of spin. It is an automatic property of all particles. No other spin value is possible. To date, no other spin values have been detected.

Using the examples in figures 2.8 and 2.11, the photon rotating in the direction given by the left-hand rule with respect to the north magnetic pole generates the negative charge found on the electron. The positive charge of the proton is generated by its photon rotating in the direction given by the "right hand rule" with respect to its north magnetic pole.

A particle such as an electron approaching two observers, as in figure 2.12, will be seen by both observers to be a spin up particle in zone A. As it passes observer 1, it becomes a spin down particle to observer 1 as shown at Ob 1.

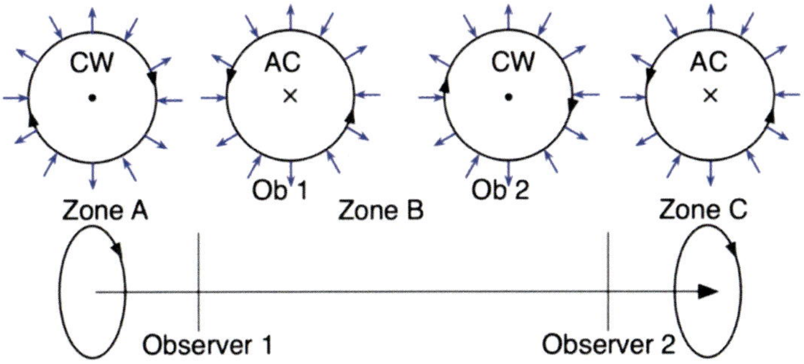

***Figure 2.12:** Schematic illustration of how the spin of a particle depends upon the direction from which it is observed.*

It remains a spin up particle to observer 2 as shown at Ob 2. When it passes observer 2 into zone C, it becomes a spin down particle to both observers. Although the electron is always spinning, it is not possible to determine in which direction it is spinning until it is measured. That has been interpreted as spin superposition.

There is one other property of particles that has troubled physicists. Their mathematics indicates that a particle with a spin half must make two turns to travel around once. This seems like an odd possibility. An evaluation of figure 2.1 shows the rotating photon making two revolutions for a particle to make one complete turn.

Under this rotating photon model, that is a property of all particles since the rotating photon must make two revolutions per wavelength to form the particle. It is difficult to conjecture any other structure that would be able to generate such a requirement.

2.9. Particle or Wave?

When he was working on his PhD thesis, de Broglie (1925) determined that, if a photon, which has a wavelength, also has momentum, a particle, which has momentum, must also have a wavelength. Known as the principle of duality, a particle can behave like a wave and a wave can behave like a particle. Scientists observed that the photon, which is an electromagnetic wave, behaves like a particle because it has fixed energy and momentum, as well as wavelength. That shows up in the photoelectric effect.

The energy of the particle is a combination of its rest mass, m_0, and its momentum, p, expressed as:

$$E^2 = p^2c^2 + m_0^2c^4 \tag{2.11}$$

In this rotating photon model, a particle is entirely an electromagnetic entity. At rest it has an energy given by $E = h\upsilon_0$, where υ_o is its rest frequency. When a particle moves at velocity, v, its energy E_v is incorporated into it by increasing its frequency. That can be expressed as

$$E_v^2 = m_v^2c^4 = (h\upsilon_v)^2 \tag{2.12}$$

Combining equations 2.3 and 2.11 gives

$$(h\upsilon_v)^2 = (h\upsilon_0)^2 + p^2c^2 \tag{2.13}$$

Equation 2.13 represents a version of Pythagoras' theorem and can be displayed as shown in figure 2.13. The arrow represents the direction of travel of the rotating photon. It shows the relationship between the energies and hence frequencies of the different components of equation 2.13. At velocity, v, its total energy is $h\upsilon_v$. Its rest energy is $h\upsilon_0$.

The kinetic energy component, KE, of its electromagnetic field, pc, is given by $Esin\theta = h\upsilon_v sin\theta = pc = h\upsilon_{KE}$, where υ_{KE} is the frequency component of the moving particle in its direction of travel. Substituting $\upsilon_{KE} = \frac{hc}{\lambda_{KE}}$ and simplifying yields $pc = \frac{hc}{\lambda_{KE}}$, where λ_{KE} is the wavelength of the rotating photon's electromagnetic field in its direction of travel. That gives:

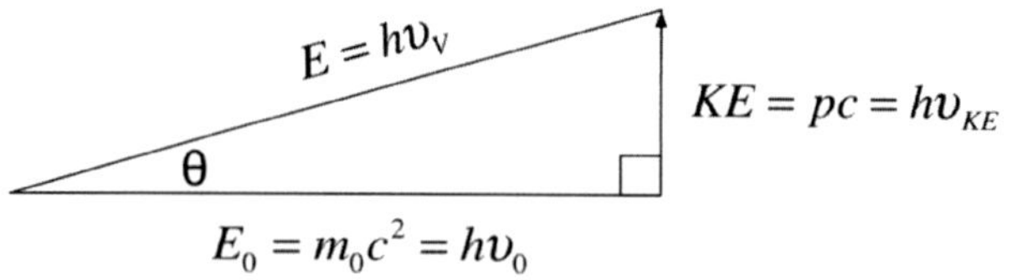

***Figure 2.13**: Illustration of the relationship between the frequencies of the rotating photon at rest and at velocity, v.*

$$\lambda_{KE} = \frac{h}{p} \qquad (2.14$$

Equation 2.14 is the expression for the de Broglie wavelength. It is simply the forward motion component of the rotating photons electromagnetic field oscillation. It is also the only part of a particle's electromagnetic wave that is available for interaction with the external world, charge interactions excluded. Experimental observations indicate that electrons behave as waves, according to de Broglie's prediction and the derivation of equation 2.14. It is perhaps evident in hindsight that equation 2.14 should be an automatic consequence of this rotating photon model.

This seems reasonable when one considers that everything is electromagnetic in nature. Every particle has a frequency associated with it, the frequency of its rotating photon. When the particle is moving, its mass, and hence its frequency, also increases. This increase in frequency introduces another wave that was not part of the particle at rest. The above mathematics shows that the new wave is exactly what de Broglie predicted.

The mathematics uses Pythagoras' theorem to show that the de Broglie wavelength is nothing more than the frequency of the extra energy given to the rotating photon in order to make it move. It is only available in the direction of motion of the particle. This frequency is always associated with a particle.

That mathematical solution does not show what physically happens. Figures 2.14 A and B show electrons, e^-, travelling to the right. 2.14 C schematically illustrates its high frequency electric field oscillations, $\approx 10^{20}$ Hz, are perpendicular to its travel direction.

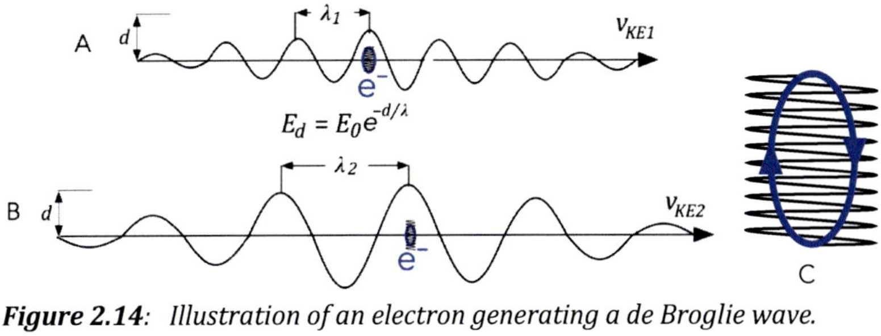

***Figure 2.14**: Illustration of an electron generating a de Broglie wave.*

High frequency electric oscillations generate a change in the electric permittivity of space (Feynman, 1963). That change surrounds it and is strongest at its position. Its influence is of limited lateral extent, terminating some distance either side of the electron.

That changing electric permittivity generates an electromagnetic wave. Like all electromagnetic waves, it travels at c. It starts out weak, builds up to a maximum at the electron's position and fades out some distance away.

2.14 A shows a faster moving electron than 2.14 B. The slower an electron moves, the greater its influence on the space surrounding it. That gives the slower moving electron the greater amplitude, and hence longer wavelength.

2.10. Photons Define Mass, Length, and Time

Some fundamental properties of the constituents of the universe are mass, length, and time. Since this model proposes that everything is composed of photons, photons should be used to define everything.

Chapter 1 pointed out that photons have mass because the electric permittivity and magnetic permeability properties of space offer some resistance to their motion through space.

Each photon has energy $E = hv$. This chapter pointed out that all individual subatomic particles were composed of a photon of the appropriate energy making two revolutions within its wavelength. The relationship between the rotating photon that is a particle and the linear photon from which it is formed is $E = mc^2$. Substituting for E gives the mass, m, of a particle as

$$m = \frac{hv}{c^2} = hv\varepsilon_0\mu_0 \qquad (2.15$$

The mass of an individual subatomic particle is determined by the frequency of the founding photon and equation 2.15. The mass of a collection of particles is determined by the sum of the individual particles mass, less their binding energies.

Length and time are related. Length is the interval between different events that occur simultaneously at separate locations. Time is the interval between different events that occur at the same location. Both are related through a space–time continuum (Minkowski, 1908).

Each requires the other for an accurate determination to be made. A standard length can be defined as either the distance travelled by a photon in a given time, or by the distance travelled by a photon of known frequency after a given number of oscillations. In the same manner, a standard unit of time can be defined as either the time taken for photons to travel a given distance, or the time taken for a photon of known frequency to make a given number of oscillations.

The standards of length and time are defined by setting the speed of light, *c*, in vacuum as $c = \frac{L}{t}$ = 299,752,458 m/sec. That defines both length and time to an accuracy of ≈ 3 parts in 10^{10}. Greater accuracy can be determined by setting that number as accurate to any number of decimal places in order to define that there are no fractions of a meter in that definition. Defining time more accurately requires the selection of an oscillation source, for example an atomic clock, which has a stable frequency that can be calibrated against a defined standard.

As section 2.4 showed, length, time, and mass are all subject to the special relativity corrections with increased speed. These mass and time corrections apply to an observer considered at rest measuring, something that is moving. The travelling observer will measure local mass and time as being the same as they would be measured by them at rest. The moving observer would notice that the distance travelled to reach a pre-determined destination would be less than the distance measured before setting out. Chapters 8 and 9 will show how time and distance are also affected by the gravitational potential. Together, chapters 2, 8, and 9 show how all measurements are relative to the observer in the theories of relativity.

Equation 1.2 shows that photon properties of wavelength, frequency and speed depend upon the product of $\varepsilon_0\mu_0$. By convention, μ_0 is defined as $4\pi \times 10^{-7}$ Newton/Ampere^2. That means mass, length and time all vary with electric permittivity. Chapter 9 shows how mass alters the electric permittivity of space. That makes all those properties relative to mass. The dependence of length and time upon mass is approximated in Einstein's gravity corrections.

2.11. Summary

The above has given a physical description of the structure that all individual matter particles will have when composed of a rotating photon. Under this model, individual particles have the following properties:

1. They consist of a photon of the appropriate energy making two revolutions within each wavelength. This gives them a radius of $r = \frac{\lambda}{4\pi} = \frac{\hbar}{2mc}$
2. They have mass, m, given by $m = \frac{hv}{c^2}$.
3. They have angular momentum $I\omega$ = ½$\hbar$. It is now called spin. It is the mass, m, of the photon rotating at its radius, r, at the speed of light, c.
4. They will all be subject to the corrections of Einstein's special relativity theory, namely

$$E = mc^2 \tag{2.2}$$

$$m_v = \frac{m_0}{\sqrt{1-\frac{v^2}{c^2}}} \tag{2.7}$$

$$t_v = \frac{t_0}{\sqrt{1-\frac{v^2}{c^2}}} \tag{2.8}$$

$$l_v = l_0\sqrt{1 - \frac{v^2}{c^2}} \tag{2.9}$$

5. The radius of each individual particle will decrease with velocity according to the relationship

$$r_v = r_0\sqrt{1 - \frac{v^2}{c^2}} \tag{2.4}$$

6. Each rotating photon continually emits and absorbs secondary or virtual photons as it travels in a circle. The emission, together with the photon's electric field, E, generates its electric charge.

7. Electric charge is generated at the circumference of the photon's two-dimensional plane of rotation.
8. There is an opposing electric field inside the rotating photon.
9. The magnetic moment for each particle is given by its charge, e, multiplied by its radius, r, namely $\mu_B = \frac{e\hbar}{2mc}$, the Bohr magneton.
10. When particles move, they orient their axis of rotation so that they travel perpendicular to their plane of rotation, with the magnetic field leading. Their constituent rotating photons thus spiral through space.
11. Only two spin states are available because the photon can only rotate clockwise or anti-clockwise. Fractional spin is not possible.
12. The two possible states of spin are opposite side of the page images of one state. They depend upon the direction from which the particle is observed.
13. The charge on a particle depends upon the direction that the photon rotates with respect to its magnetic field.
14. When circularly-polarised photons rotate, one electric field always points in outwards, generating charged particles.
15. When plane-polarised photons rotate, their electric field direction alternates. With the photon always rotating in the same direction, each half of the rotation is oppositely charged, generating a particle with no net charge.
16. An anti-particle is a mirror image of its particle.
17. When a particle meets its anti-particle, the two unlock each other's angular momentum. The two rotating photons become linear photons and travel away in opposite directions, thus conserving angular and linear momentum. Charge is automatically conserved.
18. When a sufficiently high-energy linear photon is slowed down, it will split into two rotating photons, becoming a particle and its anti-particle, thus conserving charge and angular momentum. This occurs when the photon is near the high field region of a nucleus.

19. When a particle moves, its mass increases, its charge and spin remain the same, and its magnetic moment decreases.
20. The chirality of the particle depends upon the direction of rotation of the photon with respect to its magnetic field. Positively charged particles have opposite chirality to negatively charged particles.
21. Every particle has an electromagnetic wave associated with its motion, it being equal to the de Broglie wavelength.
22. Except for exceedingly short-lived particles where the rotating photon does not complete a single wavelength, there will be no isolated non-integer charges.
23. Particle properties like electric charge, magnetic moment, angular momentum or spin, wave particle duality, and special relativity corrections exist as consequences of the structure of particles. They are not properties that are imposed upon them.

The presence of the photon as outlined above suggests that, over a very short distance, an energetic photon is indistinguishable from a particle. There is considerable evidence to suggest this has been observed (Erdmann, 1997).

Before proceeding, there is one aspect of the interaction between particles and photons that needs to be mentioned — Heisenberg's (1927) uncertainty principle. It states that it is not possible to determine the position, Δx, and momentum, Δp, of anything to an accuracy of better than Planck's (1900) constant. For photons it is expressed as

$$\Delta p.\Delta x \geq h \qquad (2.16$$

At the subatomic level, the only way to measure the position or momentum of a particle is to have it interact with a photon or another particle. As soon as that photon or other particle contacts the first particle, it interacts with it. This interaction will change either its position or its momentum, i.e., its velocity. This is a physical limitation because there are no other means to probe a subatomic particle.

There is no physical way around this limitation. More accurate results can sometimes be obtained when many measurements are made, and the results averaged. But for a single measurement, there is no way around that limitation. It is a limitation due to available measurement techniques. It is also the physical limit. The position of a photon is spread out over its wavelength, λ. A particle is spread out over $\frac{\lambda}{4\pi}$. That makes the uncertainty associated with a particle smaller than that associated with a photon. For particles it is expressed as

$$\Delta p . \Delta x \geq \frac{h}{4\pi} \qquad (2.16$$

There is no way that the uncertainty principle can cause anything to happen. It does not prevent consideration of features like the structure of an electron because its dimensions are less than the uncertainty principle allows. This does not mean that a rotation that exists in dimensions smaller than is allowed under the uncertainty principle in equation 2.16 cannot be determined. It can, and as has been shown in the above derivations. A photon making two revolutions within its wavelength is the origin of a particle's angular momentum, $\frac{1}{2}\hbar$, its spin, as well as the interchange between mass and energy through $E = mc^2$.

The above has described the properties that all particles will share as a result of their rotating photon structure. These include their mass, spin, charge, magnetic moment, and a wavelength that allows them to behave like waves under certain circumstances while remaining a discrete particle. The special relative corrections, equations 2.2, 2.4, 2.7, 2.8, and 2.9 are automatic consequences of the rotating photon structure of particles. Whenever they move, their motion will automatically be subjected to the special relativity corrections.

The proof of the correctness or otherwise of this model lies in its ability to match the known properties of matter and to predict new and unknown properties. Experiment is the only true arbiter of a theoretical prediction, and not merely the elegance of a theory. This presentation and its predictions rise and fall upon experimental evidence.

With these properties being predicted in all particles, it is now appropriate that we investigate how closely this model matches the properties of the real particles, namely the electron, proton, neutron, and neutrino, and how they incorporate these features into their structures and properties.

2.12. List of References

Bender D et al., (1984); "Tests of Mass at 29 GeV Centre-of-mass Energy", Physical Review, **D30**, pp 515 – 527

Bohr N, (1913); "On the Constitution of Atoms and Molecules", The London, Edinburgh, and Dublin Philosophical Magazine and Journal of Science, **26**, pp 151-189

Bourgoin RC, (2008); "On the Forming of Photon Toroids", Advances in Studies of Theoretical Physics, **2**, pp 215 – 227

Compton AH, (1923); "A Quantum Theory of the Scattering of X-rays by Light Elements", Physical Review, **21**, pp 483 – 502

de Broglie L, (1925); "PhD Thesis", published in Ann. Phys. (Paris), **111**, p 22

Dirac PAM, (1928); "The Quantum Theory of the Electron", Proceedings of the Royal Society of London, **A117**, pp 148 – 169

Erdmann M, (1997); "The Partonic Structure of the Photons: Photoproduction at the Lepton-Proton Collider HERA", Springer-Verlag, Berlin

Einstein A, (1905); "Zur Elektrodynamic bewegter Körper", Annalen der Physik, **17**, pp 891-921

Einstein A, (1952); "On the Electrodynamics of Moving Bodies", in "The Principle of Relativity", Einstein A et al., Dover Publications, New York, pp 35-65

Feynman RP, (1963); "The Origin of the Refractive Index": The Feynman Lectures on Physics, Vol 1. Ch 31

Feynman RP, (1985): "QED The Strange Theory of Light and Matter", Princeton University Press, Princeton, NJ, Ch 1

Heisenberg W, (1927); "Über den anschaulichen Inhalt der quantentheoretischen Kinematik und Mechanik", Zeitschrift für Physik **43**, pp 172–198,

Lorentz HA, (1904); "Electromagnetic Phenomena in a System Moving with any Velocity Smaller than that of Light", Proceedings Koninklijke Akademie van Wetenschappen, **6**, pp 809-831

Planck M, (1900); "Auf das Gesetz der Energieverteilung im Normalspektrum" Annalen der Physik, **1**, pp 719 – 736

Minkowski H, (1908); "Raum und Zeit," Jahreschericht der Deutschen Mathematiker-Vereinigung, **18**, pp. 75-88. (Address delivered at the

80th Assembly of German Natural Scientists and Physicians, Cologne, September.)
Qui-Hong H, (2007); "The Nature of the Electron" http://arxiv.org/pdf/physics/0512265v1.pdf
Robinson VNE, (2011); "A Proposal for the Structure and Properties of the Electron", http://dx.doi.org/10.4137/PPI.S7102
Sunyaev RA and Zel'dovich B, (1980); "Microwave Background Radiation as a Probe of the Contemporary Structure and History of the Universe," Annual Review of Astronomy and Astrophysics **18**, pp 537 – 560
Uhlenbeck GE and Goudsmit SA, (1925); "Spinn Elektronen und die Struktur der Spektren" Naturwissenschaften, **13**, pp 953-954
Uhlenbeck GE and Goudsmit SA, (1926); "Spinning Electrons and the Structure of the Spectrum", Nature, **117**, pp 264 – 265
Van der Mark MB and Williamson JG, (2021); "Relativistic Inversion, Invariance and Inter-Action", Symmetry **2021**, 13, 1117 https://www.mdpi.com/2073-8994/13/7/1117
Williamson JG and van der Mark MB, (1997); "Is the electron a photon with toroidal topology?" Annales de la Foundation Louis de Broglie, **22**, pp 133 – 178
Williamson JG, (2014); "A New Theory of Light and Matter", Proceedings of Science, Frontiers of Fundamental Physics, **14**, Marseille, July 15-18, pp 1 – 8
Williamson JG, (2019a); QV0007: Quicycle, Troon: https://quicycle.com/video/qv0007-john-g-williamson-unifying-the-electrons-particle-wave-natures/
Williamson JG, (2019b); "A new Linear Theory of Light and Matter" J. Phys.: Conf. Ser. **1251** 012050 https://iopscience.iop.org/article/10.1088/1742-6596/1251/1/012050/meta
Yang, L-P et al., (2020); "Quantum Spin Operator of the Photon" arXiv:2004.03771

3. Real Particles

Abstract:

The rotating photon model is applied to real particles — protons, neutrons, electrons, and neutrinos. It shows how their structure under this model gives them their various properties. It lays the foundations for all the matter known to exist in the observable universe.

3.1. Introduction

Chapter 1 described how empty space supports photons. Chapter 2 showed some of the properties that particles would have if they were all made of rotating photons. It is now necessary to show that this rotating photon model can describe the properties of real particles. In particular, it needs to match the known properties of protons, neutrons, electrons, and neutrinos, as well as seeing if it can predict previously unknown properties that can be tested experimentally. The correctness or otherwise of this model will be determined by the ability of its predictions to be verified.

3.2. The Electron

The first subatomic particle to be positively identified was the electron. Cambridge physicist, Professor JJ Thomson (1897a, b), succeeded in showing that it was a particle that had both charge and mass. Its discovery ushered in a new age for technology and the world. As scientists discovered new aspects of electrons, they realised it controlled most of the phenomena we see in the world today, gravity and nuclear properties excluded. They also found it was an enigmatic particle (MacGregor, 1992) that was difficult to describe. Standard model physicists have never proposed an adequate structure for the electron. Reasons for this include the different properties it exhibits under different circumstances. At times it behaves like a particle, at other times like a wave. Sometimes it behaves like a point particle, yet it possesses a property once described as angular momentum, which is not possible for a point particle. It has a charge of exactly the same magnitude and opposite polarity to that of a proton, even though the electron is almost two thousand times less massive.

The electron's charge does not increase when its mass increases with increasing velocity. It has a magnetic moment associated with it. The electron has been so difficult to understand that theoreticians developed quantum mechanics in order to try to explain it using wave equations. Some physicists found the electron's behaviour so "bizarre" that they asked if it really was a particle (Margeneau, 1961). Attempts at describing a structure have been reviewed by Barut (1991) and MacGregor (1992). These days most theoreticians use electron wave equations when calculating expected results. Physically it is described as a point particle with properties attached mathematically.

It is now over one hundred years since the detection and identification of electrons as particles. Although they have been well studied, very little is known about their structure. In 1990 at Antigonish, Nova Scotia, a workshop was held to try to work out the properties of and a model for the electron (Hestenes and Weingartshoffer (1991). Cambridge University hosted a conference on the electron in 1997, marking the one-hundredth anniversary of its discovery by Thomson (Kirkland and Brown, 1998). Candidates for electron structure included:

- Bohm's "Amoeba" model (Bohm, 1951; Jaynes, 1991), in which the electron can change its shape to accommodate different dimensions, depending upon its circumstances.
- Lorentz' harmonic oscillator model (Lorentz, 1952).
- The spinning sphere model and the "point particle" model, even though it sometimes behaves as if it had a dimension similar to the Compton radius (Compton, 1923).

Despite those and many other attempts, the electron's structure remained elusive. The behaviour of electrons is often predicted from many models, such as quantum mechanics (Dirac, 1928, 1938; Schrödinger, 1930, 1931), quantum electrodynamics (QED) (Jost, 1972;

Schwebber, 1994; Feynman, 1995), and spectroscopic models (MacGregor, 1992). In view of the success of the predictions from these models, some scientists doubted the need for a structure to explain the electron. There are still many unanswered questions. What is the origin of charge? What are its dimensions? How does its magnetic moment arise? What gives an electron its mass? What gives an electron its spin? Is the electron a particle always, sometimes, or never? Is it a wave always, sometimes, or never?

Before attempting an explanation, we should consider how scientists currently describe electrons. They give them a wave equation, usually following the work of Dirac or Schrödinger. Wave equations have many variations. It is a mathematical expression that extends over great distances, with a probability of finding the point particle electron somewhere within that distance. Whatever property of the electron they wish to consider, theoreticians mathematically attach it into the wave equation.

That enables them to calculate the expected outcome from a particular situation. When it matches experimental observation, they are pleased with the ability of their calculations to match experiment. Very rarely have they predicted a property before it was observed. One of the more renowned successes of this approach was when Dirac (1928) obtained an answer that had a + and a – in it. That enabled him to predict that a minus answer meant an anti-particle. Positrons were discovered four years later (Anderson, 1932, 1933).

Compton (1918) made the first attempt to give a shape and size to electrons. He deduced that it was not possible to consider the electron to be a point charge. He indicated that some of its properties could be explained if it was considered to be a "*…ring of electricity subject to rotation*."

More recently, it has been proposed that electrons are composed of toroids of electromagnetic energy (Williamson and van der Mark, 1997; Qui-Hong, 2007). That is, they are the rotating photons (Robinson, 2011) discussed in the previous chapter. Some of the implications of this structure are presented below.

Electrons have been extensively studied. Their properties are well known. Not so their structure. This presentation shows how the rotating photon model give electrons their properties.

<u>Electron Properties</u>: (Nakamura et al., 2013)

Mass (m_e):	$9.10938291 \times 10^{-31}$ kg or 0.51099891 MeV/c^2
Charge I:	$-1.60217648 \times 10^{-19}$ Coulomb
Spin (J):	$\frac{1}{2}\hbar$
Magnetic moment (μ):	$1.001159652180 \frac{e\hbar}{2mc}$

Under the rotating photon model, its mass gives the electron a frequency of 1.235598×10^{20} Hz and a wavelength of $24.26... \times 10^{-13}$ m. That frequency explains the electron's zitterbewegung (zbw) (Hestenes, 1990; Hestenes and Weingartshofer, 1991). Its frequency means that the electron seems to "shiver" at over 1.2×10^{20} times per second, when it is at rest. Under this model electrons have a radius of $1.93079... \times 10^{-13}$ m at rest.

Experiment is reality. Experimentalists have measured the diameter of electrons by accelerating them to high energies and scattering them against different targets. They found that high energy electrons scatter like point particles (Bender et al., 1984). This is where we need to look at the 'fine print.' The prediction of 1.93×10^{-13} m applies for electrons at rest. As the previous chapter stated, the radius of an electron decreases as its velocity increases. Accelerating electrons to 29 GeV increases their rest mass by over 50,000 times, giving them velocities very close to c, and according to equation 2.4, reducing their radius to $\approx 10^{-17}$ m. Particle accelerators working at many TeV, reduced the electron's radius to less than 10^{-18} m.

The reduction in radius predicted by equation 2.4 means that the rotating photon model can match the electron's measured diameter. Equation 2.4 suggests that it would be easy to use a suitable electron scattering means to measure the change in radius at lower velocity, making it easy to further verify this rotating photon model. Care needs to be taken when designing such an experiment to consider the influence of the electron's charge, which at low velocity would distort attempts to measure its physical scattering dimensions.

The electron's spin is described in the rotating photon model. Its mass predicts its dimensions, which have not been confirmed by measurement at energies less than 29 GeV. The elastic scattering work of Hofstadter (1956, 1962) suggests that 200 MeV electrons have scattering cross sections close to 10^{-16} m, also in agreement with equation 2.4.

Figure 3.1 shows a schematic "image" of an electron under this model. Figure 3.1 A illustrates an oblique view of the rotating photon and its associated electric and magnetic fields. Figure 3.1 B shows a side view of the electron.

The secondary or virtual photons, shown in figure 3.1 A as blue arrows, originate from the large "centripetal force" on the electromagnetic field that is the rotating photon. They are "thrown off" by that centripetal force. They are virtual photons because they will be reabsorbed to conserve mass if they aren't absorbed by another particle.

The centripetal force is in the plane of the rotating photon's axis. That is where the virtual photons are emitted. It means the electric charge of the particle is generated at the rotating photon's circumference and in its plane of rotation. Figure 3.1 B shows the charge field.

As mentioned with figure 2.7, electric charge, the emission and absorption of virtual photons, originates in the two dimensional circumferences of the rotating photon's axis. Its field spreads out both vertically (axially) and horizontally (equatorially) from its rotating circumference. The virtual photons become "real" photons when they are influenced by another particle. They generate the electric field.

In figure 3.1, the magnetic field is illustrated by the (green) magnetic lines of force. The electron will only move in the direction of its axis of rotation, i.e., perpendicular to the photon's plane of rotation. Figures 2.7 to 2.11 showed the rotating photon emitting and absorbing virtual photons.

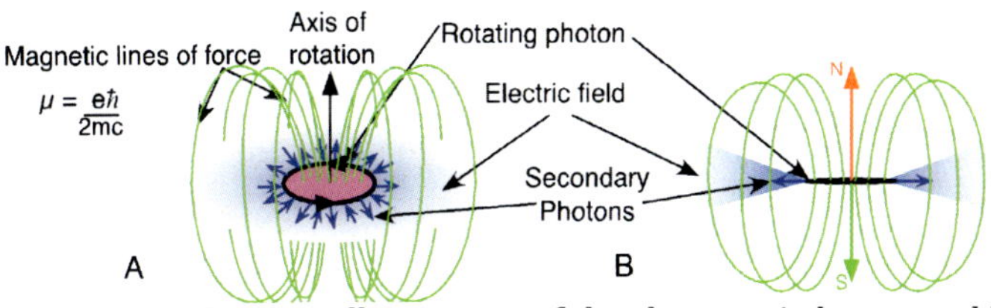

Figure 3.1: *Schematic illustrations of the electron, A shows an oblique view while B shows a sectional view through the centre.*

According to QED theory, the virtual photons are responsible for electric charge. The photon's spiralling motion means that its charge also spirals through space as it moves.

Possible experimental evidence for an electron having its charge distributed in two dimensions comes from what is known as helical Mott scattering. In normal Mott scattering, particles are scattered as if the charge is uniformly distributed around a spherical body. Helical Mott scattering implies the charge is spiralling through space. MacGregor (1992) indicated that possible evidence of this scattering was discovered with electrons in the 10 keV range. If further verified, it would add more support for this rotating photon model of matter. MacGregor also mentioned that the helical scattering effect would only occur over a small range of keV energies. For the rest of the energies, scattering would be coulombic, i.e., point like.

The rotating photon model for the structure of matter gives all particles the properties of mass, spin, electric charge, magnetic moment, de Broglie wavelength, dimensions, chirality, and parity. They apply to the electron just as they apply to all other particles. The largest difference between the predictions of this rotating photon model and experimental measurement comes from the electron's magnetic moment. This model suggests that the electron's magnetic moment, μ_B, is equal to its electric charge multiplied by its radius, namely $\mu_B = e\hbar/2m_e c$, the Bohr magneton. Experimentalists have measured the electron's magnetic moment to be 1.001159652180... μ_B, a difference of just over 0.1%. Using quantum electrodynamics (QED), theorists have calculated this to a high degree of accuracy, giving it the value 0.001159652180... μ_B. Otherwise the electron's known properties are matched by this rotating photon model.

This model also predicts several properties of the electron that are either unknown, or known but not recognized, including but not limited to:

i. Its spin of ½$\hbar$ is angular momentum. It represents the whole mass of the electron traveling in a circle of radius $\hbar/2mc$ at the speed of light. That makes it a tremendously powerful gyroscope. Like all gyroscopes, it cannot change its direction unless it has another particle with which to exchange angular momentum. This explains why electrons take every opportunity to form pairs, the driving force behind chemical reactions.

ii. Its diameter decreases with velocity according to equation 2.4. This change should be detectable.

iii. Electrons, and indeed all particles, move in the direction of their axis of rotation, with their magnetic field leading their electric field.

iv. Its electric charge is generated at its circumference, not at a point or spread throughout a sphere.

v. As an electron moves, its charge spirals through space, as indicated in figure 3.2.

This model also describes a moving electron as having a wavelength that matches the de Broglie (1925) wavelength.

Figures 3.2 A and B show the approximate paths taken by the rotating photons at keV and MeV energies respectively. Their charge is produced at the axis of travel of the rotating photon. At GeV energies and above, an electron's charge trajectory would be a straight line on this scale. The experimental results of Bender, et. al. (1984) using 29 GeV electrons show that the radius of the spiral was about 10^{-17} m. Results at TeV energies put that radius at less than 10^{-18} m. The latter is why the electron is regarded as a point particle. The reality is that an electron's shape changes with its speed, according to equation 2.4.

With so many apparently contradictory properties, the electron has long been known as an enigmatic particle. Another problem arose when physicists accelerated electrons and their mass increased, as predicted by Einstein's special theory of relativity. It was assumed that their charge would also increase because it was a property of the electron and associated with its mass.

Figure 3.2; *Schematic illustrations of the electron's charge spiralling through space. A and B show it moving at keV and MeV energies respectively.*

Physicists were perplexed when its charge did not increase. They observed that the electron's spin also did not increase when its mass increased, enabling them to postulate a relationship between spin and charge. Despite decades of research, there has been no acceptable theory linking charge and spin. It is suggested that this rotating photon model illustrates the relationship. It shows that, as electrons are accelerated to higher energies, their mass will increase, their charge and spin will remain the same, and their magnetic moment will decrease.

Under this rotating photon model, an electron's rotation axis can be described in Cartesian co-ordinates by

$$x = \hbar \frac{\sqrt{1-\frac{v^2}{c^2}}}{2m_0c}.cos\theta$$

$$y = \hbar \frac{\sqrt{1-\frac{v^2}{c^2}}}{2m_0c}.\ sin\theta \qquad (3.1$$

$$z = z_0 + \frac{vt}{\sqrt{1-\frac{v^2}{c^2}}}$$

with the provision that θ extends from 0 to 4π and t extends from 0 to $v^{-1}.\sqrt{1-\frac{v^2}{c^2}}$. Equation 3.1 shows that, at rest, the rotating photon is a hoop of radius $\frac{\hbar}{2m_ec}$, located at position, z_0. As its velocity increases, the rotating photon spirals its way through space, giving rise to its relativistic corrections of mass, length, and time. That the electron's magnetic moment is 1.001159652180 x $\frac{e\hbar}{2m_ec}$ is explained in QED and does not significantly affect the electron's dimensions. The effect of its electric charge and magnetic moment, however, extend well beyond the dimensions given in equations 3.1.

As given in equation 2.14 and figures 2.13 and 2.14 of section 2.9, an electron's motion automatically generates its de Broglie wave. An important feature of an electron's de Broglie wave is that it influences the electric and magnetic fields well beyond its position. With a de

Broglie wavelength given by $\lambda_{dB} = \frac{h}{p}$, an electron's small mass means that slow moving electrons, of a few eV, can have a wavelength of the order of 1 nm. This means that the electron influences space around it for dimensions of the order of 10^{-9} m. Is it any wonder that physicists have had difficulty understanding electrons when they display dimensions somewhere between 10^{-9} m to 10^{-18} m? This presentation has shown why the one particle can exhibit dimensions over such a large range.

Another aspect of an electron's de Broglie wave is its ability to generate diffraction and interference patterns. Diffraction is just like photon diffraction. The closer the wave's centre is to the diffracting edge, the greater will be the angle of diffraction. Electrons are also diffracted by regular crystalline surfaces. Their diffraction patterns are also seen in electron microscopes.

Quoting a comment made about Einstein's gravity: *"Mass distorts space-time. Space time tells mass how to move."* The same principle applies at the individual particle level of the universe. *"Moving particles generate waves. Waves tell particles how to move."* The Davisson and Germer and GP Thomson measurements proved conclusively that electrons behave like waves.

Interference is a different situation. The electron is a particle and can only travel through one slit. Figure 3.3 A illustrates an electron, complete with its de Broglie wave, approaching a double slit. B shows the wave passing through the double slit, in the normal manner associated with photon waves. Note the positive interference pattern.

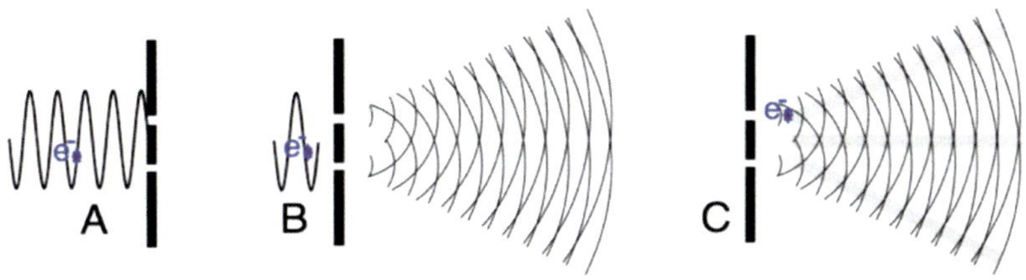

Figure 3.3: *Illustrations of a moving electron, complete with its de Broglie wave, approaching a double slit, A, with the wave passing through both slits, B, and with the electron passing through one slit, C.*

Electrons are particle and can only pass through one of the slits. In C it is shown passing through the upper slit. The reinforcement of the positive interference guides the electrons to those regions. That generates the detected interference patterns.

Consider what happens when an attempt is made to monitor one slit to find out which slit the electron passes through. For that purpose, you need to put a detector on one slit. Monitoring the slit involves placing a voltage, or voltage inducing, means on one slit. That act alters the nature of the wave that emerges from that slit. It does not matter whether the electron or its de Broglie wave passes through the slit. The presence of the detector alters the wave function. Figures 3.4 D and E show that different waves do not produce an interference pattern.

It does not matter whether the electron does or does not pass through the monitored slit. The act of doing something different, e.g., applying a voltage, an individual electron current sensor, blocking one slit, or similar, alters the phase of the portion of the wave that passes through that slit. The detector alters the wave function, preventing a regular interference pattern from being observed.

As a particle, the electron does not pass through both slits. Its de Broglie wave does! When one slit is altered, the interference pattern is not regular. No interference pattern will be generated.

Electron tunnelling is an intriguing issue, It is considered to be a peculiar quantum mechanical effect. Figure 3.5 A shows the typical quantum interpretation of electron tunnelling. An electron approaches a barrier with insufficient energy to overcome it. Under the correct circumstances, an electron emerges on the other side. That is typical quantum tunnelling.

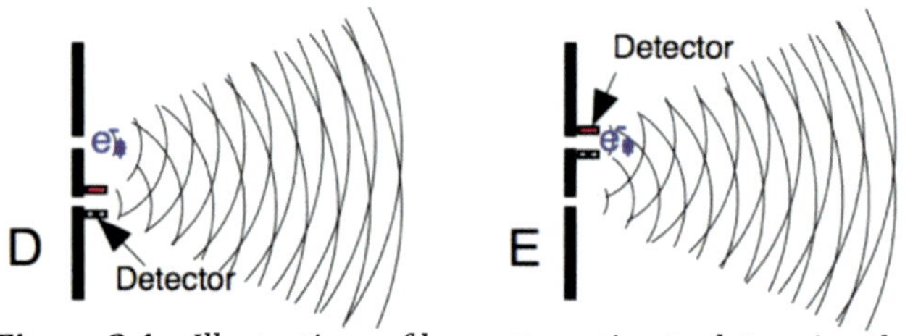

Figure 3.4: *Illustrations of how attempting to determine through which slit the electron passes alters the wave function through that slit.*

Figures 3.5 B to D show classical energy transfer, often performed as a billiard ball "party trick". B shows a ball approaching an alignment of balls. At C it strikes the row. Energy of the first ball is transferred through the middle balls to the final ball. D shows it is ejected with a decreased energy due to imperfect elasticity.

Figure 3.5 E shows the billiard balls replaced by electrons. The first electron approaches the barrier, which contains electrons. Some of its energy is transferred through the barrier. There will be losses because the barrier is not as "elastic" as billiard balls. If the transferred energy reaches the other side of the barrier with sufficient energy to overcome the work function, WF, between the two surfaces, an electron will emerge from the other side.

All barriers contain electrons! Is it electron tunnelling? Or is it energy transfer? One difference is electron tunnelling depends upon its de Broglie wavelength. Without a correction factor, the lower is an electron's energy, the greater the probability that part of its wavelength would appear on the other side of the barrier. With energy transfer, the greater is the electron's energy, the greater is the chance of knocking an electron inside the barrier past the work function.

The electron's anti particle, the positron, is a mirror image of the electron. The same photon rotates in the opposite direction with respect to its magnetic field. Apart from its opposite charge, it looks the same as

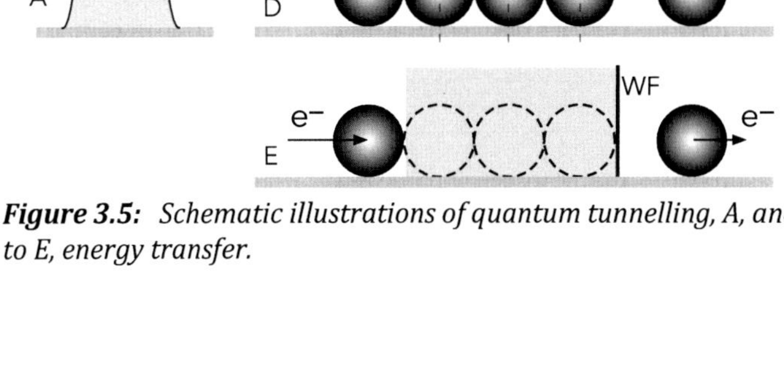

***Figure 3.5:** Schematic illustrations of quantum tunnelling, A, and classical, B to E, energy transfer.*

an electron. It has the same mass, spin, charge, and magnetic moment, as suggested in this model.

Further, when an electron and positron come close to each other, they are attracted by their opposite electric charges, move together, and unlock each other's angular momentum. The two rotating circularly-polarised photons now become linear-travelling circularly-polarised photons, and they move away in opposite directions in order to conserve momentum. Chapter 7 presents some of the ways in which an electron's structure gives it some unusual properties when it is combined with nuclei to form atoms.

Table 3.1 summarizes some of the differences between treating electrons as particles versus waves. It gives a strong indication that this rotating photon model can be used to determine electron properties in a

Treating Electrons as Particles or Waves	
Rotating Photon Particle Theory	**Wave Theory**
Electrons are detected and behave like particles.	Electrons diffract and interfere like waves.
This model shows how their structure gives them their physical properties, including their de Broglie wavelength and special relativity corrections. These result in electrons responding as if their dimensions vary between ≈ 10^{-9} m and 10^{-18} m.	Electrons are treated as point particles with a wave function defining their probability of location. Physical properties are attached mathematically. The extended limit of waves makes it easy to manipulate properties and wave equations to match observation.
All particle properties derive from their structure, which gives a physical description of the processes involved.	Waves are easily manipulated through Fourier transform and the like, and they give very little indication of the physical processes involved.

Table 3.1 *A comparison of some of the differences between treating electrons as particles or waves.*

classical manner. They are real particles that generate their de Broglie wavelength as they move. They will respond like waves when appropriate.

It becomes a matter of which approach one prefers, waves that can be manipulated to get almost any desired answer, or particles that have fixed properties and a realistic description of what is and what is not possible.

The advantage of particles is that there is only a limited number of ways they can be handled. Either they match observation, or they don't. That makes it much easier to visualize any effect of interest.

3.3. The Proton

Rutherford (1911) considered the possibility that the positive charge in the atom was concentrated in a central core. Soon thereafter Geiger and Marsden (1913) sent alpha particles through thin gold foil. Much to their surprise, they found that some of them bounced back. A comment made at the time was that: "*It was like firing shells at tissue paper and having some of them bounce back.*" Rutherford's idea was proven correct. The results could only be explained if the mass and positive charge of the atom were concentrated in a central core. The electron was known to have a very light mass compared to that of the atom. This meant that the atom's mass was concentrated with its positive charge in a central nucleus.

By sending alpha particles through nitrogen, Rutherford (1919) noticed that hydrogen ions were generated. He concluded that hydrogen ions were part of nitrogen. Further studies revealed that the positive charges were mixed with other unknown particles that accounted for about half the mass of nuclei. Those positive charges had approximately the same mass as the hydrogen atom and were called protons. Extensive studies were made of protons over the following decades and their properties are now well known.

Proton Properties: (Nakamura, 2013).

Mass (m_P):	1.674927… x 10^{-27} kg or 938.272046 MeV/c^2 (≈1836 times the electron mass)
Charge (e):	+1.60217648… x 10^{-19} Coulomb

same as electron, but opposite sign

Spin (J): ½$\hbar$

Magnetic moment (μ_{BP}): 2.793697… $\frac{e\hbar}{2m_P c}$

Root mean square radius (r_P): ≈0.87 fm

Under this model, a proton's mass gives its rotating photon a frequency of ≈ 2.268… x 10^{23} Hz and wavelength of ≈ 1.322… x 10^{-15} m. That gives the proton a theoretical radius of 0.1052 fm. Its two revolutions per wavelength mean that it has angular momentum ½$\hbar$. That is what nuclear physicists call its spin. The proton's positive charge means that its rotating photon is circularly-polarised, with an internal twist spin of $\hbar$. It also means the proton's photon is rotating in the opposite direction to the electron with respect to its magnetic field. Its mass is that of the rotating photon.

Under this model, the proton should have a radius of 0.1052 fm and a magnetic moment of $\frac{e\hbar}{2m_P c}$. At first glance, we seem to be out by a factor of ≈ 7 for the radius and just under 3 for the magnetic moment. The discovery that the proton has a magnetic moment of 2.79… $\frac{e\hbar}{2m_P c}$ led to the conclusion that it was not an elementary Dirac particle. It was considered to be composed of some other substances.

The electron's radius is approximately 1.93… x 10^{-13} m. The difference between a proton's predicted radius of ≈ 0.1 x 10^{-15} m and it measured radius of ≈ 0.8 x 10^{-15} m represents an error of about 1 in 200 compared to the electron's dimension. Similarly, the difference between the proton's predicted magnetic moment and its observed magnetic is about 1 in 600 compared to the electron's magnetic moment. From that perspective, the difference does not seem very large. This situation changes when we look at the experimentally measured charge distribution of the proton determined from electron scattering experiments at Stanford University, as shown in figure 3.6.

Curve 1 shows the expected charge distribution if all of the charge were generated at a radius of 0.105 fm, as predicted by the rotating photon model. There are practical limitations in measuring this charge distribution using elastically scattered MeV electrons. Because of their fields, charges interact at a distance. If all of the charge were concentrated

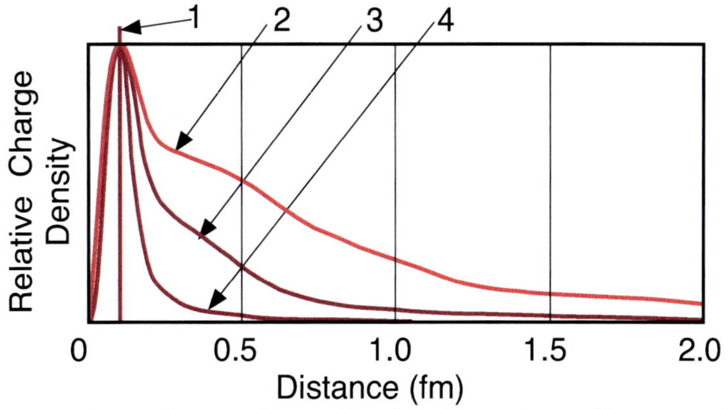

***Figure 3.6**: Theoretically predicted and experimentally measured charge densities of a proton.*

as shown in curve 1, it would be measured as having an exponential like decay with distance from the centre along a radius. This is illustrated in curve 4.

Curve 2 shows the actual, measured charge distribution of the proton. It is taken from the experimental results obtained by Hofstadter (1956) and Littauer et al. (1961 a, b). It is the integrated charge distribution, plotting all of the measured charge as if it were on one radius. Curve 3 shows the charge distribution in Curve 2 divided by the normalized radius for r > 0.105 fm. As the radius gets larger, the amount of charge on any one segment at that distance becomes smaller.

In the rotating photon model, all the charge would be at a radius of 0.105 fm, as in curve 1. The most interesting feature of the measured charge distribution in curve 2 is that, within experimental error, its maximum is also at 0.105 fm. That is not coincidence. It is accurate to ≈ 10^{-17} m, 0.01 fm. That is the very strong indication of this model's validity mentioned in § 2.5.

The other feature of note is that there is no measured charge at the centre of the proton. That can only occur if its charge is distributed in a planar or cylindrical manner. It also requires the axis of the proton's rotating photon be perpendicular to the incident electrons. Both of these predictions are made from the proton's mass and this model. They thus appear to be both accurate and significant.

The differences between prediction and observation are the differences between curves 3 and 4. Curve 4 shows what the measured radial charge distribution would be if all the charge were generated at the proton's theoretical radius of 0.105 fm. It allows for an exponential charge decay outside that radius.

Something has spread the charge out from curve 4 to curves 3 and 4. That something needs to be explained. Under this model, any explanation must be limited to the properties of the rotating photon, and this implies an underlying, non-sinusoidal frequency. Let us first look at the interaction of linear harmonics, by analogy to the proton, shown in figure 3.7.

In figure 3.7, line 1 represents the photon's travel axis. Curve 2 shows the proton's basic photon frequency, 2.2687… x 10^{23} Hz, and amplitude, X, plotted as a sine wave. Curve 3 shows a ⅓rd frequency harmonic of curve 2 with amplitude, Y. Curve 4 shows the sum of these two waves over one wavelength of the ⅓rd frequency harmonic.

The ⅓rd frequency harmonic was intentionally chosen. Figure 3.8 shows the effect of a regular, harmonic change to a sine wave. In figure 3.8, line 1 represents the base. Curve 2 represents a sine wave frequency moving through time. Curve 3 shows a regular interruption to curve 2. The peaks and troughs of the curve 2 sine wave are removed. Curve 4 shows a ⅓rd frequency harmonic oscillation.

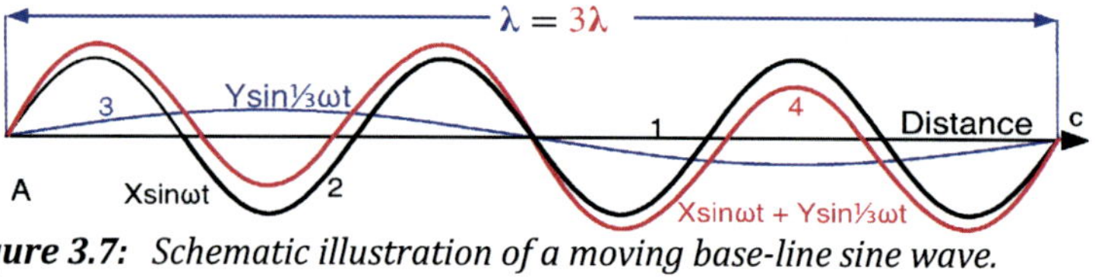

Figure 3.7: *Schematic illustration of a moving base-line sine wave.*

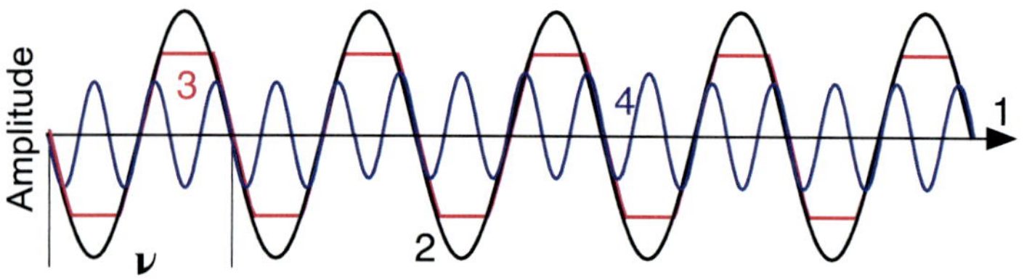

Figure 3.8: *Schematic illustration of how non-sinusoidal components of a wave lead to one-third harmonics.*

Whenever curve 3 has part of its peak or trough removed, curve 4 is moving in the opposite direction. The same representation applies to a frequency or wavelength travelling through time or space.

This serves to illustrate that adding a ⅓rd frequency harmonic provides a good first approximation to the regular interruption of a sine wave. In the same manner, a series of ⅓rd frequency harmonics (i.e. ⅓rd, 1/9th, 1/27th, etc.) give an accurate representation of a regular distortion to a sine wave.

This shows up in other aspects of electromagnetic radiation. The suppression of ⅓rd harmonics in electromagnetic transmitters can be a serious problem since they are so easily generated when the waveform is not purely sinusoidal (Langford-Smith, 1955). Although Fourier transforms can use other harmonics to obtain the original waveforms, ⅓rd frequency harmonics form significant components in many electromagnetic fields. That is why only they are considered in this study. ⅓rd wavelength harmonics require higher energy, and hence higher mass, properties the proton does not possess.

Figure 3.9 shows what happens when the linear wave, shown in figure 3.7, is converted to a rotating wave. 3.10 A represents the original wave from figure 3.7. Figure 3.9 B shows the two waves travelling in circular paths. The result is two rotations.

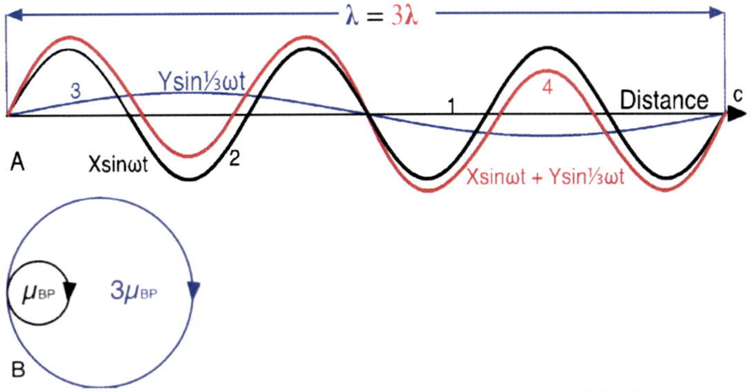

***Figure 3.9:** An illustration of the linear wave, 4, composed of the basic wave, 2, superimposed on the one third frequency harmonic, 3.*

Each exists for a fraction of time. The ⅓rd frequency harmonics radii will be larger than that of the original photon's radius.

The principles of the conservation of mass, charge, and angular momentum mean they cannot exist at the same time. Each exists for a fraction of time and change to the other as determined by conservation principles.

Figures 2.1 and 2.6 showed what happens when a sinusoidal wave goes from a linear trajectory to a circular trajectory. In a similar manner, figure 3.9 B shows what happens when a non-sinusoidal wave goes from a linear format to a rotating format. The sinusoidal wave, 2, forms the central core. Its ⅓rd frequency harmonic, 3, forms a circle with three times the core radius. If the rotating photon is circularly-polarised, both segments will have an electric charge. For protons, that charge is positive.

For the fraction of time each exists, they generate a magnetic field of strength equal to its charge multiplied by its radius, i.e., $\mu = e.r$. In that manner, the proton's charge remains the same while its magnetic moment increases because its radius increases.

The mass, m_{os}, of each oscillation, is determined by:

$m_{os} = h\upsilon_{os}/c^2$. With one third the frequency, the mass of the $3\mu_{BP}$ oscillation is one third that of the μ_{BP} oscillation.

The mathematics and physics required to establish the rotation was mentioned in section 2.3. Greater detail of the mathematics can be found in Williamson (2019b) and van der Mark and Williamson (2021).

As mentioned earlier, one ninth and one twenty seventh frequency harmonics are also possible. Collectively they are called the one third frequency harmonic series. They can occur in electromagnetic radiation. They show themselves in the rotating photon model by the existence of other oscillations of 3, 9, and 27 times that of the primary photon's wavelength and (therefore) radius. It is shown later that the predicted ⅓rd frequency harmonics in the series have been detected. They have the properties as predicted under this model. It is suggested that the matches with experiment, the number of known properties this model can match, along with the number of predictions it makes, are sufficient to demonstrate the accuracy of this assumption.

In a pure sine wave application, the rotating photon would travel in a circle of radius 0.1052 fm, making ≈ 4.554... x 10^{23} revolutions per second. This is herein called the basic or P/1 oscillation. As mentioned in chapter 2, the act of rotation causes the particle to emit and absorb secondary photons. Their emission and absorption will give the particle the same charge as that of the electric field on the outside of the rotation. For a proton, the rotating photon will be circularly-polarised and will have a positive electric field exposed to the outside world. It will therefore be positively charged.

As it rotates the existence of one-third frequency harmonics means that the photon is not always the P/1 cycle rotating about its central core. When the influence of a one-third harmonic dominates, the photon no longer traverses its basic oscillation P/1 oscillation. Instead, it rotates at a lower frequency over a larger radius.

Applying the principle of only a series of one-third harmonics being possible, means that the only curved trajectories in which the photons can now travel will be those with frequencies of ⅓rd, $^1/_9{}^{th}$, and $^1/_{27}{}^{th}$ of the fundamental proton's photon frequency. These have the same form as the ⅓rd frequency oscillation shown in figure 3.10. For visual clarity they are not illustrated. For the purposes of this discussion, they will be denoted P/3, P/9, and P/27 respectively.

Figure 3.10 illustrates the oscillations considered for the P/1 and P/3 harmonics. Figure 3.10 A shows the basic P/1 oscillation. It has a radius of ≈ 0.1052 fm and is shown in this case as rotating anti-clockwise, denoted as spin positive, with its north magnetic pole pointing out of the page.

Table 3.2 gives the values of these oscillations. It shows that as the harmonic increases, the radius and magnetic moment increases, spin and charge remain the same and mass decreases. That gives it right-handed chirality with respect to the direction of its north magnetic pole. It is therefore a positive particle. While it exists, it has mass m_P = 938.2723... MeV/c^2, charge +e, magnetic moment, $\mu_{BP} = \frac{e\hbar}{2m_P c}$, and spin ½$\hbar$.

Figure 3.10 B shows a possible combination of P/3 frequency harmonics when the P/1 oscillation generates them. Conservation of

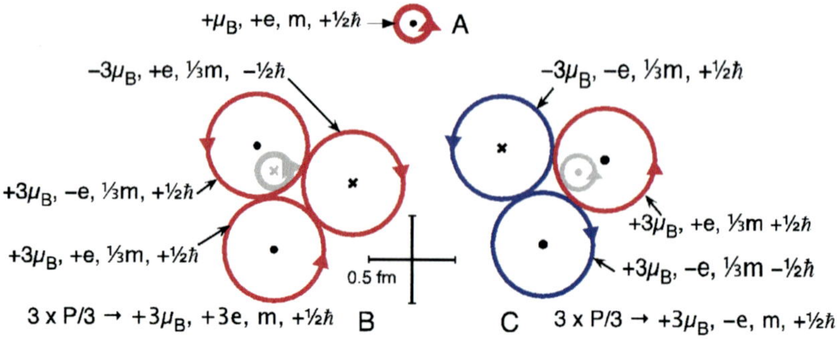

Figure 3.10: *Illustrations of the P and P/3 harmonic oscillations of the proton. A shows the core proton oscillation. B and C show a possible P/3 combination that conserves electric charge, mass, and spin.*

Oscillation	Radius (fm)	Magnetic Moment (μ_{BP})	Spin	Charge	Mass (MeV/c^2)
P/1	0.10515	1	½ħ	+e	938.272
P/3	0.31545	3	½ ħ	+e	312.904
P/9	0.95635	9	½ ħ	+e	104.313
P/27	2.86896	27	½ ħ	+e	34.767

Table 3.2: *Some properties of the different proton one-third frequency harmonics considered under this model.*

mass requires three P/3 oscillations to be generated. That combination conserves mass and spin and generates $+3\mu_{BP}$ and +3e electric charge. That is too much positive charge for that to form all oscillations.

3.10 C shows another combination that also conserves mass and spin, generates $+3\mu_{BP}$ charge. It generates −1e, negative charge. When the two are averaged, the result is that mass, charge, and spin are conserved, with the total magnetic moment now being $+3\mu_{BP}$. The combination of figure 3.10 B and C is referred to as a P/3* harmonic oscillation. It has the averaged properties of $+3\mu_{BP}$, +e, m, and +½ħ.

It can be linearly proportioned with the P/1 oscillation to get an answer that retains the proton's positive electric charge, mass, and spin, while increasing its magnetic moment from μ_{BP} to $3\mu_{BP}$.

In figure 3.10 and other representations of the harmonic oscillations, the chirality rule of figures 2.8 and 2.11 is used. The dot represents the north magnetic pole pointing out of the page. That gives an anti-clockwise rotating oscillation a positive electric charge. The cross represents the north pole pointing into the page, giving a clockwise oscillation a negative electric charge. Each closed loop generates a unit electric charge. There is no possibility of fractional charge being generated for longer than about 10^{-25} sec.

Adding the P/1 and P/3* oscillations in the proportion of 10.35% P/1 to 89.65% P/3* gives a proton a magnetic moment of $\approx 2.793\ \mu_{BP}$, while still retaining its mass, an electric charge of +e, and its spin of $+\frac{1}{2}\hbar$. It does not match the charge distribution given in curve 2 of figure 3.6. To do that it is necessary to introduce P/9 harmonic oscillations.

One such example is given in figure 3.11. It shows one of the positive P/3 oscillations replaced by two positive P/9 oscillations and one negative P/9 oscillation. It shows that this combination gives a structure with the proton's mass and positive charge. It now has a magnetic moment of $+9\mu_{BP}$ and a spin of $\frac{1}{2}\hbar$. That is referred to as a P/9* oscillation.

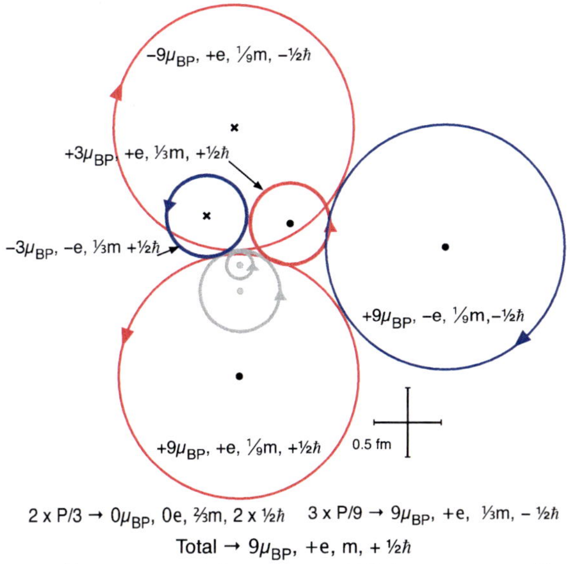

Figure 3.11: *Illustrations of three proton P/9 harmonic oscillations relative to its P and P/3 harmonic oscillations.*

As was the case with the P/3 oscillations, other combinations of P/9 oscillations from a positive P/3 oscillation can also give a P/9* oscillation that has an added average of $+9\mu_{BP}$ +e, m, and $+½\hbar$. They are not shown here because it is a simple exercise to demonstrate that possibility.

Table 3.3 lists the values of the P/1, P/3* and P/9* oscillations derived under these illustrations. It is now a matter of getting a combination of these three that matches both the proton's observed magnetic moment and its charge distribution.

Oscillation	Radius (fm)	Magnetic Moment (μ_{BP})	Electric Charge	Mass	Spin
P/1	0.1052	1	e	m_P	$½\hbar$
3xP/3 = P/3*	0.3156	3	e	m_P	$½\hbar$
2xP/3+3xP/9 = P/9*	0.9470	9	e	m_P	$½\hbar$

Table 3.3: *A listing of the properties of the series of one-third harmonic oscillations that will be used in the calculations of the magnetic moment and charge distribution of the proton.*

Any combination that matches these properties will automatically have the correct mass, electric charge, and spin. One such combination is given in table 3.4. They are needed to give a total value close to the proton's experimentally observed magnetic moment of 2.793697 μ_{BP}.

The results in table 3.4 represent good first approximations of values for the P/1, P/3*, and P/9* oscillations. Two thirds of the P/9* oscillation is made up of P/3 oscillations. That means that about 34% of the proton is composed of P/1 oscillation, about 63% is composed of the P/3 oscillations and about 3% is made up of the P/9 oscillations. There is a variable range of those values that will match the observed properties Note that, although mass, charge, and spin are conserved, magnetic moments are not conserved. They are a derived value given by the Bohr magneton, $\mu_{BP} = e.r = \frac{e\hbar}{2m_P c}$.

Oscillation	**Radius** (fm)	**Percent** †	**Magnetic Moment** (μ_{BP})
P/1	0.105	34.003	0.34003
P/3*	0.315	58.101	1.74303
P/9*	0.947	7.896	0.71064
Total:		100	2.7937

Table 3.4 *Percentages of the charge and hence mass of the rotating photon associated with the three oscillations indicated.*
† Note that other oscillations within a small range of P/9, typically 7% to 10% can also give the required magnetic moment, subject to changing the ratios of the P/1 and P/3* oscillations as well.*

The above study shows it is possible for this rotating photon model of matter to match the observed proton properties. An important feature of that match is the exact prediction of its measured central core charge radius. That is the highest peak shown in curve 2 of figure 3.6. Equally important, curve 2's bulge between ≈ 0.2 fm and ≈ 0.8 fm corresponds to the 0.6 fm diameter of the P/3 oscillations. They also make up over half of the proton's mass, charge, and spin. Their charge decays exponentially from there. It blends in with the P/9 oscillations that extend the positive charge out well beyond 2 fm. Together they create curve 3 (in figure 3.6), and when integrated over 360°, they create curve 2.

This presentation shows that it is possible to match this model to observed proton properties. The proton's root mean square (rms) radius is extended from the central core's 0.105 fm by the generation of rotating P/3 and P/9 harmonic oscillation. Two possible origins of this have been suggested. One is that the proton's wave is not purely sinusoidal. This would generate the series of one third frequency oscillations shown in figures 3.10 and 3.11 and quantified in tables 3.3 and 3.4. The other is that the primary oscillation is rotating so fast over such a short distance that it vibrates, producing the above mentioned one third frequency harmonic oscillations.

In either event, these oscillations are governed by the conservation of mass, charge, and angular momentum. The proton's measured magnetic moment is an average of its varying instantaneous magnetic moments. Figure 3.12 shows different ways of visualizing protons. These can be used for modelling protons in reactions. Figure

3.12 A shows a side elevation of the proton's positive charge plane surrounded by its magnetic field.

The central charge originates in the circumference of the photon's rotation plane. Its field spreads out into the "z" direction from there. Charge originating from the P/3 and P/9 oscillations also originate at the circumference and in the rotation plane. The virtual photons spread the field out from there. Figure 3.13 B shows an oblique view of the charge distribution as it is generated in the plane of rotation. The height in this plot represents intensity, not spatial distribution.

Figure 3.12 D shows a plan view of the P/1. P/3 and P/9 harmonic oscillations. The P/1 central core is entirely positively charged. The P/3 and P/9 oscillations average two positive oscillations and one negative oscillation for each set of three similar harmonics. The positive oscillations therefore dominate. This leaves the whole structure embedded in a positive charge. It is strongest at the P/1 radius and has a bump at about the 0.5 fm radius. From there, it diminishes to zero just beyond 2.5 fm.

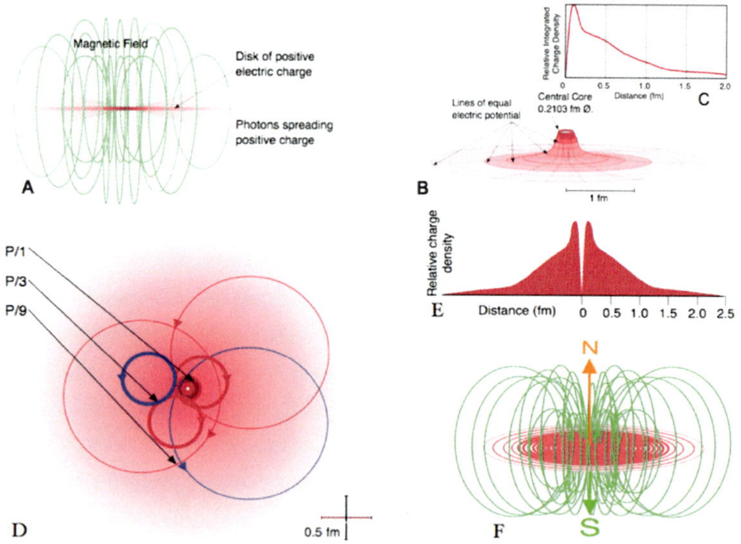

Figure 3.12: *Schematic illustrations of different representations of a proton. B, C, D and E show only the planar distribution of the electric charge. A and F include representations of the magnetic field.*

This indicates that the proton is a dynamic structure in which the P/1 oscillation generates a series of P/3 and P/9 harmonic frequency oscillations that rapidly interchange.

Mass, charge, and spin are conserved. Charge is spread out over a greater distance. The proton's magnetic moment increases from 1 μ_{BP} to 2.7937... μ_{BP}.

Figure 3.12 E shows a plot of the integrated charge distribution on either side of the centre. The horizontal axis represents distance from the centre and the vertical axis represents relative charge intensity. Figure 3.12 F shows an oblique view of the proton's planar charge distribution. The red circles represent the changing charge distribution that is generated in the photon's rotational plane. They do not represent the rotating photons. The green curves represent the proton's magnetic field.

There is proof of the existence of these harmonics. The primary P/1 oscillation is the measured diameter of its central core being 0.105 fm, as shown in figure 3.6. The P/3 oscillations have been detected by the presence of three features close to the central core. Two of them are positively charged and the other is negatively charged. They are generally considered to be proof of the existence of quarks. The P/9 oscillations have a rest mass of one-ninth that of the proton, or 104.3 MeV/c^2. That is very close to the mass of a muon. The next chapter will show that, when separated from a proton, these harmonic oscillations exhibit all the properties of muons.

As mentioned in the previous chapter, the circular motion of the photons within particles automatically gives them their special relativity corrections for mass, length, and time when they are accelerated. When moving, they automatically orient themselves so that their plane of charge rotation is perpendicular to their direction of motion. This is why it is possible to measure their charge distribution, shown in figure 3.6. Their motion also generates their de Broglie wave. There remains plenty of scope to expand this work.

All three of these oscillations, P/1, P/3 and P/9, "pop" into and out of existence as required in order to conserve the various principles. Their lifetimes are unknown. The muon, which has the properties of a P/9 oscillation, has a half-life (outside the proton) of ≈ 2.2 microseconds. They may last longer within the proton. It is suggested that, if a proton's magnetic moment could be determined in under 1 pico second, it would not give the same value. The proton's magnetic moment is an accumulated average, not a fixed value.

These oscillations do not simply remain in the plane of rotation, as indicated in figure 3.12 E. Their charge distributions will buffet them into different tilt angles. If they were in a structure rotating at 10^9 rpm, the change in orientation required to get the proton to orient in their direction of travel would be less than the internal buffeting angle to which they are subjected.

Each of these representations can be used to model proton behaviour, as required to suit the situation. An important feature of the illustrations is that there is no place within the structures for any 'solid matter' to exist. Everything about the proton is electromagnetic. Its mass is inherent in the rotating photon that is the particle. Its electric charge, spin, and magnetic field are generated by the photon's rotation. The proton's magnetic field and rms charge distribution are caused by the circular rotation of ⅓rd frequency harmonic oscillations. There is nothing else present.

From discussions in chapter 2, as illustrated in figure 2.8, there would be a negative field in the proton's central core. Using sufficiently accurate techniques, it could be possible to detect it.

It should be noted at this stage that the "standard model" of subatomic particles considers protons to be composed of two positive up quarks and one negative down quark. One of the problems associated with that interpretation has been the proton's "spin anomaly" (European Muon Collaboration, 1988). Their findings suggest that the spin attributed to the quark constituents of the proton was not sufficient to explain its measured spin. Suggestions for accounting for the remainder of the spin include the possibility that relativistic orbital angular momentum of quarks could explain it (Thomas, 2015). It would be an interesting exercise to evaluate this model, in which every component is traveling at *c* and contributing to the angular momentum, to see if it could match the observed results.

The anti-proton is a similar photon rotating in the opposite direction with respect to the magnetic field, giving it negative charge. The rotating photon model predicts that all other properties would be the same, as observed. Both the proton and anti-proton still exhibit the two states of spin, have the same mass and wavelength, and obey the same rules of parity, duality, and chirality.

3.4. The Neutron

The discovery of positive nuclei and protons by Rutherford and the β rays emitted from radioactive nuclei led to the initial consideration that the positive charge repulsion of the protons in the nucleus was somehow reduced by electrons in the nucleus. This was aided by the observation that the weight of many nuclei was approximately double the weight of the number of protons known to exist in the nucleus. It soon became apparent that this was not the case, and other explanations were sought for how the protons could be bound in the nucleus. The situation was somewhat improved with Chadwick's (1932) discovery of the neutron. While it was not known how neutrons bound the protons together, they at least provided an opportunity for theories to be developed.

Over the decades, neutrons have been extensively studied and determined to have the following major properties:

Neutron Properties: (Nakamura, 2013)

Mass (m_N):	1.6749… x 10^{-27} kg or 939.56379… MeV/c^2 1838.6836... times electron mass
Spin (J):	$\frac{1}{2}\hbar$
Charge (*e*):	0
Magnetic moment (μ_N):	-1.9130427... $\mu_{BN} = e\hbar/2m_N c$
Root mean square radius:	≈ 0.87 fm

Under this model, these neutron properties would give its rotating photon a frequency of 2.269... x 10^{23} Hz and a wavelength of 1.320... x 10^{-15} m. This would give the neutron a theoretical radius of 0.1050 fm and no magnetic moment. Its zero charge means it is composed of a rotating plane-polarised photon. That makes a neutron significantly less stable than a proton.

An isolated neutron, not bound to a proton to form a nucleus, will spontaneously decay to a proton and an electron, emitting an anti-electron neutrino at the same time. Free neutrons have a half-life of about 10 minutes. This appears to indicate there must be a proton and an electron inside a neutron. The neutron is often referred to as a neutral proton because it has many similar properties. To make things a little more complicated, it is also possible for a neutron to absorb a positron and become a proton. So, a proton must also contain a neutron and a positron.

The standard model explains the interchange between protons and neutrons by introducing quarks. They were originally proposed by Gell-Mann (1961, 1962) and Zweig (1964). Quarks were proposed as fundamental particles. Up quarks (u) had spin $\frac{1}{2}\hbar$ and charge $+\frac{2}{3}e$. Down quarks (d) had spin $\frac{1}{2}\hbar$ and charge $-\frac{1}{3}e$. A proton contained two up quarks and one down quark. A neutron contained two down quarks and one up quark. When a neutron decayed to a proton, one down quark changed to an up quark, emitting an electron and anti-electron neutrino at the same time. It was also possible for an up quark in a proton to absorb an electron and emit a neutrino to become a down quark and change the proton to a neutron. That and other possibilities meant the up and down quarks must either contain electrons, positrons, and both types of neutrinos, or else it must generate them during interchange. That would not, however, classify them as fundamental particles that could not be split.

One way this complication can be resolved is for the proton, neutron, electron, and neutrino to be composed of the same substance. That substance is re-arranged to accommodate these particles within their structure. In this rotating photon model, that substance is photons. Chapter 2 showed that the proton's photon must be circularly-polarised while the neutron's photon must be plane-polarised. This difference in polarisation means there will be a difference in something called intrinsic spin, and it will have a value of $\hbar$. Twist spin (intrinsic twist) is an internal angular momentum that is present in one nucleon and absent in the other. In this case, the polarization of the rotating photon is the difference.

The proton's circularly-polarised photon has a twist spin, that is different from that of the neutron's plane-polarised photon. Scientists describe the twist spin as having two values, usually indicated as 1 and 0 (multiplied by $\hbar$). They can only determine that there is a difference between them, but not which one has the twist spin. That leaves us free to describe the neutron in the rotating photon model, knowing that this difference is explained by it.

Figure 3.13 shows the neutron's measured charge distribution superimposed upon that of the proton (Hofstadter 1956, 1962 and Littauer et al. 1961 a, b). Curve 1 represents the neutron's integrated charge distribution superimposed on the proton's integrated charge distribution in curve 2. The main feature to note is that the neutron has an internal charge distribution, even though its total external charge is zero.

Figures 2.6 C and D show that a rotating plane-polarised photon exposes a positive electric field during its first half wavelength and a negative electric field during its second half wavelength.

If the whole oscillation contained only the first harmonic, herein referred to as N/1, neutrons would have spin and mass, but no charge distribution and no magnetic moment. We can now apply the method used in deriving the proton's magnetic moment to the neutron.

Both the proton and neutron have a positive core of radius 0.105 fm. Where the proton's P/3 oscillations occur, the neutron has a negative charge. That suggests the neutron's N/3 oscillation is negative, while the proton's P/3 oscillation is positive. The proton's P/9 oscillations have their greatest influence beyond about 1 fm. From there they extend out to about 2 fm. That is the same region where the neutron's internal charge is also positive.

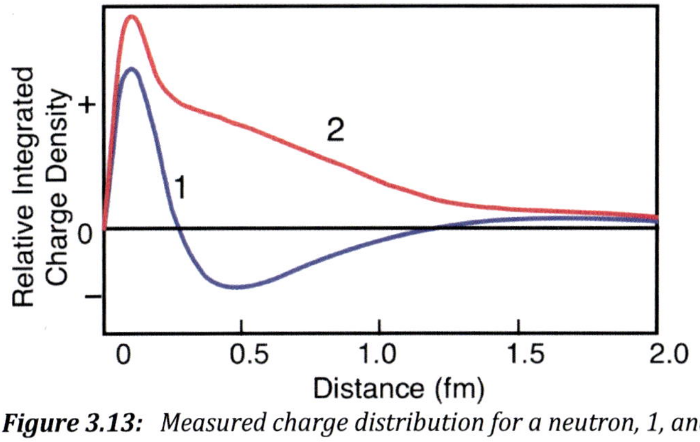

Figure 3.13: *Measured charge distribution for a neutron, 1, and proton, 2.*

It is apparent that the plane-polarised nature of the rotating photon makes the N/1 and N/9 oscillations positive and the N/3 oscillations negative. As in the case of the proton, it is now necessary to find a combination of harmonic oscillations that give the total spin, charge, magnetic moment, and mass to the neutron.

The first feature to note about such a quest is that there is no way a neutron's N/1 and N/3 oscillations can give the neutron a large negative magnetic moment. The only way that can occur is if there is another negative oscillation outside the neutron's N/9 positive oscillation. The only possibility is a negative N/27 oscillation. That requires the first half cycle to generate the positive N/1 and N/9 oscillations while the second half cycle generates the negative N/3 and N/27 oscillations. The properties of those oscillations are given in table 3.5. An indication of an instantaneous plan view of a neutron with such a structure is given in figure 3.14.

Figure 3.14 shows the central N/1 core with a radius of 0.105 fm. It generates a lot of positive charge that extends out to about 0.25 fm. It is surrounded by two negative and one positive N/3 oscillations, giving the neutron a negative region that extends out to a radius of about 1.2 fm.

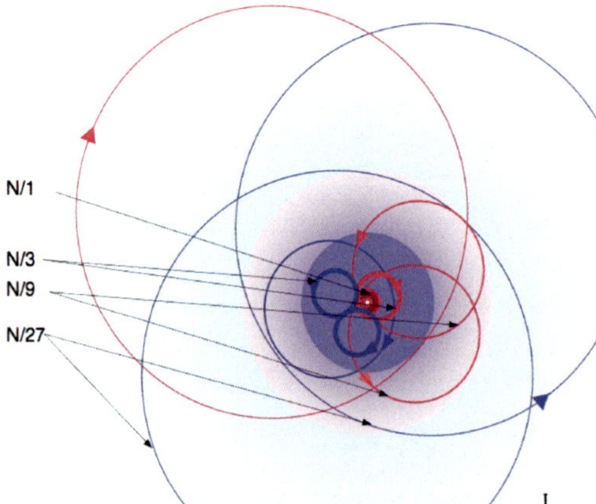

Figure 3.14: *Schematic illustration of the various harmonics and the charge distribution they generate in a neutron under this model.*

Oscillation	Radius (fm)	Magnetic Moment (μ_{BP})	Spin	Charge	Mass (MeV/c^2)
N/1	0.1050	+ 1	½$\hbar$	+ e	939.564
N/2	0.3150	– 3	½$\hbar$	– e	313.188
N/9	0.9450	+ 9	½$\hbar$	+ e	104.396
N/27	2.8350	– 27	½$\hbar$	– e	34.799

Table 3.5: *Calculated values of the properties of the different neutron harmonics of the rotating photon that is the neutron, based upon its mass and this model.*

In turn, that is surrounded by two positive and one negative N/9 oscillations that extend positive charge out to about 2.5 fm. That is surrounded by two negative N/27 oscillations and one positive N/27 oscillation, which extend a weak negative charge out to about 5 fm radius. The alignment of spins and magnetic moments are not shown here. Interested readers should be able to derive them.

Using the same approach that was used to analyse the proton, it is possible to get a set of oscillations that can be added to yield the conservation of mass, charge, and spin, and to match the neutron's observed magnetic moment. One such list is shown in table 3.6.

Figure 3.14 and table 3.6 combine to show that the neutron's inner core is positive. That makes the neutron a neutral proton. It only requires a small change to go from a positively charged proton to a neutral proton. Just like the proton, there is a narrow range of ratios of the N/1 to N/27 oscillations that can match the neutron's observed magnetic moment and charge distribution. That spread will have the following properties:

i. There is a positive central core with zero charge at its centre and maximum intensity at $\approx$ 0.105 fm, with total charge $\approx$ +0.37e.

ii. There is a negative ring of charge about the central core, which will extend from $\approx$ 0.2 fm to $\approx$ 0.8 fm, with charge of $\approx$ – 0.42e.

iii. There will be a weak positive ring outside the negative ring, starting at $\approx$ 1.2 fm and extending to $\approx$ 2.5 fm, with a charge of $\approx$ +0.13e.

iv. There will be a very weak negative ring outside the positive ring that extends from $\approx$ 2.5 fm to $\approx$ 5 fm, with a charge $\approx$ −0.08e.

Oscillation	**Radius** (fm)	**Magnetic Moment** (μ_{BP})	**Percent Charge**	**Magnetic Moment**
N/1	0.105	1	+37.1	+0.371 μ_{BN}
N/3*	0.315	3	-41.9	–1.257 μ_{BN}
N/9*	0.945	9	+12.9	+1.161 μ_{BN}
N/27*	2.835	27	-8.1	–2.187 μ_{BN}
Total:			(+50–50) =0	**–1.913 μ_{BN}**

Table 3.6: *Calculated values of the charge and fractions of the neutron oscillation to give the observed magnetic moment and charge distribution of the neutron.*

As in the case of the proton, there is a small range of harmonic oscillations that can match the neutron's measured magnetic moment. This suggests the structures are inherently stable. They do not rely upon fine-tuning to obtain a match.

There is one feature in figures 3.11 and 3.14 that is worth mentioning. The P/3 and N/3 oscillations contain the greatest charge, mass, and spin fractions of their respective particles. They consist of three oscillations, two of which are positive in protons and negative in neutrons.

Deep inelastic scattering experiments show these features distributed near the centre of the nucleons. They have the varying charges indicated. They could be interpreted as the existence of quarks. The P/1 and P/3, as well as N/1 and N/3 oscillations are concentrated in the nucleons, centres. Their "popping into and out of existence" would confuse results from measuring spin component within nucleons No matter what their contribution to total nucleon spin will not give the total.

Although neutrons have almost the same mass as protons, they are physically larger because of their negative N/27 outer oscillations. These form a very weak negative charge ring on the outside of the neutron, containing less than –0.1*e*. The inner core positive ring, its surrounding negative ring, and the second positive ring have all been detected, as shown in figure 3.13. The outer negative ring has not been directly detected, although there is some experimental evidence that suggests it exists. This evidence comes from the dimensions of nuclei measured by charged particle scattering and from neutron scattering.

Charged particle scattering gives nuclear dimensions that are slightly smaller than those obtained from neutron scattering. The standard model suggests that this is due to the protons, and hence the positive charge, being concentrated near the centre of the nucleus. Because neutrons are neutral, they will not experience the proton's charge and thus they will measure the mass radius of the nucleus. This model suggested that an incoming neutron's weak negative outer ring will experience repulsion from the same weak outer rings associated with the neutrons in the nucleus, and this will yield slightly larger dimensions in scattering.

Other representations of the neutron are given in figure 3.15. In these and other figures, red represents positive charge and blue represents negative charge. Figure 3.15 A illustrates a side view of the neutron's planar charge distribution, surrounded by its magnetic field in the third dimension.

Even though neutrons have no external residual charge, their internal charge distribution under this rotating photon model gives them a magnetic field. Table 3.5 shows that the overwhelming majority of a neutron's magnetic field is generated by the N/27 oscillations. They are generated over an area 729 times larger than the N/1 oscillation and 81 times larger than the N/3 oscillation. A measurement of a neutron's magnetic field distribution would show that it is strongest in the centre.

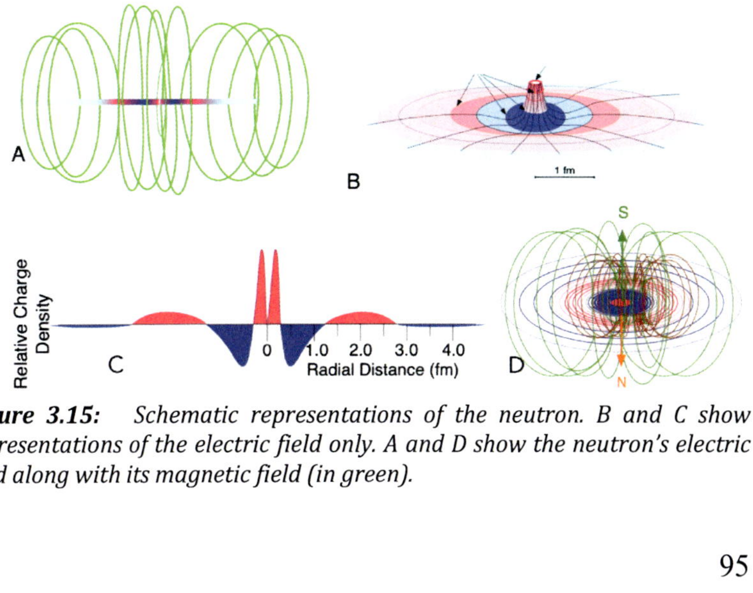

Figure 3.15: *Schematic representations of the neutron. B and C show representations of the electric field only. A and D show the neutron's electric field along with its magnetic field (in green).*

The charge planes in both the neutron and proton have been represented as stable structures. The dynamic nature of the changing harmonic oscillations means they will vibrate at high frequencies, typically ≈ 10^{24} Hz.

3.5. The Neutrino

The existence of neutrinos was first postulated by Pauli to explain differences in the products that result when neutrons decay into protons and electrons. The neutron has zero charge and spin ½$\hbar$. The proton has charge $+e$ and spin ½$\hbar$. The electron has charge $-e$ and spin ½$\hbar$. While there has been conservation of charge, there are two ½$\hbar$s arising from only 1 spin ½$\hbar$ particle. That required another particle to conserve the extra ½$\hbar$ spin. Pauli postulated the necessity for another particle to account for that difference. That required a property of no charge and spin –½$\hbar$. Fermi called the particle a neutrino, Italian for little neutron. The reaction for a neutron spontaneously decaying into a proton, electron, and neutrino is summarized in figure 3.16.

The important features are given in the descriptions below each image. The neutrino has no charge, spin ½$\hbar$, mass as indicated, and no twist spin. It is a rotating plane-polarised photon. The proton has positive charge, spin ½$\hbar$, mass as indicated, and twist spin $\hbar$. The electron has spin ½$\hbar$, negative charge, mass as indicated, and twist spin $\hbar$. It was the missing ½$\hbar$ that led to Pauli's postulation of another

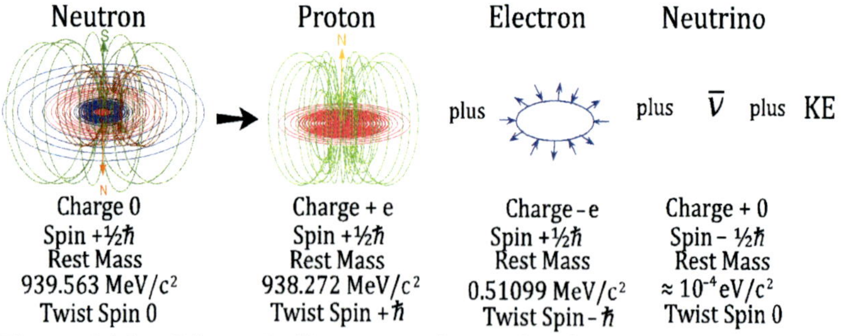

Figure 3.16: *Schematic illustration of an isolated neutron spontaneously decaying to a proton, electron, and anti-electron neutrino.*

particle to balance spin. In this case, it would have to have a spin of −½ħ, opposite to the electron's spin. That is why it is an anti-electron neutrino.

Everything is balanced. The mass difference between the neutron and the proton-plus-electron is the energy imparted to the electron and neutrino.

Neutrinos were first discovered by Reines and Cowan (1956), working at the Savannah River Site. They are known to have very unusual properties. Their major property of interest is that they have the same ½ħ spin as electrons and protons. They have no charge and no magnetic moment. Attempts to measure their mass have only succeeded in applying an upper limit to it — currently, less than 0.1 eV/c^2. This is one of the reasons why spin was regarded as a quantum mechanical property that had no physical basis. Neutrinos are also very difficult to detect.

Under this model, the neutrino's zero charge means it is made of a rotating plane-polarised photon. That it has no detectable magnetic moment implies that it has a pure sine wave form. Because neutrinos are rotating plane-polarised photons, they do not have the ħ twist spin associated with the circularly-polarised photon that makes up a proton. This gives them the same twist spin as a neutron. As mentioned earlier, magnetic moments are a derived property. There is no 'conservation of magnetism.'

The reaction in figure 3.16 also occurs in neutron-rich nuclei, as neutron's decay into protons in order to make the structure more stable.

From an historical perspective, the original suggestion by Uhlenbeck and Goudsmit (1925, 1926) that electrons possess an intrinsic angular momentum seemed reasonable. The later detection of electrons as point particles seemed to rule that idea out, though. The detection of the same property in protons and neutrons, which have a rest mass of almost 1840 times that of an electron, added further to the idea that the property, now called 'spin,' could not be angular momentum. The suggestion that neutrinos, suspected to have a mass less than *one billionth* that of a nucleon, could have the same angular momentum seemed absurd. For that reason, spin was considered a quantum mechanical property with no known mechanical analogue. This presentation shows that a neutrino's spin is indeed angular momentum, and that this understanding allows many unknowns about them to be resolved.

The rotating photon that is the neutron has zero charge, no twist spin, intrinsic spin as angular momentum ½$\hbar$, mass ≈ 939.3 MeV/c^2, and a residual magnetic moment. An isolated neutron's plane-polarised photon rotation is unstable. It self-unravels to become the circularly-polarised photon known as a proton, with a mass of 938.5 MeV/c^2. Conservation of charge means that another circularly-polarised photon of opposite charge must also be generated. The only stable charged particle within the appropriate mass range is an electron. Its rotating photon is circularly-polarised, taking away the intrinsic spin in the proton's rotating photon structure. This conserves both intrinsic spin and electric charge.

The combined rest mass of the proton, 938.5 MeV/c^2, and the electron, 0.511 MeV/c^2, are less than the initial rest mass of the neutron. Also, both the proton and electron have opposite spin ½$\hbar$, cancelling each other out. This leaves some residual energy and the neutron's initial ½$\hbar$ angular momentum to be conserved. This is the role of the neutrino. It is a rotating plane-polarised photon that makes two revolutions within its wavelength. It is ejected when the neutron's oscillation converts from plane-polarised to circularly-polarised.

To a first approximation, the neutrino's mass, may be calculated from the time it takes a neutron to decay into a proton and electron once the decay process has started. This process is otherwise known as the "weak" interaction. It occurs in ≈ 10^{-10} sec (Hughes, 1991). Under this rotating photon model, that is the time it takes for a neutrino to be emitted once the neutron has started to decay. This gives neutrinos a rest mass of about 4 x 10^{-5} eV/c^2, a wavelength of about 30 mm, and a radius of ≈2.5 mm. It needs to be that large in order to have the same angular momentum as the proton.

A neutrino as a spiralling plane-polarised radio wave. Their spiralling motion means they still have the same angular momentum as a proton, while still having a very small mass. Figure 3.17 A shows that neutrinos are emitted with, and exist at, different energies. It also shows that the greatest number of neutrinos detected by far, cosmological neutrinos, have an energy in the vicinity of ≈ 10^{-4} eV, similar to their rest mass. The cosmological neutrino flux has not been measured. It was estimated from the Big Bang residual. It may be much higher.

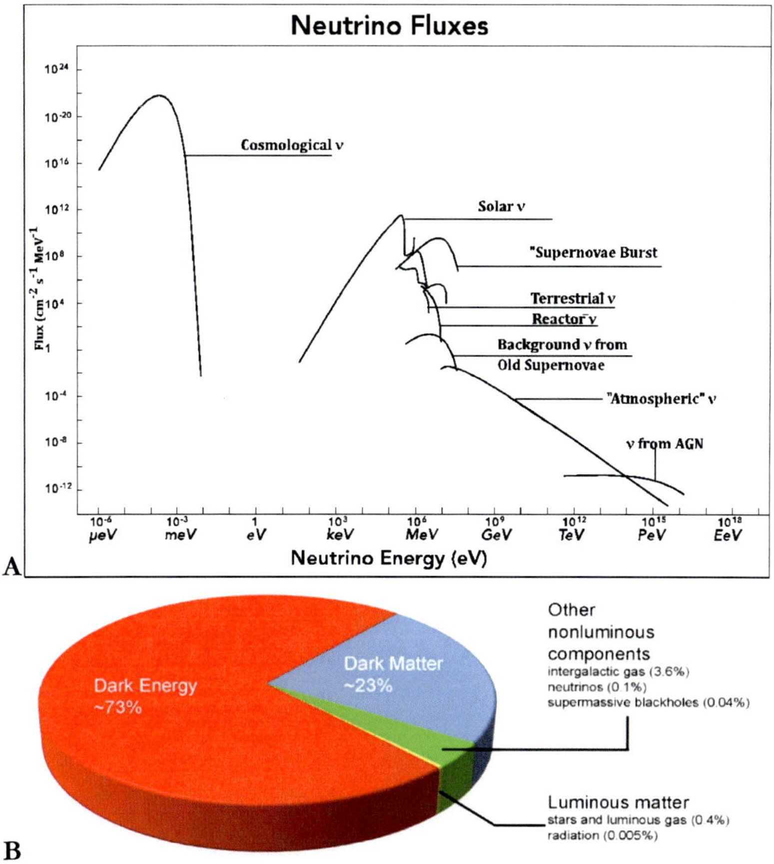

Figure 3.17: *Information about neutrinos.* ***A*** *Neutrino flux versus energy for different sources. Copied from:*
https://masterclass.icecube.wisc.edu/en/learn/detecting-neutrinos
B *Ratio of luminous and non-luminous matter to dark matter and dark energy required to match experimental observations under standard models. (Source: https://www.osti.gov/servlets/purl/1410794)*

Most solar neutrinos are emitted with energies in the vicinity of 0.8 MeV. That approximately corresponds to the mass difference between a neutron and the combined mass of a proton and electron, which is ≈ 0.78085... MeV/c². Neutrinos have a very small mass and carry away much of the energy difference between a neutron and it decay products. The nucleons and electrons are constrained by the high density solar mass and can't accept much more energy.

Rearranging equation 2.7 for relativistic mass correction, to express velocity in terms of mass, gives

$$v = c.\sqrt{1 - \frac{m_0^2}{m_v^2}} \qquad (3.2$$

Equation 3.2 means a 1 eV neutrino of rest mass 4 x 10^{-5} eV/c^2 will travel at ≈99.999% of the speed of light. For all practical purposes, that approximates to considering neutrinos as travelling at the speed of light. This suggests another way to determine the rest mass of a neutrino. Experimental observations have shown that neutrinos from supernovae within our galaxy arrive within seconds of the detection of photons from the same event. Figure 3.17 indicates that supernovae neutrinos are emitted with an energy of ≈ 10 MeV. Using a neutrino mass of ≈ 10^{-4} eV/c^2, equation 3.2 gives a 10 MeV neutrino a speed of ≈ $(1 - 10^{-11})c$.

After travelling for ≈ 30,000 light years (LYs), photons would arrive ahead of neutrinos by a few seconds. That is complicated by suggestions that neutrinos can escape from the dense mass of a supernova explosion more quickly than can photons.

All those factors indicate that neutrinos have a low rest mass, in the range 10^{-5} to 10^{-4} eV/c^2. Additional support for this rest mass comes from the measured neutrino flux versus energy spectrum shown in figure 3.17. The peak of cosmological neutrino flux is at about 10^{-4} eV. That energy approximately corresponds to the peak energy of the cosmic microwave background radiation temperature of space, at ≈ 2.7°K. It is suggested that these neutrinos have been redshifted until they have no energy left for motion.

They then absorb background photons, giving them their observed energy in the vicinity of 10^{-4} eV. It could be argued that neutrinos of any mass could pick up these photons and produce that spectrum. It is herein suggested that neutral neutrinos can only absorb those photons and pick up their energy if they had similar dimensions. Any proposed point-particle model for the neutrino would not have any mechanism for absorbing these low energy photons.

Figure 3.17 B suggests that neutrinos, which A suggests are mostly cosmological neutrinos, make up about 2.5% of the total baryonic mass. An estimated baryonic mass density of $\approx 10^{-26}$ kg/m^3 means about 6 hydrogen atoms per cubic metre. Protons have a mass equivalence of $\approx$ 1 GeV/c^2. Neutrinos have a mass equivalence of $\approx 10^{-4}$ eV/c^2. Each baryon has a mass equal to $\approx 10^{13}$ neutrinos. 2.5% of the baryonic mass density gives cosmological neutrinos a density of over 10^{12} neutrinos/m^3. They are by far the universe's most common components.

The above enables us to draw the following conclusions about neutrinos:

i. They are rotating plane-polarised photons that make two revolutions per wavelength. That gives them the properties of zero charge and spin as angular momentum, $I\omega = ½\hbar$. If they were a sinusoidal oscillation, they would have no magnetic moment. None has been detected.

ii. They have a rest mass of $\approx 5 \times 10^{-5}$ eV/c^2. That gives them a wavelength of about 30 mm and a radius of about 2.5 mm if they are a single wavelength oscillation.

iii. Their low mass means that neutrinos with an energy of about 10^{-2} eV will travel at about 0.99c. That is why neutrinos are considered to travel at the speed of light.

This raises the question: If neutrinos are so large, how can they travel through everything and be so hard to detect? The first thing to realize is that any hoop with its whole mass travelling in a radius at the speed of light is a very powerful gyroscope. Mass for mass, neutrinos are the most powerful gyroscopes in the universe. The only thing that could be more powerful is a circularly-polarised photon with energy of $\approx 10^{-5}$ eV/c^2 or less. There are not many of them.

Trying to get a neutrino to change its direction is difficult. Electrons have no way of absorbing them. For a nucleon to absorb a neutrino, it must strike the nucleon at exactly the right angle with its leading point. Its small mass means that point is very small and if it misses the spot, it will pass the nucleon with the rest of it unable to react. This leaves neutrinos with almost nothing with which to interact.

The higher a neutrinos energy, the smaller its radius. A GeV neutrino has a similar radius to the central core of a nucleon. That makes it easier for GeV neutrinos to react with nucleons. Their increasing radius diminishing with energy makes it more difficult for lower energy neutrinos to react with nucleons.

An isolated neutron is a rotating plane-polarised photon. It is unstable. Why should an isolated neutrino, also a rotating plane-polarised photon, be stable? At 2.7°K, the background temperature of space corresponds to an energy of about 10^{-4} eV. The background temperature equivalent for neutrinos is about 1.95°K. There are no energy levels lower than that to which a neutrino can decay.

The above has treated the quantum property called spin as angular momentum. There is about a 10^{13} difference between the masses of a nucleon, $\approx 10^3$ MeV/c^2, and those of a neutrino, $\approx 10^{-4}$ eV/c^2. This is possible because the radius of the neutrino's rotating photon, $\approx$ 2 mm, is $\approx 10^{13}$ times larger than that of a nucleon's diameter $\approx 10^{-16}$ m.

These calculations for the rest mass of an electron neutrino are only approximations. They suggest a mass about 10^4 times smaller than the current best measurements. At the same time, it matches known neutrino properties, such as being detected a few seconds after photons emitted from supernovae within our galaxy. It also matches the known neutron property of the time taken for a neutron to decay to a proton, electron, and neutrino.

An important feature that could be overlooked is the cosmological neutrino flux. The results from Lincoln (2017) suggest that neutrinos, which are mostly cosmological neutrinos, make up about 2.5% of the total detected mass of the universe. In approximate figures, baryons have a mass of $\approx 10^9$ eV. Neutrinos have a mass of $\approx 10^{-5}$ eV. There are $\approx 10^{14}$ neutrinos per baryon. It is suggested the universe has a baryonic matter density of 1×10^{-26} kg/m^3. That corresponds to a density of $\approx$ 6 protons per m^3. That gives neutrinos a density of $\approx 5 \times 10^{14}/m^3$, or $5 \times 10^5/mm^3$. Allowing for neutrinos at near rest speeds occupying several mm^3, the density of neutrinos is typically $10^9/cm^3$.

It was mentioned earlier that rotating plane polarized photons were not as stable as rotating circularly polarize particles. Why don't neutrinos, which are rotating plane polarized photons, decay? The answer is found in their low rest mass, $\approx 10^{-5}$ eV/c^2. There is no lower state to which they can decay. They have no option but to rotate as a plane polarized photon of very low energy, with a wavelength a little longer than that of cosmic microwave background radiation.

3.6. Nucleon Interchange

A neutron can become a proton by emitting an electron or by capturing a positron. A proton can become a neutron by emitting a positron or by capturing an electron. Both require the emission of a neutrino. Figure 3.16 illustrated a neutron decaying into a proton, electron, neutrino, and additional energy. Examples of other interchanges are given in figure 3.18.

Figure 3.18 A shows a neutron capturing a positron to become a proton. It also emits a neutrino. Figure 3.18 B shows an excited proton capturing an electron to become a neutron, also emitting a neutrino.

Figure 3.18 C shows a neutron capturing a neutrino and becoming a proton by ejecting an electron. Figure 3.18 D shows an energetic proton emitting a positron and neutrino to become a neutron.

Figures 3.16 and 3.18 suggest that a proton contains a neutron, positron, and neutrino, while a neutron contains a proton, electron, and anti-electron neutrino. The easiest way for this to occur is if they are all made of the same substance that simply rearranges itself according to the circumstances. When all particles are made of rotating photons, a simple rearrangement of the oscillations produces all the variations necessary.

Consider figure 3.16. A neutron is composed of a rotating plane-polarised photon. For the reasons mentioned in chapter 2, isolated neutrons are not as stable as isolated protons.

A neutron's opposing charges and magnetic poles overlap within its structure. In addition, its photon does not have the internal $\hbar$ intrinsic twist spin to help keep the particle stable.

Under this model, the properties of charge, spin, and intrinsic (photon twist) spin are conserved during nucleon interchange. Mass/energy is also conserved, with the additional mass imparting energy via $KE = \Delta mc^2$ to the residual particles.

The physical process in figure 3.18 A involves a neutron's plane-polarised photon converting to a more stable circularly-polarised photon, a proton, by absorbing a positron. In doing, so it has to emit a plane-polarised photon rotating in the opposite direction to that of the proton. That is an electron neutrino. This conserves electric charge and

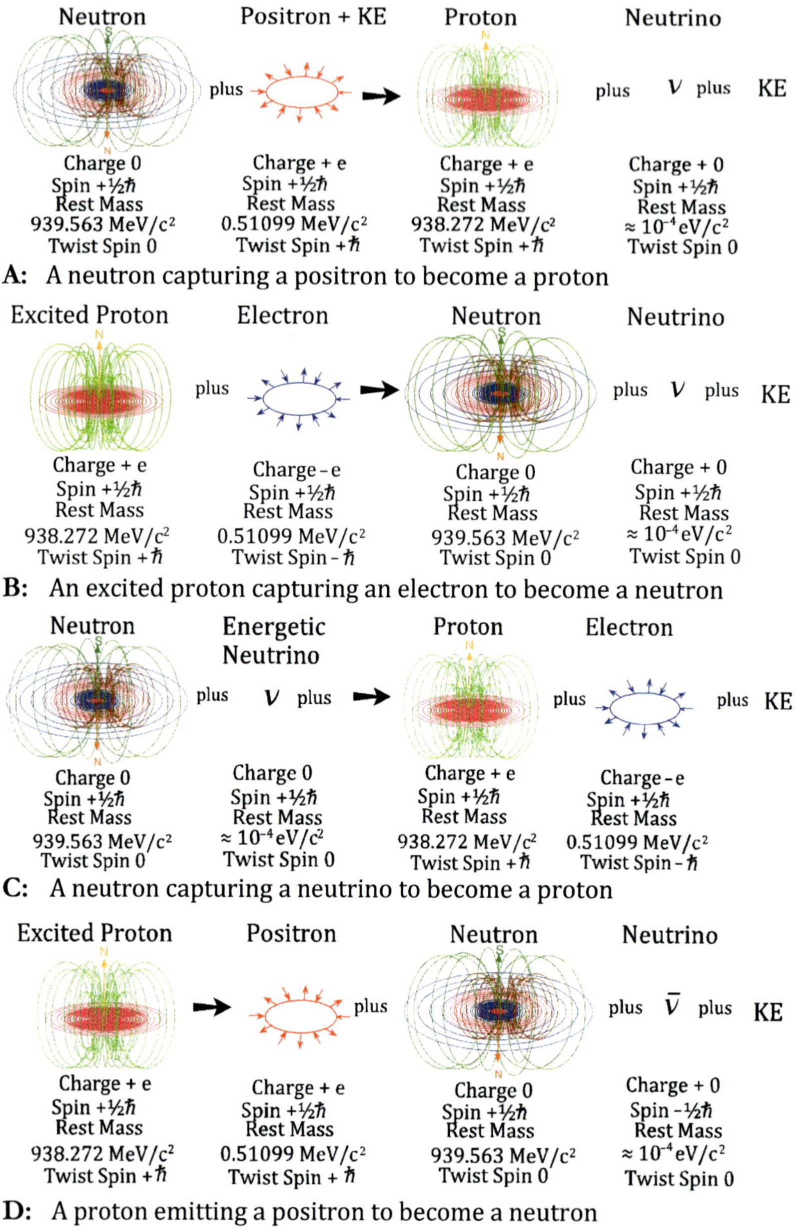

Figure 3.18: *Schematic illustrations of some interchanges between nucleons.*

angular momentum. Excess energy is consumed by the escaping neutrino.

The process is reversible. Figure 3.18 B shows an energetic proton capturing an electron. It is equivalent to the case of an energetic electron being captured by a proton, which can occur when a nucleus has too many protons for it to be stable.

An electron, which is significantly larger than a nucleus can, as part of its motion around a nucleus, pass close to an excited proton. The proton captures the electron, converting its circularly-polarised photon motion to the plane-polarised rotation of a neutron. In order to balance the $\frac{1}{2}\hbar$ additional spin angular momentum the electron brought with it, it emits an electron neutrino, one that rotates in the same direction as the electron. All properties are therefore conserved during rearrangement.

Figure 3.18 C shows an energetic neutrino impinging on a neutron. The neutron absorbs its oscillation. Conservation of angular momentum requires it to emit an electron and rearrange its rotating plane-polarised photon to become a rotating circularly-polarised photon, i.e., a proton. Charge and angular momenta are conserved in the rearrangement.

Figure 3.18 D shows what can happen to a proton in the presence of a high positive field. The proton takes energy from the high positive field and emits a positron. The positron carries away the circular polarisation associated with the proton's positive charge and $\hbar$ intrinsic twist, leaving it as a plane-polarised rotating photon, whereby it becomes a neutron. Conservation of angular momentum means it must also emit a positron neutrino, otherwise known as an anti-electron neutrino.

In all of these reactions, properties are conserved. No additional "anythings" are needed. Magnetic fields are not conserved because they are a derived property that depends upon the electric charge and radius of the rotating photon.

Nucleons are not pure sine waves, Dirac particles, like electrons and neutrinos. The latter are single oscillation sinusoidal waves. Sections 3.2 and 3.3 showed that the nucleons have harmonics associated with them. They manifest as $\frac{1}{3}$rd frequency harmonics of the fundamental nucleon frequency. A neutr½on emitting an electron could thus occur by emitting the electron, possibly from one of its N/3 harmonics. It then becomes a positive P/3 harmonic. The oscillations in the remainder of the nucleon simply re-arrange under the conservation of angular momentum, and the neutron becomes a proton.

Neutrons have in the past been regarded as protons onto which a negative charge has been imposed. In other words, it is a neutral proton. That concept was continued in this rotating photon model. Both have a central positive core that accounts for approximately one-third of its mass, charge, and spin. Apart from the slight mass increase and a notable absence of external charge on the neutron, a key difference between the proton and neutron is the difference between their $\hbar$ intrinsic twists.

They are the difference in polarisation states of the two rotating photons. It is herein referred to as intrinsic twist to differentiate it from the ½$\hbar$ angular momentum, spin, of the particles. In this model, charged particles have circularly-polarised photons and neutral particles have plane-polarised photons.

One feature to note here is that it is not possible to force an electron into a neutron to convert it into a negative proton. That would require the electron to rearrange the rotational direction of the positive N/1 oscillation to a negative N/1 oscillation. It also requires the polarity of the other oscillations to be changed. Consider the chirality of the particles. Both the proton and neutron have the same right-handed chirality, while the electron has left-handed chirality, see figure 2.11. To maintain chirality, the neutrino must be right-handed, opposite to that of the electron. Changing the chirality of a proton or neutron requires a loot of energy. It is not easy.

Under this model of matter, particles consisting of rotating circularly-polarised photons have their like electric and magnetic fields superimposed upon each other during every half-wavelength revolution. That means their fields lock into and reinforce each other, adding to the stability of the structure. Once established, their ½$\hbar$ intrinsic spin angular momentum of their rotating circularly-polarised photon, and the $\hbar$ intrinsic twist represent very strong forces. They are the origins of the electric charge, and hence, also of the direction of the magnetic fields. The $\hbar$ intrinsic twist, ½$\hbar$ intrinsic spin and overlapping magnetic fields are the powerful forces responsible for the stability of the proton and electron.

When plane-polarised photons rotate, their opposing electric and magnetic fields overlap, repelling each other. That combines the lack of $\hbar$ intrinsic spin, i.e., angular momentum, to make neutral particles inherently less stable than charged particles. It helps to explain the half-life of free neutrons compared to the infinite lives of protons and electrons.

A free neutron is unstable and decays, with an average half-life of about 10 minutes, producing a proton, electron, and anti-electron neutrino. The neutrino, which also consists of a rotating plane-polarised photon, has no energy level below which it can decay. It has no option but to remain a neutrino. When a neutron is bound to one or more protons, it becomes as stable as a proton. It is suggested that this is because stable proton oscillations overlap the neutron's unstable oscillations, holding them in place. How that overlap occurs is discussed in chapter 5.

3.7. From Linear to Rotating Photons

Some of the properties of linear photons have been studied and are somewhat understood. There is still more to learn about them. Not many attempts have been made to understand the properties of the rotating photons that make up matter particles. Some aspects of linear photons were discussed in chapter 1. The properties that particles would have if they were composed of a photon rotating twice within its wavelength were discussed in chapter 2. This chapter has focused specifically on the only four known stable particles — electrons, protons, neutrons, and neutrinos.

An important aspect of this work centres on how and why linear and rotating photons can interchange. The interchange from rotating to linear photons was discussed earlier in the annihilation of particle/anti-particle pairs. Their opposing charges cause them to be attracted to each other. When they interact directly, they unlock each other's angular momentum. The two rotating photons become linear photons and travel away from each other in opposite directions in order to conserve linear momentum. In the process, each particle releases energy given by $E = mc^2$, where m is the mass of the particle.

It is now time to look at the reverse process — the conversion of linear photons into rotating photons. Experimentally, this is achieved by aiming a beam of high energy linear (gamma ray) photons at either an opposing beam of photons, or a solid target. When a photon impinges on another photon, both are compressed. Their compression brings their magnetic fields closer to overlapping. The magnetic field interacts with the photons ħ intrinsic twist to pull it into making two revolutions per wavelength. That converts the linear photon into a rotating photon with intrinsic spin ½ ħ. Conservation of momentum principles require that to occur with both photons rotating in opposite directions with respect to their magnetic fields.

A similar process occurs when a high energy linear photon impinges upon a nucleus that does not absorb it. The compressed photon's magnetic field pulls it towards the two rotations per wavelength state. If it goes into a single double rotating photon, it violates conservation principles. To avoid that, it splits into two equal photons that rotate in opposite directions. Their opposite directions give them opposing charges, a particle/anti-particle pair.

That still leaves a deficiency of ħ intrinsic twist. The original photon had ħ intrinsic twist. Each of the particles' rotating photons now have ħ intrinsic twist. That difference has to be obtained from somewhere. That can come from the ½ħ intrinsic spin of one nucleon within the nucleus being reversed. The process of going from + ½ ħ to – ½ ħ, or vice versa, releases the required ħ angular momentum in a single step that reverses the orientation of one nucleon as it provides the angular momentum for the second photon's intrinsic twist to form a charged particle.

In both cases, the ends of the photons will not join. The particles magnetic field forces the rotating photon to make exactly two revolutions per wavelength. There is no other possibility. As mentioned earlier, photons prefer the double rotation structure over the linear structure because the former gives a magnetic loop, a structure that is lower energy than the open ended magnetic field of a linear photon.

This describes the physical mechanism by which circularly-polarised linear photons form double-looped rotating photons known as particles or Fermions. What is not known is why only the electron and proton frequencies are stable. The neutron frequency depends upon the proton. Electron neutrinos are stable because their frequency is so low there are no lower energy levels available. The neutrino frequency is stable because its energy level is so low there is nothing below it

It is well known that there is a wide range of frequencies of linear photons, from wavelengths of less than 1 fm to several km, as shown in figure 1.4. There is nothing in the above to suggest that there should not be a wide range of rotating photons to give particles of a variety of different masses. To date, the only stable particles detected are the proton, electron, neutrino, and bound neutron. The wide range of frequencies among rotating photons is found in the relativistic mass increase as the speed of those stable particles is increased.

Neutrinos cover the widest range. Figure 3.17 shows they have energies between 10^{-5} eV to 10^{15} eV. That gives them frequencies between $\approx 10^{10}$ Hz and 10^{30} Hz. Electrons have frequencies above $\approx 2 \times 10^{20}$ Hz. Nucleons have frequencies above about 4×10^{23} Hz. Linear photons cover approximately the same frequency range as neutrinos. High-energy linear photons, gamma rays, have been detected at approximately the same energy levels, and hence frequency range, as that of high-energy cosmic rays. There does not appear to be any limitation on the energy of photons that can be converted to particles with relativistic mass correction.

3.8. Summary

Under this rotating photon model, all particles are composed of photons of the appropriate energy rotating at c in a circle of circumference equal to half its wavelength. They are 100% electromagnetic in nature. Their equivalence of energy and mass, given by $E = mc^2$, is that energy is a photon of the appropriate frequency traveling at c in a straight line. Mass is a similar photon traveling in a circle of circumference equal to half its wavelength. Applied to electrons, it gives them a rest radius of 193... fm. Applied to protons, it gives a core radius of 0.1052... fm, and to neutrons, a core radius of 0.1050... fm.

Experimental measurements of the charge distribution of protons and neutrons have shown that the central core of both nucleons matches these values within experimental error. The proton's rms radius is larger by a factor of about 7. This study has attributed that larger radius to a series of ⅓rd frequency harmonics spreading out its charge and increasing its magnetic moment.

It has long been known that the neutron is a proton that has no charge because a negative charge has been added into it. Under this rotating photon model of matter, that occurs because the proton is composed of a circularly-polarised photon while a neutron is composed of a plane-polarised photon.

With nucleons composed of rotating photons, it would not be surprising if high-energy linear photons had properties similar to the high energy rotating photons that make up nucleons and other particles. It is believed this has been observed (Erdmann, 1997). It is suggested that Erdmann was describing "The Photonic Structure of Partons", not "The Partonic Structure of Photons", the title of his publication.

In addition to the properties of all particles described in chapter 2, the following describes some unknown properties of particles that can be explained or predicted from this model.

An electron is a torus of electromagnetic energy with a central rest radius ≈ 1.93 x 10^{-13} m. Its radius will diminish with velocity according to the relationship $r_v = r_0\sqrt{1 - \frac{v^2}{c^2}}$. The otherwise enigmatic properties of electrons are due to their structure. When moving, they orient themselves with their magnetic field leading their rotating charge plane.

Electric charge is generated at the rotating photon's circumference. It is two dimensional at the circumference from where its secondary or virtual photons are emitted. Exactly the same single electron experiment will give rise to different results because the leading edge of the electron's charge plane is at a different orientation when it travels the same trajectory.

It does not appear possible to orient electrons in any manner such that their leading-edge charge would always have the same orientation. Its mass rotating at the speed of light in a circle of radius 193 fm means it is a very strong gyroscope, making it difficult for it to change its direction when moving. This makes it behave in a way that would not be expected for "ordinary" particles. It also means that it will seek out another particle, preferably another electron, with which it can exchange photons, and thus move more easily in non-linear paths. This is the reason why electrons attach in pairs, which is the foundation of chemical bonding that shapes the materials around us.

The structures of the nucleons make them more complex. They are still entirely electromagnetic, with their charge being generated in the two dimensions of the rotating photon's circumference. Their entirely electromagnetic nature makes them essentially indistinguishable from photons of similar frequency over short time periods.

The charge distributions and magnetic moments of the nucleons are both greater than those predicted from their mass using equation 2.2b. This is attributed to a series of ⅓rd frequency harmonics that spread out their charge and increase their magnetic moment.

This also explains why the neutron has a magnetic moment when it has no net charge. It has the charge distribution as shown in figure 3.13 plus an additional weak outer negative charge ring extending out for a few fm. The neutron's neutral charge is due to having 50% positive charge and 50% negative charge. Figure 3.14 provides a way of visualizing the reasons for that charge distribution.

As a 'neutral proton', a neutron, interchanges with a proton by emitting or absorbing other appropriate photons. Left to its own devices, an isolated neutron will decay to a proton, electron, and anti-electron neutrino with all four particles being distinct, rotating electromagnetic oscillations. The neutrino carries away the additional angular momentum as spin via its large radius and low mass. It can carry off the neutron's intrinsic spin as its own intrinsic spin because it is also a rotating plane-polarised photon.

With the charge distribution of individual nucleons being two-dimensional their thickness is due to their magnetic field being in the third dimension. The charge distribution is generated by the series of ⅓rd frequency harmonics that appear and disappear as they extend the nucleon's radial dimension beyond that of its central core. The charge is not instantaneously spread out over that increased dimension. It is generated at the circumferences of the rotating frequency harmonics, being strong at those circumferences. It is the averaged charge distribution that is measured in figures 3.6 and 3.13. The significant implications for the binding of protons to neutrons to form nuclei, is discussed in chapter 5.

As seen from tables 3.2 and 3.5 and figures 3.13 and 3.14, ⅓rd frequency harmonics — the P/3 and N/3 oscillations — represent the largest fraction of both protons and neutrons. They are located near the centre of the nucleons and have varying charge, with both positive and negative oscillations present in both nucleons. Protons have two positive and one negative P/3 oscillations. Neutrons have two negative and one positive N/3 oscillations.

It is possible that deep inelastic scattered electrons could highlight them, giving images similar to those expected from quarks in similar positions. It is suggested that if measurements were made of the spins of those structures, they would show that their combined spins were not enough to account for the spin of the nucleons because they would not include the spins of the other oscillations.

The properties of neutrinos, as determined from the weak interaction time constant, suggest that they have a rest mass of the order of 5×10^{-5} eV/c^2. This exceedingly small mass enables even low energy (such as 1 eV) neutrinos to travel at almost the speed of light. Having no electric charge, no magnetic field, and a very large angular momentum for their mass, it is difficult for them to interact with other particles.

The above has shown how the rotating photon model can explain the properties of real particles. This model matches the known properties of electrons, protons, neutrons, and neutrinos. It makes several predictions of unknown properties of these particles. As shall be shown in chapters 5 and 6, some of these properties are very useful in explaining certain unknowns in the current standard model.

Equally important, it provides detail about neutrinos. It shows how they can have a very small mass and yet still have the same angular momentum as a nucleon with 10^{13} times their mass. It is herein suggested that the different "flavours" of neutrinos — the electron, muon, and tau — are the same rotating plane-polarised photons that simply have different rest masses. The neutrino's structure enables them to travel so close to c as to be experimentally indistinguishable from c for almost all energies. 10 MeV energy neutrinos emitted by supernovae within the Milky Way galaxy will arrive within a few seconds of photons emitted at the same time from the same event.

It is also possible to appreciate why photons from electrons and nuclei are circularly-polarised. The particles emitting them are all rotating photons. It would be difficult for a rotating structure to emit a plane-polarised photon.

This presentation suggests that spin is not a peculiar quantum mechanical property with no physical analogy. It is angular momentum, as was originally proposed by Uhlenbeck and Goudsmit (1926). The stability of rotating photon structures come from the conservation of their two sources of angular momentum, the reinforcing nature of their interlocking electric and magnetic fields, and that they have no lower energy levels to which they can decay.

That gives us the five fundamental building blocks of the universe — protons, neutrons, electrons, neutrinos, and photons. They are the universe's building blocks, the main ingredients needed to understand it.

However, they are not the only particles detected in the universe. There is also an array of sub nuclear particles that have been detected from collisions between high energy and "target" nuclei. These come from cosmic rays and particle accelerators. They are distinguished by their very short half-lives. The two hundred or so separate particles that have been identified far outnumber the five fundamental particles mentioned above. Those two hundred generated particles will be referred to as "elementary particles," and the next chapter describes their origins, along with the properties of some of them.

3.9. List of References

Anderson CD, (1932); "The Apparent Existence of Easily Deflectable Positives", Science, **76**, pp 238-239

Anderson CD, (1933); "The Positive Electron", Physical Review, **43**, pp 491 – 494

Barut AO, (1991); "History of the Electron", In "The Electron, New Theory and Experiment", Ed. Hestenes D. and Weingartshofer A., Kluwer Academic Publishers, Dordrecht NL, pp 105 – 148

Bender D, et al., (1984); "Tests of QED at 29 GeV Energy", Physical Review, **D30**, pp 515 – 527

Bohm D, (1951); "Quantum Theory", Prentice Hall Inc., Engelwood Cliffs, N.J.

Chadwick J, (1932); "Possible Existence of a Neutron", Nature, **129**, p 312

Compton AH, (1918); "The Size and Shape of the Electron", Washington Academy of Sciences, **VIII**, pp 1 – 11

Compton AH, (1923); "A Quantum Theory of the Scattering of X-rays by Light Elements", Physical Review, **21**, pp 483 – 502

de Broglie L, (1925); "PhD Thesis", published in Ann. Phys. (Paris), **111**, p 22

Dirac PAM, (1928); "The Quantum Theory of the Electron", Proceedings of the Royal Society (London) **A117**, pp 610-624

Dirac PAM, (1938); "Classical Theory of Radiating Electrons" Proceedings of the Royal Society of London **A167**, pp 148-169

Erdmann M, (1997); "The Partonic Structure of Photons" Springer Verlag, Berlin

European Muon Collaboration, (1988); Ashman J, Badelek B, Baum G, et al., "A measurement of the spin asymmetry and determination of the structure function g1 in deep inelastic muon-proton scattering", Physics Letters B, 206, pp 364 – 370

Feynman RP, (1995); "QED The Strange Theory of Light and Matter", Princeton University Press, Princeton NJ, Ch 1

Geiger H and Marsden E, (1913); "The Laws of Deflexion of a Particle Through Large Angles", Philosophical Magazine, **25**, pp 604 – 623

Gell-Mann M, (1961); "The Eight-Fold Way: A Theory of Interaction Symmetry", Caltech Synchrotron Laboratory Report CTSL-20

Gell-Mann M, (1962) "Symmetries of Baryons and Mesons", Physical Review, **125**, pp 1067 – 1084

Hestenes D, (1990); "The Zitterbewegung Interpretation of Quantum Mechanics", Foundations of Physics, **20**, pp 1213 – 1232

Hestenes D and Weingartshofer A (1991), Ed., The Electron, New Theory and Experiment, Proceedings of Antigonish Conference, Kluwer Academic Publishers, Dordrecht

Hofstadter R, (1956); "Electron Scattering and Nuclear Structure", Reviews of Modern Physics, **28**, p 214

Hofstadter R, (1962); "The Electron Scattering Method and its Application to the Structure of Nuclei and Nucleons", Nobel Lecture Physics, 1942-1962, Elsevier Publishing Company, Amsterdam, pp 560 – 581

Hughes IS, (1991); "Elementary Particles", Cambridge University Press, Cambridge UK, Ch 7

Jaynes ET, (1991); "Scattering of Light by Free Electrons as a Test of Quantum Theory", In "The Electron, New Theory and Experiment", Ed. Hestenes D. and Weingrtshofer A., Kluwer Academic Publishers, Dordrecht NL, pp 1 – 20

Jost R, (1972); "Foundations of Quantum field Theory" in "Aspects of Quantum Theory", Ed., Salam A. and Wigner EP, Cambridge University Press, Cambridge, UK

Kirkland A and Brown PD, Eds., (1998); "The Electron, Proc. Int. Cent. Symp. On the Electron", IOM Communications, London

Langford-Smith F, (1955); "Radiotron Designers Handbook", AWA Valve Co, Sydney

Lincoln D, (2017); "God's Thoughts, Practical Steps Towards a Theory of Everything", https://www.osti.gov/biblio/1410794

Littauer RM, Schopper HF and Wilson RR, (1961a); "The Structure of the Proton and Neutron", Physical Review Letters, **6**, p 286

Littauer RM, Schopper HF and Wilson RR, (1961b); "The Structure of the Proton and Neutron", Physical Review Letters, **7**, pp 144 – 147

Lorentz HA, (1952); "Theory of Electrons", Dover Books, NY

MacGregor MH, (1992); "The Enigmatic Electron", Kluwer Academic Press, Dordrecht, NL

Margeneau H, (1961); "Preliminaries" In Quantum Theory, 1, Elements, Ed Bates D.R., Academic Press, New York

Nakamura et al., (2013); (Particle Data Group) http://pdg.lbl.gov/2013/tables/contents_tables.html

Qui-Hong H, (2007); "The Nature of the Electron" http://arxiv.org/pdf/physics/0512265v1.pdf

Reines F and Cowan CL, (1956); "The Neutrino", Nature, **178**, p 446

Robinson VNE, (2011); "A Proposal on the Structure and Properties of the Electron", Particle Physics Insights, **4**, pp 1-18 http://dx.doi.org/10.4137/PPI.S7102

Rutherford E, (1911); "The Scattering of α and β particles by Matter and the Structure of the Atom," Philosophical Magazine, **21**, pp 669 – 688

Rutherford, E. (1919). "Collision of α particles with light atoms. IV. An anomalous effect in nitrogen". The London, Edinburgh, and Dublin Philosophical Magazine and Journal of Science. Series 6. **37** (222), pp 581–587. doi:10.1080/14786440608635919.

Schrödinger E, (1930); "Über die kräftefreie Bewegung in der relativistischen Quantenmechanik", Sitzungsberichte der Preußischen Akademie der Wissenschaften. Physikalisch-mathematische Klasse, pp 418 – 428

Schrödinger E, (1931); "Zur Quantendynamik des Elektrons", Sitzungsberichte der Preußischen Akademie der Wissenschaften. Physikalisch-mathematische Klasse, pp 63 – 72

Schweber SS, (1994); "QED and the Men who Made it, Dyson, Feynman, Schwinger and Tomonaga", Princeton University Press, Princeton NJ

Thomas AW, (2015); "Origin of the spin of the proton", Australian Physics, **52**, pp 91 – 95

Thomson JJ, (1897 a); "On the Cathode Rays", Proceedings of the Cambridge Philosophical Society, **9**, pp 243 – 244

Thomson JJ, (1897 b); "Cathode Rays", Philosophical Magazine, **4**, pp 293

Uhlenbeck GE and Goudsmit SA, (1925); "Spinn Elektronen und die Struktur der Spektren" Naturwissenschaften, **13**, pp 953-954

Uhlenbeck GE and Goudsmit SA, (1926); "Spinning Electrons and the Structure of the Spectrum", Nature, **117**, pp 264 – 265

van der Mark M and Williamson JG (2021); "Relativistic Inversion, Invariance and Inter-Action", Symmetry, 2021, **13**, 1117. https:// doi.org/10.3390/sym13071117

Williamson JG, (2019a); QV0007: Quicycle, Troon: https://quicycle.com/video/qv0007-john-g-williamson-unifying-the-electrons-particle-wave-natures/

Williamson JG (2019b); "A New Theory of Light and Matter", Journal of. Physics: Conference. Series. **1251** 012050 doi:10.1088/1742-6596/1251/1/012050

Williamson JG and van der Mark MB, (1997); "Is the electron a photon with toroidal topology?" Annales de la Foundation Louis de Broglie, **22**, pp 133 – 178

Zweig G, (1964); CERN Reports 8182/TH, 401 and 8419/TH, p 412

4. Elementary Particles

Abstract:

The nature of elementary particles is presented. It is shown that muons and pions are an integral part of the structure of nucleons. The remaining baryons are induced into the nucleons when they are accelerated and produced when they are decelerated by collisions with other particles. They are not part of the structure of nucleons at rest.

4.1. Introduction

Chapters 2 and 3 showed how the common particles were formed from rotating photons. From studies of cosmic ray bombardments and particle accelerators, other particles are known to exist. As they are part of the universe, their presence, and roles in it need to be described and understood. When they were first detected they were called "elementary particles." The term was originally used to describe the array of sub-nuclear particles that were detected from cosmic ray sources and particle accelerators (Hughes, 1991). That terminology is continued herein. The elementary particles are divided into two main groups, leptons, and hadrons.

Leptons, meaning "light," are spin half particles that have properties similar to the electron and its neutrino. They are considered to be Dirac particles, i.e., single constituent particles. They are regarded as not taking part in the strong force interaction that holds nucleons together in the nucleus.

Hadrons are regarded as particles that do take part in the strong force interactions. They are considered to be composed of more fundamental particles, called quarks. Hadrons are divided into two groups, mesons, and baryons. Mesons are considered to be composed of two quarks and have unit spin of 1 or 0 (x $\hbar$). Baryons are regarded as particles composed of three or more quarks. They are either protons or particles more massive than protons.

Bosons, which theoreticians use to describe force-carrying particles, are sometimes included as elementary particles. They all have spin 1 (x $\hbar$). The photon, being the force-carrying particle for electromagnetism, is regarded as a boson.

Altogether, over two hundred separate particles have been detected. This chapter discusses their origins and the role they play in the structure and properties of the universe. Chapter 2 showed how the rotating photon model could be used to explain the properties of particles. Every individual subatomic particle consists of a photon of the appropriate energy rotating in a circle, making two revolutions within its wavelength. The mass of the particle is set by the rotating photon's frequency. The higher its frequency, the greater the particle's mass. The relationship is linear through the equations $E = h\upsilon$ and $m = h\upsilon/c^2$.

Electric charge is explained by the electric field, E, of the rotating photon, and the emission and absorption of virtual photons that are generated by the centripetal force as the rotating photon travel in a circle. Spin is the angular momentum of the photon. The magnetic moment is generated by the photon's magnetic field, B. Charged particles consist of rotating circularly-polarised photons. Neutral particles consist of rotating plane-polarised photons.

The discovery of elementary particles resulted in a theory brought forward to explain them. It is called the "standard model." The theory and the story of their discovery are interconnected.

4.2. Their Origins

After the discovery of the electron and proton, the next particle discovered was the neutron. Isolated neutrons would spontaneously decay into a proton and electron with a half-life of about 10 minutes. The energy difference between the mass of a proton and that of a neutron generated an electron-strength oscillation to conserve charge. The neutron decay therefore yields two circularly-polarised rotating photons, the proton, and the electron. The neutron's original spin, ½$\hbar$, and its plane-polarised component remained. There is also an intrinsic spin, $\hbar$, difference between the proton and the neutron. It was hypothesized that the difference could be carried off as a neutrino, a 'little neutron.' Neutrinos were subsequently discovered.

All other elementary particles have two features in common. Firstly, they are produced by energetic particles colliding with other particles, either stationary or moving in the opposite direction. None are produced by spontaneous decay.

Secondly, they are all short lived and decay by emitting linear photons or particles that decay to linear photons, the original particle, and possibly lepton-anti-lepton pairs. To understand the origins and role of these 200 or so elementary particles, we must consider what happens to particles when they are accelerated.

When an accelerating electric potential is applied to a charged particle, it imparts energy to it, increasing its speed. Through the special relativity corrections, several events occur. Its increased speed applies a relativistic velocity correction, increasing its mass according to Equation 2.7. It also slows down its time according to Equation 2.8. The energy, E, imparted to the proton is divided between the mass of the particle, through $E = mc^2$, and its speed, v, through the relationship:

$$E^2 = p^2c^2 + m_0^2c^4 = m_v^2c^4 \qquad (4.1$$

The higher the speed of a particle, the more its mass increases and its time slows down. The origins of the elementary particles can be best considered by what happens to protons when they are accelerated to high energies. A rest proton's fundamental frequency is 2.2688... x 10^{23} Hz. Protons also have $1/3^{rd}$ and $1/9^{th}$ frequency harmonics. As far as accelerated protons are concerned, they are at rest. Their mass and time frames are in keeping with their rest properties. They can travel indefinitely as a stable particle. The existence of high energy cosmic rays with energies of 100 GeV, that have travelled for hundreds of thousands of years, testifies to their stability.

Their increased frequency is accompanied by increased mass and slowing down of time compared to observers at rest, as per the special relativity corrections discussed in chapter 2. In its own time frame, the nucleon's oscillations are at the same frequency as their rest frequency oscillations. Only external observers will be aware of their increased frequencies through their increased mass. As far as the protons are concerned, nothing has changed. While travelling at those high energies, they have no reason to (nor will they) emit any radiation, particles or otherwise.

By way of example, a 1 GeV proton has a mass approximately double its rest mass. Its time scale is approximately 70% of its rest time scale. A 1 TeV proton has a mass approximately 1,000 times its rest mass and its time scale is approximately 0.1% of its rest time scale.

Consider, then, what happens when they are stopped by collision with other particles at rest. Their mass and frequency are now no longer in equilibrium with their new rest time reference scale. Their additional mass makes their frequencies higher than the stable proton frequencies. They must therefore rid themselves of the extra energy contained within them.

4.3. Muons and Pions

Stopped particles can rid themselves of excess energy in one of two ways. They can emit linear photons through a process called bremsstrahlung, or breaking radiation, or they can emit rotating photons. The outer most stable oscillations in a proton are its P/9 harmonics. They are an integral part of the proton and exist within it. The stopping process jolts the outermost proton oscillation from the structure. Its stable frequency is ⅑th of the frequency of the central proton oscillation. Its rest mass would therefore be about ⅑th of a proton's mass.

A P/9 oscillation jolted out of a proton would have properties like that of a muon. Table 4.1 shows a comparison of their properties. The only difference between the two is that the muon has a mass approximately 1.4% greater than that of a P/9 oscillation. That is attributed to the mass release from the muon's binding energy within the proton.

Muons are unstable. Their half-life is typically 2.2 x 10^{-6} sec. A μ^- decays to an electron. When electrons are emitted, they are accompanied by an electron neutrino. That gives a spin ½$\hbar$ imbalance. The electron neutrino generates an anti-muon neutrino to conserve angular momentum.

Property	P/9 Oscillation	Muon
Rest Mass (MeV/c^2)	104.3	105.6
Spin	$\frac{1}{2}\hbar$	$\frac{1}{2}\hbar$
Charge	e^+ or e^-	e^+ or e^-
Intrinsic Spin	$+\hbar$ or $-\hbar$	$+\hbar$ or $-\hbar$
Decay Products	$\mu^+ \rightarrow e^+ + \nu_e + \nu_{\mu}^-$; $\mu^- \rightarrow e^- + \nu_e^- + \nu_\mu$	

Table 4.1: *Comparison between a proton's 1/9th frequency harmonic oscillation and a muon.*

The reverse occurs when a μ^+ emits a positron. An anti-positron neutrino, which is an electron neutrino, is also released. It is accompanied by an anti-muon neutrino necessary to conserve angular momentum.

The excess energy between the muon's mass and that of the electron and neutrino is carried away by the kinetic energy of the particles. They will have considerable excess energy because of the high energy difference between the accelerated and rest energy. Muons have a rest energy of $\approx$ 105 MeV/c^2, while accelerated particles have energies of GeV and greater.

Muons are not fundamental oscillations like electrons. They are unstable, decaying with a half-life of about 2.2 x 10^{-6} sec. They were the first "elementary particles" discovered.

At that time there was considerable interest in what enabled the positive protons to be held in the nucleus by the neutral neutrons. Yukawa (1935, 1937) proposed that protons and neutrons could be held together by the mutual exchange of a particle that had a mass of about 250 times the mass of an electron. If those particles were exchanged quickly enough, his mathematics indicated they could hold a nucleus together against the proton's repulsive force. He postulated that the particle needed to have a mass of about 135 MeV/c^2. It was soon after that tracks of a 105.6 MeV/c^2 particle were identified by Anderson and Neddermeyer (1936). With the only tracks previously identified being due to alpha particles, protons, positrons (previously predicted by Dirac, 1928) and electrons, the surprise discovery of the new particle prompted Ravi to ask: "*Who ordered that?*" The newly discovered particle was called a mesotron, from the Greek word for "middle". The title was later shortened to mesons. Yukawa claimed the mesotron as his predicted particle, even though it was in error by over ten percent. Further analysis showed that particle did not have the right properties to be Yukawa's predicted particle.

A decade later, a particle with a mass of 139.6 MeV/c^2 was identified (Occhialini and Powell, 1947; Lattes, Occhialini and Powell, 1947). The charged pion had a mass closer to Yukawa's prediction and it was recognized as Yukawa's predicted particle. Its exchange was considered to form the basis of the strong nuclear force for some time. Other particles were also identified from cosmic rays, the charged and neutral kaons, and the neutral lambda.

Nuclear physics was an important topic. Experimentalists realized they could learn more about sub-nuclear particles if they could build their own particle accelerators and smash them into other particles under controlled conditions. That justification for particle accelerators soon led to the discovery of over 200 known separate particles (Close, Marten and Sutton, 1987; Beringer et al., 2012). They were all produced by smashing high-energy charged particles into selected targets. All had very short half-lives, varying between $\approx 2 \times 10^{-6}$ second and $\approx 10^{-24}$ sec. Over the following decades, theoreticians sought to give order to them.

Going back to the decelerating proton, the above showed that a P/9 oscillation can be jolted out of a high-energy proton when it is stopped. The reverse process can also occur. A high-energy proton can jolt an outer oscillation from another nucleon, proton, or neutron, when it is struck.

Figure 3.18 D showed that when a proton emits a positron, it becomes a neutron. To do that, it must generate an N/27 oscillation as well as a neutrino. The same would occur when a proton emits a positive P/9 oscillation. When protons shed P/9 oscillations, they would generate negative ***N***/*27* oscillations. The ***N*** notation refers to nucleons in general. That separates them from neutron oscillations, which are referred to as N. A negative ***N***/27 oscillation would be attracted by the jolted positive P/9 oscillation. An N/27 oscillation has a negative charge. An ***N***/27 oscillation comes from either nucleon and can have either a positive, neutral, or negative charge. All have a mass of $\approx$34.8 MeV/c^2.

The flurry of the rapid time frame change does not lead to an orderly decay of particles. There are other possibilities beyond a P/9 oscillation generating a negative **N**/27 oscillation. A proton's negative P/9 oscillation could generate a positive **N**/27, which would also be attracted to it. There is also the possibility that, in the flurry of activity, a neutral ***N/27*** ^{0} oscillation could be generated. Its close proximity to the positive or negative P/9 oscillations results in it attaching itself to one of them.

These possibilities also exist if there are neutrons in either the target or projectile nuclei. A proton losing a +ve P/9 oscillation becomes a neutral nucleon, automatically generating an **N**/27. The possibilities include **N**/9^{+} and **N**/27^{-}; **N**/9^{+} and **N**/27^{0}; **N**/9^{-} and **N**/27^{+}; **N**/9^{-} and **N**/27^{0}. To a first approximation, the masses of the N/9 and P/9 oscillations, collectively referred to as **N**/9, are ≈ 104.4 MeV/c^2. In the same manner, the masses of the **N**/27 oscillations are 34.8 MeV/c^2.

When they are added, they have a mass of ≈139.2 MeV/c^2. As in the case of a muon, their release from a nucleon would add an extra ≈1.5% to their theoretically expected mass. That would give them a theoretically expected mass of 141.2 MeV/c^2. They all consist of two spin ½ $\hbar$ particles. Theoretically these could be either added or subtracted to give them a spin of 1 or 0.

In practice only spin 0 pions are detected. With both generated at the same time in the same place, spin conservation suggests they will have opposing spins. When combined their combination gives spin 0 particles.

Their charge possibilities are:

N/9^{+} and **N**/27^{-} and **N**/9^{-} and **N**/27^{+} gives 0 charge,
N/9^{+} and **N**/27^{0} gives + ve charge,
N/9^{-} and **N**/27^{0} gives – ve charge.

When positive and negative charge particles unite, they have a stronger binding energy than when a charged and neutral particle unite. The charged particles will have a higher rest mass than the neutral particles. Not being a fundamental particle, they would be unstable. Charged particles would decay to a charged muon, an **N**/9 oscillation, plus a neutral **N**/27^{0} oscillation. The neutral particle is composed of two opposite charged particles of approximately equal size. That would enable them to unlock each other's angular momentum. Decaying to two gamma rays is a high possibility.

The nearest detected particles to a mass of 141.2 MeV/c^2 and with spins 0, are pions. Charged pions, $\pi^+$and π^-, have a rest mass of 139.6 MeV/c^2. Their binding energy is ≈ 1.5 MeV. Neutral pion, π^0, have a rest mass of 135 MeV/c^2. Their opposing charges gives them a binding energy of ≈ 6 MeV.

Charged pions, $\pi^+$and π^-, decay to μ^+ and μ^- respectively, and neutrinos. Neutral pions, π^0, mainly decay to two gamma rays.

Together, the above is a good indication that the pions are indeed composed of a combination of the **N**/9 and **N**/27 nucleon harmonic oscillations. They are jolted out of energetic nucleons when their time reference frame is slowed down by collision with another nucleon. The energetic nucleon has excess mass for that time reference frame and must get rid of it. As mentioned in section 3.7, rotating photon structures have a lower energy level. They are preferred over linear photons when sufficient energy (and possibly other circumstances) allow.

When those pions and muons are jolted out of the nucleons, they carry part of the original nucleons' energy with them. Through equation 4.1 they have a velocity and higher frequency than their rest frequency. Between them the pions and muons ultimately decay to electrons and neutrinos and/or their anti-particles. Except for neutral pions, all others decay to particles with the excess energy imparted as kinetic energy to those particles. The fact that charged pion and muon decay rarely involve photons is another indication that double rotation circularly-polarised photons are more stable than their linear counterparts. Neutral pions decay to photons because their similar size enables them to unlock each other's angular momentum.

Table 4.2 summarizes the properties of the various harmonics available to be jolted out during impact. For ease of calculation, the protons have been normalized around a nucleon (**N**) mass of ≈ 939 MeV/c^2, an error of less than 0.5%.

Figure 4.1 summarizes the dimensions of these different pions. The oscillations are shown with the muon inside the N/27 oscillation. There is no reason why their paths could not cross, or why the oscillations could not join up outside both oscillations.

The important feature here is that this model gives an approximate dimension for muons and pions at rest.

Oscillation	**Radius** (fm)	**Mass** (MeV/c^2)
N/1	0.105	939
N /3	0.315	313
N /9	0.945	104.3
N /27	2.835	34.8

Table 4.2: *The radii and masses of different one-third frequency harmonics of the nucleons, using the normalized nucleon (**N**) mass of 939 MeV/c^2 for ease of calculation.*

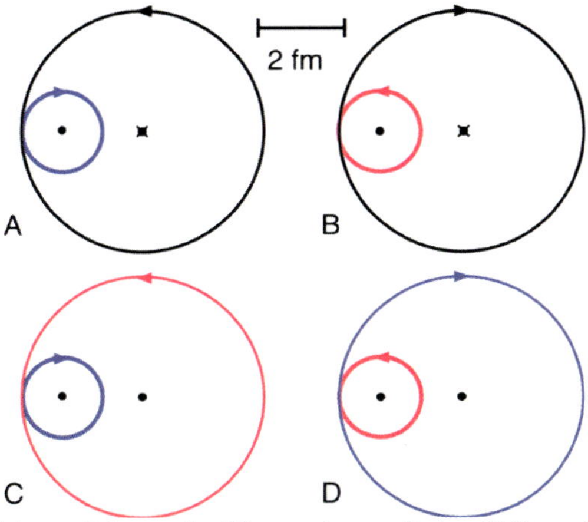

Figure 4.1: *Schematic illustrations of the N/9 and N/27 oscillation combinations that make up the charged and neutral pions. Red represents positive, blue negative and black neutral harmonic oscillations.*

If their dimensions were measured by inelastic electron scattering when they were accelerated, they would be reduced by equation 2.4. Note that the pions and muons are emitted with additional energy. When they subsequently decay to electrons and gamma rays, those particles will contain the additional energy.

That raises the question: "Why haven't **N**/27 oscillations been detected independent of the **N**/9 oscillation?" ***N***/27 oscillations are only activated when an ***N***/9 oscillation is ejected from the nucleon. They will accompany it. That explains why pions and not muons are emitted in particle accelerator collisions. It was the short half-life of pions that led to the earlier discovery of muons when cosmic rays were the only source of accelerated particles.

It is perhaps noteworthy that when Littauer et al. (1961 a, b) noticed the charge distributions in the nucleons, as illustrated in figures 3.6 and 3.13, they speculated about pion resonance clouds surrounding the core of the nucleons. This suggests that their speculations had some merit.

4.4. Some Other Elementary Particles

Since the discovery of muons and pions, experimentalists have identified over 200 elementary particles. The uniqueness of the muons and charged pions lies in their long half-lives, 2×10^{-6} sec and 1.2×10^{-8} sec respectively. Although they are created in the upper atmosphere by incoming cosmic rays, the relativistic corrections mean that some could reach ground level and be detected in cloud chambers or on photographic emulsion plates. Scientists realized they could detect more particles if they detected them at higher altitudes. High altitude balloons carrying various detection methods were launched. In addition to detecting more muons and pions, they also detected tracks from other particles.

The early success of the quark model, first forwarded by Gell-Mann (1961, 1962) and Zweig (1964), led to the belief that those particles were also yielding information about the structures inside the nucleons. Rather than wait for the availability of cosmic rays, particle physicists requested funds for particle accelerators. They could smash charged particles into other particles at known energies and detect all the products generated. It was then that the particle explosion really began. Within a few decades, the number of particles known to exist went from eight (the electron, proton, neutron, and neutrino, as well as their anti-particles) to over 200.

To make sense of it all, the range of quarks and gluons was increased. The up and down quarks were accompanied by strange, charm, top, and bottom quarks. Gluons came in 8 colours. The present model suggests alternative explanations. Look at what remains after some $\underline{\boldsymbol{N}}/9$ and $\underline{\boldsymbol{N}}/27$ oscillations have been jolted out of accelerated nucleons.

The following uses the example of a proton of rest energy ≈ 1 GeV accelerated to many GeV. Upon impact, its time reference rapidly and dramatically changes. It must shed the excess energy that was put into it to achieve the high energy. It can only do that by shedding photons, either linear or rotating.

The emission of linear photons is known as bremsstrahlung, breaking radiation. The process is straight forward. Excess mass energy is converted directly to linear photons.

Rotating photons are emitted by the P/9 and N/9 oscillations being jolted out of the nucleon. They are charged. Just as when a proton emits a positron to become a neutron, it must also generate an N/27 oscillation. As mentioned above, different combinations are available.

Considering protons only, P/9 oscillations are the outermost oscillations and the easiest to jolt out. They cause N/27 oscillations to be generated in the same place and time. Being emitted together with opposing spins, they will form pions. Neutrinos are also necessary.

As soon as the pions are ejected, new P/9 and N/27 oscillations are formed to replace them. That process continues, interspersed with linear photon emission, until the proton's mass matches its new rest time reference frame. It explains why pions are the most numerous particles emitted in high energy collisions.

What happens to the original proton? Each time a pion or linear photon is jolted out, it loses some of its energy and possibly charge. It cascades through a series of quasi stable states until it reaches its new rest time reference mass.

Table 4.3 indicates some of the particles that have been encountered in the final stages of an accelerated proton returning back to the stable state of either a proton or a neutron.

Each of those stages have been identified and labelled as a separate particle. They aren't! They are just protons cascading through a series of quasi-stable states as they continue to rid themselves of excess mass until they reach the stable rest mass of either a proton or neutron.

Many other possibilities exist. The higher the energy to which the nucleons are accelerated, the greater the number of combinations through which they would transition back to rest mass.

Particle	**Mass** (MeV/c^2)	**Decay Mode**†	**Lifetime** (sec)
Ξ^-	1321.3	$\Lambda^0 + \pi^-$	1.64 x 10^{-10}
Ξ^0	1314.9	$\Lambda^0 + \pi^0$	2.9 x 10^{-10}
Σ^-	1197.4	$n + \pi^-$	1.48 x 10^{-10}
Σ^0	1192.5	$\Lambda^0 + \gamma$	7.4 x 10^{-20}
Σ^+	1189.4	$n + \pi^+$, or $p + \pi^0$	0.80 x 10^{-10}
Λ^0	1115.6	$p + \pi^-$, or $n + \pi^0$	2.63 x 10^{-10}

Table 4.3: *Selected hadrons and some properties. The masses are expressed as MeV/c^2. (†Indicates the most common decay modes.)*

The particles listed in table 4.3 that decay to nucleons are not nucleons to which pions have been attached. That would give the particle a different spin, just as attaching a **N**/27 to a muon with spin ½ ($\hbar$) to produce a pion changes its spin to 0. The accelerated nucleons have higher frequency oscillations they must get rid of in their new rest time frame.

During the stopping process the excess energy is divested as either linear or rotating photons. The same effect occurs when cosmic ray protons strike upper atmosphere nuclei. The energy excess and the next lower quasi-stable proton state determine the radiation that will be emitted.

Protons with remnant excess mass form the majority of the so-called "elementary particles" that are more massive than the protons at rest. The have spins of ½, 1½ ($\hbar$) or higher. They are called hadrons and have at least three oscillations. Of those, the proton and bound neutron are nucleons that have higher frequency oscillations they must get rid of when they are stopped, and their time frame reference is changed. They are not separate particles that need separate explanations. They are different stages of energetic nucleons shedding their excess mass as energy in the form of linear photons or particles in the form of rotating photons when they return to their stable rest mass oscillation.

The reason for the easy generation of muons and pions can be seen from the illustrations of the proton, figure 3.13 D, and neutron, figure 3.15. The P/9, N/9 and N/27 oscillations are the outermost. They would be shed first. There are many other particles called mesons. They have spin 0 or 1, indicating they are a combination of two oscillations.

This presentation has not made an exhaustive study of all the elementary particles detected so far. The only oscillations studied so far are the outer P/9, N/9, and N/27 harmonics. This study has shown that these harmonic oscillations can explain the high rate of generation of muons and pions and their mass, spin, charge, and decay products. Other combinations are possible.

The rate at which accelerated nucleons smash into other nucleons would not be expected to give a smooth transition. The **N**/9 and **N**/27, as harmonics of the basic nucleon oscillation, would be expected as by-products because they are the outermost oscillations associated with nucleons at rest. In the chaos of high energy collisions, many different frequency harmonics could be expected.

However, as they near rest energy, the one-third frequency harmonics would show through. The products of particles with a mass less than nucleon rest mass would be expected to transition through pions and muons on their way to electrons/positrons, neutrinos, and gamma rays. Most of those with rest mass greater than nucleon rest mass would also be expected to emit pions and muons on their way to the stable nucleons, P^+ or N^0, and other stable particles e^+, e^-, ν, and γ. It should not be surprising that there should be some order in the way the energetic oscillations reach stability.

Many other particles have been detected, including W, Z, and Higgs bosons. Bosons are all spin $1(\hbar)$ particles. This means they are composed of at least two spin $\frac{1}{2}\hbar$ particles. They have masses of ≈ 81, 92 and 126 GeV/c^2 respectively. Their lifetimes are less than ≈ 10^{-21} sec. Their decay modes vary. But decay they all do because they are unstable. Their end products are photons, neutrinos, and electrons and/or protons and/or neutrons. There is nothing else.

Nucleons have a rest mass of ≈1 GeV. Accelerating them to over 1,000 times their rest mass gives plenty of opportunity for different combinations of particle oscillations to be generated and detected. It should not be surprising that some order is found in these combinations. Placing them in perspective, a proton at rest mass has a frequency of ≈ 2.7×10^{23} Hz. Particles with a fraction mass of a rest mass approximately 100 times more massive than a proton would have a frequency of ≈ 10^{25} Hz. Under this model, their detection is nothing more than a combination of energetic rotating photons making a few rotations before decaying.

4.5. Nucleon and Anti-Nucleon Annihilation

Electrons and positrons are rotating sine wave oscillations. When they come together, they unlock each other's angular momentum and release photons corresponding to the rotating photons that were the particles.

The situation with proton/anti-proton annihilation is more complex. Protons have three harmonic oscillations, P/1, P/3 and P/9. As has been shown earlier these can be separated from the proton to give muons (and pions). They are not tightly bound to the proton's core.

When a proton and anti-proton meet, they will unlock each other's angular momentum and generate photons as expected. The three harmonics are separate rotating photons that do not exist continuously. This leads to many possibilities. As well as photons, muons, pions and other mesons, other possibilities include the energy of some annihilated particles being captured by other particles, releasing high energy particles.

These have been observed, with various combinations of mesons, including pions, being found in their annihilation products (Klempt et al., 2008).

The reason for this is quite apparent from figures 3.12 and 3.15. The different oscillations only exist for fractions of time. When one oscillation meets it anti-oscillation, it releases photons, leaving insufficient energy to maintain the proton's complete structure. The loosely bound P/9 harmonics still travel their circular trajectories. As such, proton/anti-proton annihilation releases the outer oscillations, muons. That automatically generates a loosely bound ***N***/27 oscillation. Being at rest and in close proximity, they combine to form a pion.

Neutron/anti-neutron annihilation is similar. The question is often asked: "How can a neutral particle have an anti-particle of opposite charge?" The answer can best be seen in figures 3.16 B and C. An anti-neutron (AN) would have an AN/1 negative core, surrounded by an AN/3 positive oscillation, a further AN/9 negative oscillation, all surrounded by a weak positive AN/27 oscillation. A neutron/anti-neutron pair would interact in the same manner as a proton/anti-proton pair.

They are the results for proton/antiproton annihilations at rest energy. Higher energy particles give greater opportunity for even more combinations.

Just as a neutron is like a neutral proton, an anti-proton is similar to an anti-neutron. An anti-proton would annihilate a neutron. An anti-neutron would also annihilate a proton in a similar manner. The ever changing array of harmonic oscillations in the nucleons gives a considerable array of possibilities when they annihilate. That is what is detected.

4.6. Elementary Particles Under this Model

This rotating photon model of matter is one in which everything is composed of electro-magnetic radiation, that is, photons. Photons are supported by the electric permittivity, ε_0, and magnetic permeability, μ_0, of free space. Photons come in an array of wavelengths ranging from many kilometres to fractions of a femto-meter, as shown in figure 1.4.

They come with two polarisation states, circular and plane, and two travel modes, linear and circular in a circumference of half their wavelength. Linear photons constitute energy. Those travelling in a circular path are called particles. Particles composed of rotating circularly-polarised photons are electrically charged. Those composed of rotating plane-polarised photons have no residual charge.

Both photons and particles have an ability to transmit inertia and resist force. This comes about because of the ability of ε_0 and μ_0 to both transmit and resist the passage of photons through free space. If they did not resist the passage of photons, at least one of ε_0 or μ_0 would be 0 and photons would move at infinite speed through space.

The values of ε_0 and μ_0 both provides the means for photons to move and resists that movement, giving photons the property of (inertial) mass. This also shows up in the famous equation, $E = mc^2$ (equation 2.2). Energy, E, is photons travelling in a straight line. Mass, m, is the same photon travelling in a circle, making two revolutions per wavelength.

The chirality of the rotating photon determines the charge of the particle. Positively charged particles rotate in one direction with respect to their magnetic field. Negatively charged particles rotate in the other direction. This makes an anti-particle the mirror image of its corresponding particle. A particle's property of spin is only available in two states. A spin up particle is the "other side of the page" image of a spin down particle. All these and more are natural properties of the rotating photon model of matter.

The rotating photon model also shows that neither photons nor particles are point objects to which properties such as charge, magnetic moment, wavelength, spin, including intrinsic spin, are attached. Under this model they all have a physical reason for existing and possessing the values they have. It is the physical act of the photon rotating that generates the electric charge and magnetic field. The magnetic field is automatically perpendicular to the photon's rotation plane. Unlock that rotation and the charge and magnetic field disappear.

This explains the quandary of why electric charge remains the same when a particle is accelerated. After all, electric charge is attached to the mass of the particle. As an accelerated particle's mass increases through the special relativity corrections, it may seem that charge, which is attached to it, should also increase. It does not. The reason is simple. As the particle is accelerated, its spin angular momentum remains the same while its velocity and mass increase. Its radius decreases according to equation 2.4. Because its spin is the same, the same momentum of virtual photons is emitted as the photon travels in its circle.

Equation 2.10 indicates that, under this model, the Bohr magneton, the magnetic moment of a particle, is equal to its charge multiplied by its radius. As the particle is accelerated to higher velocities, its mass increases while its radius decreases. A decreasing radius means its magnetic moment also decreases. Its electric charge remains the same because its $\frac{1}{2}\hbar$ angular momentum does not change.

Experimentalists have measured the radius of electrons and found them to behave like point particles. This is true. But all these measurements have been made at energies of the order of tens of GeV to TeV. Equation 2.4 shows that, at such energies, electrons will have a radius $\approx 10^{-17}$ to 10^{-18} m and scatter like point particles. This is because they were accelerated to high energies. In the same manner the special relativity corrections for matter are due to the rotating photon nature of matter and the constancy of the speed of light for all observers, as Einstein predicted. These and other properties are all automatic consequences of particles under this rotating photon model of matter. In this model, no other form of matter is possible.

Table 4.4 lists the fundamental relationships between the different types of photon states and the particles they generate under this rotating photon model of matter. The important feature is that everything is composed of photons.

That makes it easy to consider interchanges between protons and neutrons. They are photon rearrangements in which electrons and neutrinos are generated, as required under conservation principles.

Figure 4.2 illustrates these relationships. It shows there are two polarised states of photons that generate two types of particles. There are three resonance states, electrons, nucleons (protons and neutrons) and neutrinos. Within the nucleons, there are sub- resonances of one third, one ninth, and one twenty seventh of the fundamental nucleon photon frequencies. An anti-particle is the matter particle's photon rotating in opposite direction with respect to its magnetic field.

Rotating plane-polarised photons generate neutral particles. Rotating circularly-polarised photons generate charged particles. It should be noted that circularly-polarised photons have their magnetic fields reinforcing each other, north over north, south over south. Those properties combine with the intrinsic $\hbar$ twist angular momentum of a circularly-polarised photon's 360° twist through one wavelength and the ½$\hbar$ angular momentum of its double rotation per wavelength to make charged particles a more stable and hence preferred option over linear photons of the appropriate energy.

Types:	
Linear - Energy	Rotating – Matter Particles
Polarization States:	
Elliptical Extremes are plane and circular	Circular – Charged Particles Plane – Neutral Particles
Resonances:	
Nil – thought to be continuous	Electron, Nucleon, Neutrino
Not applicable	None, Proton/Neutron, e, μ, tau

Table 4.4: *Fundamental relationships under the rotating photon model of matter.*

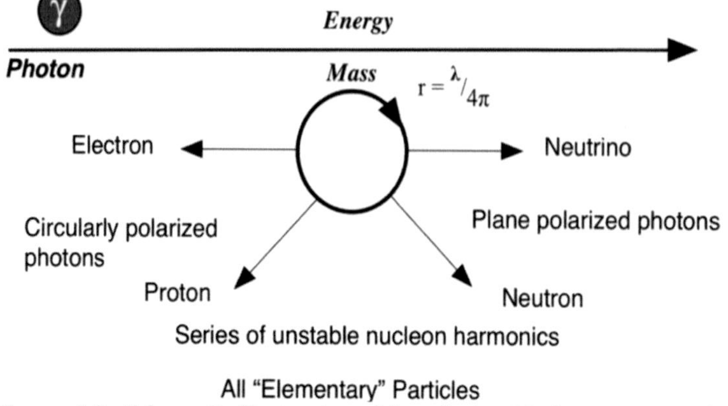

Figure 4.2: *Schematic illustration of the relationship between particles under the rotating photon model.*

A photon travelling in a straight line is energy, denoted γ. The same photon travelling in a circle that makes two revolutions per wavelength is a particle. Two and only two revolutions are necessary because that number superimposes charge and magnetic field upon each other. Any other number, whole or fraction, will have the electric and magnetic fields interfering with each other, and it will not be stable.

By far the most common particles emitted in particle collisions are pions and neutrinos. As mentioned earlier, pions and muons are structures within nucleons. When high energy nucleons, in which they are part, are brought to rest, they are jolted out as the nucleon's time reference frame changes with their decreased speed. The nucleon has excess mass and must shed some. They do that be emitting linear (γ) and rotating (μ, π and ν, and others) rotating photons.

Where possible, photons prefer rotating to linear states. Their closed loop magnetic fields gives them a lower energy state than the open fields of linear photons. This does not necessarily happen instantaneously. It leaves the original nucleons with additional energy in quasi equilibrium states. They continue to emit further γ, μ, π, ν and other psrticles as they cascade through a series of quasi equilibriums until they reach a stable state. The μ, π and other particles are not stable either. They continue to emit γ and ν until they reach the next lowest stable state, electrons as either e^+ or e^-.

As for the remaining 200 or so detected elementary particles, they do not exist in nucleons at rest. They are generated when particles are given additional energy. Their increased energy increases their mass and hence their frequencies. Their increased frequencies are countered by time slowing down within the nucleon, due to the relativistic corrections.

In its own time frame reference, a particle is at rest. High energy nucleons remain stable at high energy. When they are slowed down or stopped their frequencies are no longer in tune with their new time reference frame. They must get rid of the additional mass that accompanies their increased frequencies. It is not surprising that some order is found in this process. That does not necessarily mean their structures or properties have any influence on the behaviour of nucleons at rest.

That should be compared to the standard model of particles. Table 4.5 shows a list of the fundamental particles under the standard model. It shows that standard model physicists require 61 particles to explain the observed properties of matter.

Of the required sixty-one particles, the quarks and gluons, which total 44, have never been separately isolated and detected. The W, Z, and Higgs bosons have very short lives. Figure 4.3 presents the way those various particles are meant to interact.

Particle	**Types**	**Generations**	**Anti-particle s**	**Colours**	**Total**
Quarks	2	3	Pair	3	36
Leptons	2	3	Pair	None	12
Gluons	1	1	Own	8	8
W	1	1	Pair	None	2
Z	1	1	Own	None	1
Photon	1	1	Own	None	1
Higgs	1	1	Own	None	1
				Total:	**61**

Table 4.5: *List of fundamental particles under the standard model.*

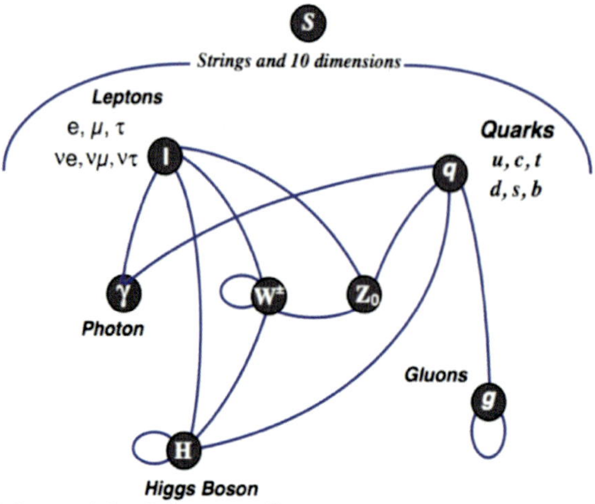

Figure 4.3: *Schematic illustration of the reactions between the fundamental particles under the standard model.*

Despite its complexity and claimed successes, it has not helped to accomplish its main task, an understanding of nuclear binding and the structure and properties of nucleons and nuclei. It is a mathematical construct that has not shown to bear any relationship to the physical world.

So limited has been the progress that it has spawned a branch of physics called string theory. It uses 10 spatial dimensions, 7 of which have never been detected. It has particles too small to ever be detected and uses mathematics that mere mortals can't understand. Its predictions don't match observation. Other mathematical attempts at going beyond the standard model, have been made, also with limited success.

This rotating photon approach of going beyond the standard model is founded on known physical principles. One of its features is that accelerating particles to high velocities will, through equation 2.4, cause the radius of the various oscillations to decrease. Rotating photon protons are not "billiard ball" like structures that can be considered in "conventional" interpretations of scattering.

Nor do they have some solid surface on which elastic scattering can occur. All collisions occur through their electric and magnetic fields. The electric field is essentially two dimensional near the proton. The interactions of passing electric fields are strong enough to dislodge P/9 and other oscillations.

The central core's rest radius of ≈0.105 fm is already exceedingly small. The P/3 oscillation is a little larger at ≈ 0.315 fm. Together they contain over 90% of the proton's charge and mass. Their field is very strong and, to a first approximation, diminishes linearly with the distance from the rotating photon's circumference. It makes little difference to electric field-based collisions whether the radius is ≈ 0.4 fm, or 0.004 fm. The electric field is intense. As such there will not be as great a reduction in proton – proton collision cross sections with velocity as suggested by equation 2.4.

There is also the added complication of the "loosely" bound P/9 oscillations, μ^+ , being knocked out of the protons. This presentation makes no attempt to model proton – proton collisions.

One overwhelmingly dominant finding comes from all the particle accelerator reactions studied. The original particles gain mass from the electromagnetic fields used to accelerate them. When they are stopped, they emit that added energy and revert to a stable state, usually their original form. They do so by emitting either stable linear photons or unstable rotating photons. The latter decay to linear photons and/or to one or more of the stable particles — proton, neutron, electron, or neutrino. No other stable particles have ever been detected.

At some stage it must dawn on particle physicists that this may be due to there being nothing else but photons, electrons, protons, neutrons, and neutrinos. If they have been unable to extract non-photonic materials out of protons, neutrons, and electrons, perhaps it is because they are also composed of photons.

4.7. Electron Stability

Electrons have a rest mass of ≈ 0.51 MeV/c^2. Modern particle accelerators have accelerated them to many TeV, increasing their mass by a factor of several million. When they are smashed into different targets, they emit an array of gamma and X-rays. The original electron remains intact. This raises the question: "Why are electrons so stable?"

There are two aspects to the answer. The first is that, for their mass, the electron's ½$\hbar$ angular momentum, as well as its photon's twist intrinsic spin $\hbar$, are almost 2,000 times stronger than the equivalent effect of protons. The second is that there are no charged stable oscillation frequencies below that of an electron's rest frequency. Even if their angular momentum could be broken, they have no oscillation to which they can decay.

Neutrinos have a much lower frequency and no charge. Electrons cannot decay to them. The only way of removing the stability of an electron is to bring it close to a positron. The two will unlock each other's $\frac{1}{2}\hbar$ angular momentum and the two linear photons will move away in opposite directions, conserving both linear and angular momentum.

4.8. Summary

Over two hundred different sub-nuclear particles have been discovered. The important ones are the proton and neutron that make up the nuclei of all atoms, and the neutrino, which enables those nucleons to interchange. Together they form stable nuclei around which electrons gather to form atoms. They are the basis of our world, the other planets, celestial objects, galactic gas and dust clouds. They last for many tens of billions of years, if not forever. They all have a structure, as was demonstrated in chapter 3. Included in their rest structure are sub-nucleon oscillations that can be knocked out of nucleons to form pions and muons. They decay rapidly to electrons, neutrinos, and photons because they are the only stable oscillations with masses below them. They are part of the structure of protons and neutrons at rest. When isolated, muons last for only a few microseconds. The smallest voluntary action we can do is to blink, which lasts for about 20 milli-seconds. That is 10,000 times longer than the half-life of a muon and about 1,000,000 times longer than the half-life of a charged pion.

Using particle accelerators, physicists have detected a myriad of different identifiable particles. The main feature of all those is their incredibly short lives. Apart from the muons and the charged pions, all other so-called elementary particles have rest half-lives somewhere between $\approx 3 \times 10^{-10}$ second and $\approx 10^{-24}$ second. Their high velocity slows their time down through the special relativity corrections, making them much easier to detect.

The above has demonstrated the physics behind the contribution the pions and muons make to the structures of the nucleons. What contribution can these other short-lived particles make to the stability of particles that last for tens of billions of years? They do not exist in nucleons or nuclei at rest. They are generated by the acceleration of the nucleons to higher energy. Their higher energy adds energy to each of the rotating photon harmonic oscillations.

When nucleons travel at high velocities, their oscillations are stable because, despite their increased mass and hence (rest) frequency, the special relativity correction for time decrease is the same as that for mass increase. The accelerated nucleon does not "experience" a different frequency and, as far as it is concerned, it is still a stable structure. If the nucleon or nucleus is stable at rest, it will not emit any "elementary" particles while it is traveling at high velocity. It is only when the accelerated nucleons are stopped that the harmonic oscillations return to their rest energy, reverting to stable nucleons.

Along the way the excess energy is liberated through the release of a series of photons and oscillations, principally containing pions. Other mesons are also emitted, particularly from higher energy nucleons.

To suggest that these can give us an indication of the structure of a proton and neutron at rest is not feasible. Apart muons and pions, and some forms of N/3 oscillations, they do not exist in particles at rest.

It is suggested that this model gives an opportunity for the experimental results obtained to reveal properties of the various ***N***/1, ***N***/3, ***N***/9 and ***N***/27 harmonic oscillations. In turn this can give greater insight into the nucleons at rest.

Suffice it to say that some of these elementary particles, particularly the muons and pions, are part of the structure of nucleons at rest. The others play no role in the structure or properties of the universe. They were included in this study because they have been detected and should be explained.

The field of elementary particle physics developed out of a desire to find out what the constituents of protons and neutrons are, and to some extent electrons. It was believed that if they were smashed into pieces, those pieces would tell them something about their structure. It is not possible to smash nucleons together without accelerating them to significant energies. As mentioned above, this has the disadvantage that it adds additional energy. The only interpretation is that the particles released were added to them during the acceleration process.

The predominant finding of all the work has been that only photon, or particles that decay to photons and the original particle, are ever detected. Particle/anti-particle pairs formed will annihilate each other to form photons.

There is an alternative to unlocking the inner structure of particles without adding energy to them. Bring a particle and its anti-particle together. They will annihilate each other and, in the process, release all particles that are present when they are at rest.

The most common by-product of an electron/positron interaction is two linear photons, each with frequency $\upsilon = \frac{\sqrt{p^2c^2+m_0^2c^4}}{h}$. The particle's momentum is contained in the increased frequency of its oscillation. There is no residual charge, spin, magnetic moment, or anything else. The reverse is also true, that a sufficiently energetic photon can, when slowed down by proximity to a nucleus, generate an electron and a positron, both of which will have the properties given in Chapters 2 and 3.

As discussed above, the by-product of nucleon/anti-nucleon interactions can produce pions, muons, and photons, and may include more than one per nucleon. In turn the pions and muons decay to produce electrons or positrons, as well as neutrinos and photons. The reverse is also true, that a sufficiently energetic photon striking a nucleus can also generate a nucleon/anti-nucleon pair.

Together, these observations represent the best experimental indication that protons, neutrons, and electrons are made of rotating photons, as presented in chapters 2 and 3. Further, the other detected elementary particles are induced into them by the energy added to accelerate them, and released when they are stopped, as mentioned in section 4.5.

Looking for and assigning special properties to some of the myriad possibilities with half-lives of less than 10^{-10} sec, and assigning important roles to them through mathematics, is not sound science. Physics makes things happen, not mathematics. Mathematics describes the magnitude of an effect, not its cause. Given that the source of elementary particles is the energy put into rest particles in order to accelerate them to high energies, there is little to be gained from going to higher and higher voltages to get shorter and shorter wavelength particles.

The standard model's major success has been predicting the array of particles detected from high energy nucleon collisions. It has not been useful for predicting the behaviour of nucleons within nuclei. The next chapter shows how this rotating photon explains nuclear binding. Those myriad of detected sub-nuclear particles do not play any role, muons and pions excluded.

4.9. List of References

Anderson CD and Niedermeyer, (1936); "Cloud Chamber Observations of Cosmic Rays at 4300 Metres Elevation and Near Sea-Level", Physical Review, **50**, pp 263 – 271

Beringer J, et al., (2012); "(Particle Data Group)" Physical Review **D86**,

Close, F, Marten M and Sutton C, (1987); "The Particle Explosion" Oxford University Press, New York

Gell-Mann M, (1961); "The Eight-Fold Way: A Theory of Interaction Symmetry", Caltech Synchrotron Laboratory Report CTSL-20

Gell-Mann M, (1962) "Symmetries of Baryons and Mesons", Physical Review, 125, pp 1067 – 1084

Hughes IS, (1991); "Elementary Particles", Cambridge University Press, Cambridge, UK,

Klempt E, et al., (2008); "The Anti-Nucleon-Nucleon Interactions at Low Energy: Annihilation Dynamics", http://arxiv.org/pdf/hep-ex/0501020v1

Lattes, CMG, Occhialini GPS and Powell CF, (1947); "Observations of the Tracks of Slow Mesons in Photographic Emulsions", Nature, 159, pp 453 – 456

Littauer RM, Schopper HF and Wilson RR, (1961a); "The Structure of the Proton and Neutron", Physical Review Letters, 6, p 286

Littauer RM, Schopper HF and Wilson RR, (1961b); "The Structure of the Proton and Neutron", Physical Review Letters, 7, pp 144 – 147

Occhialini GPS and Powell CF, (1947); "Nuclear Disintegrations Produced by Slow Charged Particles of Small Mass", Nature, **159**, pp 186 – 190

Yukawa H, (1935) "On the Interaction of Elementary Particles I", Proceedings of the Physico-Mathematical Society of Japan, 17, pp 48 – 57

Yukawa H, (1937); "On the Interaction of Elementary Particles II", Proceedings of the Physico-Mathematical Society of Japan, 19, pp 1084 – 1093

Zweig G, (1964); CERN Reports 8182/TH, 401 and 8419/TH, p 412

5. Nuclear Binding and the Structure of Nuclei

Abstract:

It is shown how the rotating photon structure of nucleons allows them to bind together to form nuclei. They bind in layers in which the nucleons remain intact. The strongest binding between the nucleons within a layer is proton to neutron only. Within each layer, nucleons form strongly bound hexagonal close packed alpha particle structures. Those layers are held together by a weaker force. This presentation enables the structure and many properties of any nucleus of any Z and A to be determined.

5.1. Introduction

We now have the basic building blocks of the universe. It is time to start understanding the universe that exists around us. This is best done by understanding atomic nuclei. They form most of the mass of Earth, the other planets, the sun, as well as other stars and indeed the whole universe. They are the platform that holds the electrons. In turn, the electrons give materials their properties. An understanding of nuclear properties developed in earlier work is presented to show how this study builds upon their knowledge.

Rutherford (1911) proposed that the positive charge of atoms is concentrated in a central nucleus. Geiger and Marsden (1913) verified it. The nature of the nucleus and the binding between the nucleons has been the subject of much debate since then. Chadwick's (1932) discovery of neutrons within nuclei assisted that debate. How could neutral particles hold positive particles together in such a small volume? Over the next few decades many different models of the nature of nuclear binding and the structure of nuclei were forwarded. Many are successful in explaining some of the observed properties of nuclei. None are applicable to all nuclei.

The following is an introductory background to the standard model description of the structure of nuclei and the binding force that holds them together. It serves to introduce some of the properties of nuclei that need to be explained. Bohr (1936) and Bohr and Wheeler (1939) developed a liquid drop model as the first proposed model for the structure of nuclei (Myers (1977). Weizsacker (1935) was one of the first to develop a semi-empirical mass formula that enabled some properties of nuclear binding to be calculated.

Deviations from the theoretical liquid drop model calculations occurred at periodic intervals, leading to the idea of "magic numbers" for nucleons (Bohr and Wheeler 1939). The liquid drop model was successful in explaining the general trends of binding energies and the sizes of nuclei (Weizsacker 1935). It was not successful in explaining the mechanisms of nucleon-nucleon binding that enabled positive and neutral charges to be concentrated in such small volumes.

Some doubted that it was feasible to have individual nucleons forming a "structure-less fluid" in order that properties such as surface tension were meaningful (Hasse and Myers, 1988). A variation of the liquid drop model included a cluster model, in which nucleons clustered together to form alpha particles, ^{4}He. That helped explain the frequency with which alpha particles were dislodged from nuclei through bombardment or radioactive decay (Cook, 2010). The inability of a single structure model to explain the observed nuclear properties has resulted in over 30 nuclear models being currently employed in nuclear physics (Hodgson, 1982, Greiner and Maruhn, 1996).

Wigner (1937) noted that a face centred cubic lattice structure could model some of the quantum mechanical properties of nuclei. Experimental measurements of quadrupole moments, displayed by Townes et al. (1949) and later updated by Segre (1965), suggested certain numbers had special properties, like the "closed shell" numbers of electrons in atoms. Myers and Swiatecki (1966) plotted the differences between the experimentally measured nuclei masses and those predicted by liquid drop models. This gave further credence to the idea of nuclear shells with magic numbers, showing that some nuclei with Z, the number of protons in a nucleus, or N, the number of neutrons in a nucleus, equal to 2, 8, 20, 28, 50, 82 and N = 126 had special characteristics because of their higher binding energies.

MacGregor (1976) offered a variation of this, in which the alpha particle structures were arranged in layers. It was known as the Ising model. This model was useful in explaining nuclear fission. Another variation proposed to account for the "magic numbers" was the Pauling (1965) Spheron model (Pauling and Robinson, 1975). Nucleons were considered to form small clusters of 2, 3, 4 etc, nucleons, and that these clusters were tightly packed within the nucleus.

Such models were moving away from the liquid drop model towards independent particle models, in which the nucleons retained their integrity as individual particles. A shell model for nuclear structure was proposed by Mayer and Jensen (1955). It was based upon experimental evidence that suggested their existence (Brown and Jackson, 1976).

Variations of the shell model are still favoured in nuclear studies. Studies of variations of a lattice type structure, particularly involving face cantered cubic and close packed structures, have also given some useful insight into the structure of some nuclei (Moszkowski, 1955; Santiago and Chung, 1993; Cook 2010). Moszkowski summarized some of the many different nuclear models that had been developed by 1955.

Those models included ideas based upon independent particles, strong interaction, liquid drop, Fermi gas, potential wells, compound nuclei, alpha particle, shell single particle, shell many particles, spin coupling, unified, collective rotational and spheroidal core, among others. Those models tended to consider the detected nuclear structures and quantum principles, rather than the mechanism of nucleon binding. No one model was able to explain all the observed nuclear properties.

It is possible that a better understanding of the structure of nuclei would be achieved if the nature of nuclear binding were considered. Most of the models mentioned do not offer any mechanism for nucleon-nucleon binding. Various ideas for nucleon-nucleon binding were forwarded, some of which depended upon their existence inside a potential well. Satisfactory ideas could be calculated by the addition of different parameters, but there was no clear indication of the mechanism of such binding.

A particle exchange mechanism proposed by Yukawa (1935) suggested that binding could occur through the exchange of pions. Subsequently the exchange was proposed between quarks, the fundamental particles then considered to be the major constituents of the nucleons (Robson, 1978; Petry, 1984; Miller, 1984; Bleuler, 1984).

To date they have not had a major influence upon the ideas of nuclear structure (Cook, 2010). Some authors have gone as far as to suggest that quarks are simply not the correct constituents to be used in the discussion of nuclear structures (Elliot, 1990).

That still leaves the nature of the nuclear force largely unanswered. It is generally believed that nucleons are held together by a strong nuclear force, with radioactivity being caused by the weak nuclear force. Quantum chromodynamics (QCD) has been used to calculate aspects of the nuclear force, which is based upon the exchange of gluons altering the nature of quark interactions (Glashow, 1961: Weinberg, 1967).

Attempts were made to describe properties of the strong nuclear force by QCD theory. It is known to have a short range of the order of 2 - 3 fm. It goes from being attractive at nucleon distances greater than ≈1 fm to being strongly repulsive at distances of less than ≈ 1 fm (Goldman et al, 1988; Moretto and Wozniak, 1993). It is said to be typically 100 times stronger than the electromagnetic force, based upon the distribution of charge in nucleons being spread throughout their volume.

Even with that short range, there are still some nuclei that have loosely bound nucleons. They attach to nuclei for a few milliseconds at radii up to about 5 fm. They are sometimes referred to as satellite nucleons. Their existence seems to suggest the strong force may have a slightly longer range, although with a weaker magnitude. Despite about forty different models of the structure of nuclei and the use of some 61 fundamental particles and bosons (Braibant et al., 2009), there are still many unknowns about the nature of the binding force and the structure of nuclei.

The topics of the structure and properties of nuclei is an extensive one covering many different models and thousands of known nuclei. It has not been possible for one presentation to cover all properties of all nuclei. The major objective of this chapter is to present a new model for the arrangements of nucleons within a nucleus, showing how they bind together, as well as how the model matches some of the observed properties of nuclei. There are too many properties and too many nuclei to cover in this brief presentation, which is based upon the structure of the proton and neutron derived earlier in chapters 2 and 3.

5.2. Background - The Structure of Nucleons in the Rotating Photon Model

Chapters 2 and 3 showed that the individual subatomic particles, the proton, neutron electron and neutrino, were composed of rotating photons of energy $E = hv = mc^2$, where m is the mass of the individual particle, c is the speed of light, v is its frequency, and h is Planck's constant. Under this rotating-photon model, both the proton and neutron are considered to consist of oscillating, self-sustaining closed loops of electromagnetic energy, each consisting of a central positive core of radius $r = \hbar/2m_N c$, where m_N is the nucleon's mass. This central core has a radius of ≈ 0.105 fm, a frequency greater than 10^{23} Hz and accounts for approximately one third of the charge, mass, and spin of the nucleon.

The proton has a series of one-third and one-ninth frequency harmonics outside its central core, generating the remaining charge and the majority of its magnetic moment. The neutron has a series of one-third, one-ninth, and one-twenty seventh frequency harmonics outside its central core. The neutron's core and one-ninth harmonic oscillations are positively charged to a total of $+½e$, where e is the electric charge on a proton or electron. Its one-third and one-twenty seventh harmonic oscillations are negatively charged to a total of $-½e$.

The sum of both charges is 0, making the neutron a neutral particle. The rotation of the series of one-third harmonics generates the electric charge of the nucleons in the two dimensions of their plane of rotation and at the circumference of each individual rotation. The nucleon's magnetic field is in the third dimension (see chapter 3).

Chapter 2 showed that as rotating photon particles move, they are automatically subject to the special relativity corrections of mass, length, and time. It also showed how a component of their frequency in their direction of motion is equal to that of their de Broglie wavelength. These properties are built into the nucleon's structure. There is no need for them to be considered separately.

Under the rotating photon model of subatomic particles, the photon's rotation results in it continually emitting and absorbing secondary or virtual photons as it travels in a closed loop trajectory. It was postulated that this gives these particles the property of electric charge, e. The Bohr magneton magnetic moment of a particle is e multiplied by its radius.

A consequence of this model is that the electric charge originates at the circumference of the rotating photons that are the nucleons. The electric charge in an individual particle originates in two-dimensions and not in three dimensions as indicated in the standard model. Its secondary photons spread the charge's influence into the third dimension as they move away from the rotating photon that generated them.

Associated with each nucleon is an array of rotating photons that spread the charge out from its central core. That charge is entirely positive for the proton but is both positive and negative for the neutron. When these particles are placed side by side, positive charge segments of the proton overlap negative charge segments of the neutron, with the distance between these charges being effectively zero.

The force of electric attraction between two opposing charges varies according to the inverse of the square of their distance. If the distance between two opposing charges is very small, the attractive force between them will be very large. The proton's positive charge overlaps the neutron's negative charge with very little separation. That enables the attraction between a proton and neutron to be described as a "division by zero" force. The following sections indicate how this "division by zero" attractive force can be used to explain many of the known properties of nuclei.

All sorts of mathematical arguments can be used to suggest that this division by zero attraction is strong enough to bind protons together against their mutual repulsion; or is not strong enough because the positive charges of the protons also overlap in this same plane. Therefore, their repulsion when positive charges overlap will be strong enough to overcome the negative/positive overlap between the neutron and proton. These objections can be overcome because "like charges repel" and "opposite charges attract". Where the rotating photons are in the same plane, the negative charge rings from the neutron would attract the positive charges of the proton and the two would pass through each other at what becomes zero distance. The positive charges of the proton and neutron will repel each other by slightly bending their trajectory and passing over one another, forming an interleaved structure.

That would have the effect of distorting the circle and as such would alter the magnetic field produced by the rotating photons. It will be shown later in this chapter that there is a slight distortion of the magnetic fields of the nucleons, which it is suggested is caused by this effect.

The two-dimensional nature of charge at the circumference means that the extent of the positive repulsion between them is lessened. In this manner the physics can be used to show that this binding mechanism is possible. The final arbiter of this model should be its ability to match experimental measurements and predict new and previously unobserved phenomena.

It should be noted that the standard model indicates that the charge distribution on a proton and neutron is three-dimensional. In that situation, the attraction caused by the overlap between negative sections of a neutron and the positive charge of a proton would be far too small to counteract the repulsion between neighbouring positive protons.

When it comes to nuclear binding, the two-dimensional nature of electric charge distribution can generate a strong force when opposing charge loops overlap in the same plane. Add to that the magnetic field in the third dimension. The nucleons would orient their spin axes so that there was maximum magnetic binding between adjacent and local nucleons. It will be shown later why this means that both the spins and magnetic moments of almost all even Z and A nuclei are zero. That marks a distinct difference between this model and those that represent nucleons with charge uniformly distributed throughout their volume.

It should also be pointed out that the dimensions of the individual nucleons are determined from their observed masses and magnetic moments plus this rotating photon model. No other mechanisms or postulates about their dimensions are required. All dimensions determined from this study will come from this rotating photon model.

Figure 5.1 shows the integrated projection of the electric charge distribution of the nucleons on the vertical scale, versus radius from the centre of the nucleon on the horizontal scale. Figure 5.1 A was taken from curve 2 of figure 3.7, a proton; B was taken from curve 1 of figure 3.14, a neutron. Red is positive, blue is negative. Heights above and below the baseline represent charge densities, not vertical height.

Together they give an approximate representation of the amount of charge that would overlap when protons and neutrons are side by side and close to each other. Chapter 3 indicates the distribution of charge is due to a series of one-third frequency harmonics of the nucleon's photon oscillation.

For protons these are denoted P/3 and P/9 for the ⅓rd and ⅑th harmonics, corresponding to masses of 312.9 and 104.3 MeV/c^2 respectively. For the neutron, these are denoted N/3, N/9 and N/27, corresponding to masses of 313.2, 104.5, and 34.8 MeV/c^2 respectively.

Figure 5.2 shows a plan view of the proton, A, and neutron, B, illustrating these harmonic oscillations as circles within a background of the electric fields they generate. Red represents positive charge and blue represents negative charge for both the oscillations and the field. The intensity of the colour is an approximate indication of the field strengths.

Each of the one-third frequency harmonics consists of two oscillations of one charge and one of the other. That gives the overall segment the charge of the dominant harmonic oscillation. The P/3 oscillation are positive. The N/3 oscillations are negative. The "pop" into and out of existence, as required to conserve spin, charge and mass.

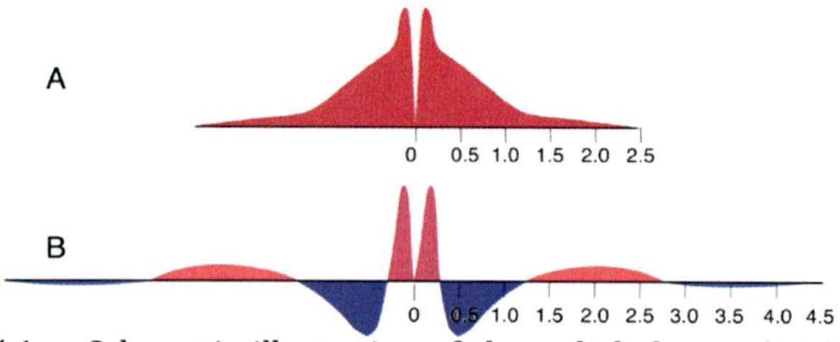

Figure 5.1: *Schematic illustration of the radial charge distribution of a proton, A, and neutron B.*

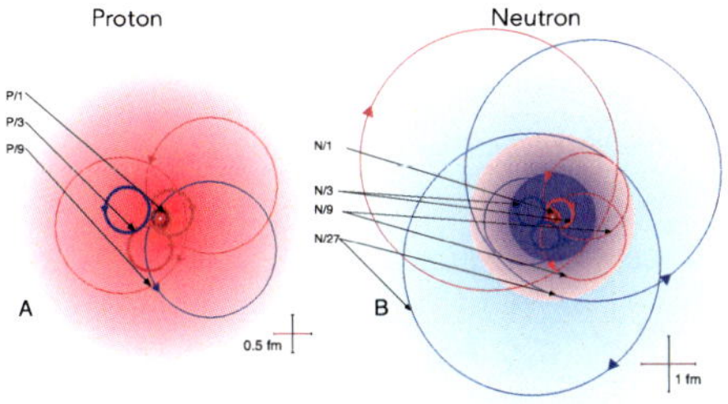

Figure 5.2: *Schematic illustrations of a proton, A, and neutron, B, showing an instantaneous view of the different harmonics and the planar nature of the electric fields they generate.*

Figure 5.3 shows two illustrations of both nucleons. A and C illustrate oblique views of the proton and neutron respectively. They show the charge density curves with the magnetic field lines superimposed upon them. They represent the best way of visualizing nucleons. Figures 5.3 B and D show plan views of their charge distributions respectively. Lines are used instead of shades to indicate intensities.

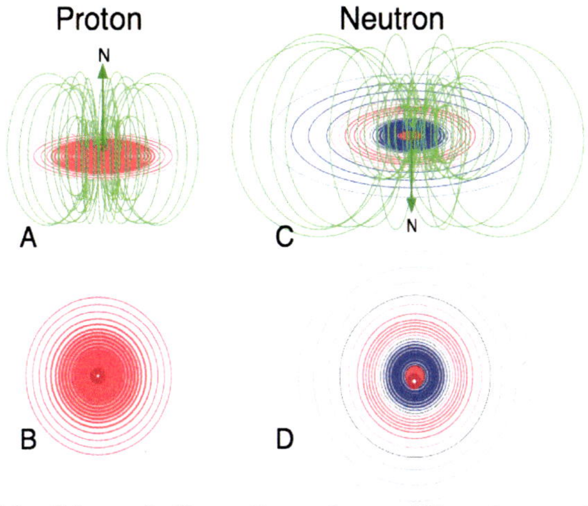

Figure 5.3: *Schematic illustrations of some different ways of visualizing the proton (A, B) and the neutron (C, D)*

The proton has unit positive electric charge. Its oscillations all generate a magnetic field in the one direction. The neutron has segments of opposing charge that rotate in the same direction and hence its magnetic field also has opposing segments.

In figure 5.3, both nucleons are shown as spin up particles. That gives the proton its north magnetic field pointing up. The neutron's north magnetic field is pointing down. When they align in the same plane, their same spin will give them a closed loop magnetic field, see figure 5.5 A and B. Opposing nucleons with the same spin generate magnetic binding.

Figures 5.1 to 5.3 were taken from chapter 3. They can be used for modelling the behaviour of nucleons within nuclei. In doing so there are some points that need to be remembered.

Charge, which is distributed in a planar array, is generated at the circumferences of the circles shown in figure 5.2. The planar nature of the charge distribution means the lowest energy state is when the proton's positive charge plane is aligned with the neutron's negative charge plane.

When positively charged circumferences in the proton overlap with negatively charged circumferences in the neutron, the effective distance between them is zero. That creates the conditions for a strong binding between the two nucleons. The radial distribution of the charge is indicated in figure 5.1. The magnetic field, which is perpendicular to the charge plane, adds to the effect of overlapping electric charge oscillations.

The effect of electric binding only applies over the short distances when protons' positive charge segments directly overlap neutrons' negative charge segments. Magnetic field strengths diminish as the inverse cube of the distance between the poles. These effects therefore combine to give a strong force that acts over a short distance.

With these models as background, it is now time to look at what happens when nucleons come together and form nuclei.

5.3. Nuclear Binding Under the Rotating Photon Model

The inverse square law that governs the electric force is given by

$$F_E = \frac{e_1 e_2}{4\pi\varepsilon_0 d^2} \qquad (5.1$$

where F_E is the force experienced between the two electric charges, e_1 and e_2, and separated by distance, d, and ε_0 is the electric permittivity of space. The first feature to note from equation 5.1 and this model is that, with electric charge generated in a planar disc, opposing charges will attract each other and form a structure in which the two discs will be in the same plane.

This gives the opportunity for the opposing charges to overlap when a proton is close to a neutron and in the same plane. They effectively have "zero" distance between their opposing charge segments. With d being very close to zero, the presence of two small opposing charges in the same plane suggests the possibility that their attraction may overcome the electric repulsion between two larger charges at greater distance.

For example, fractional charge overlaps of $+0.1e$ and $-0.1e$ at a distance of 0.1 fm can overcome the repulsion between 2 x $+1e$ charges at a distance of 1.1 fm. Neutrons have positive and negative charge segments. That allows for the possibility that they can hold protons together in a nucleus, relying only upon the electromagnetic forces.

Electric attraction between like nucleons is not possible in the same plane. While neutrons have no net charge, their most exterior harmonic is negative. If they were to overlap, the overlapping of like charge segments would exceed the overlapping of opposing charge segments, resulting in repulsion. This prevents neutrons from getting too close to each other.

Protons have entirely positive charge. If two protons were to align their charge discs in the same plane, as shown in figure 5.4A, they would repel each other. Repulsion between protons would be severe, preventing any possibility of proton-to-proton binding in the same plane. That does not mean that binding between protons is not possible.

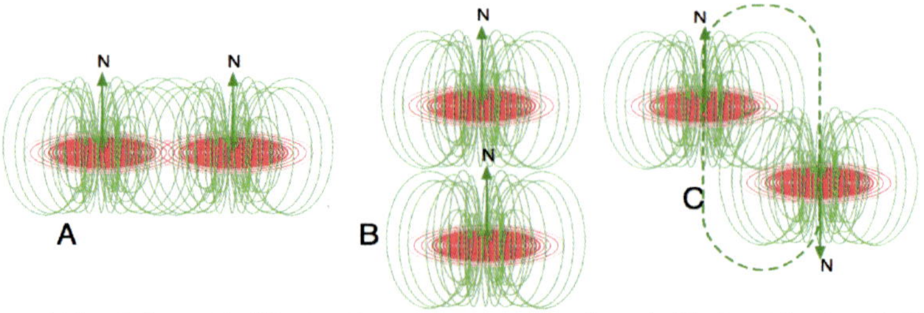

Figure 5.4: *Schematic illustrations two protons placed side by side, A, axially aligned with the same spin, B, off axis with opposite spins, C.*

If two protons of the same spin were to be axially aligned, as shown in figure 5.4 B, their magnetic fields would align, giving magnetic attraction between them. Magnetic attraction is the only mechanism by which proton to proton binding can occur.

Figure 5.4 C shows off axis protons with opposing spins. Binding would be by a magnetic loop indicated by the dashed line. A similar situation applies for neutron-to-neutron bonding.

Next consider a single proton and neutron, a deuterium nucleus. Two examples are given in figure 5.5 A and C illustrate the situation where the two nucleons' centres are separated by 5 fm.

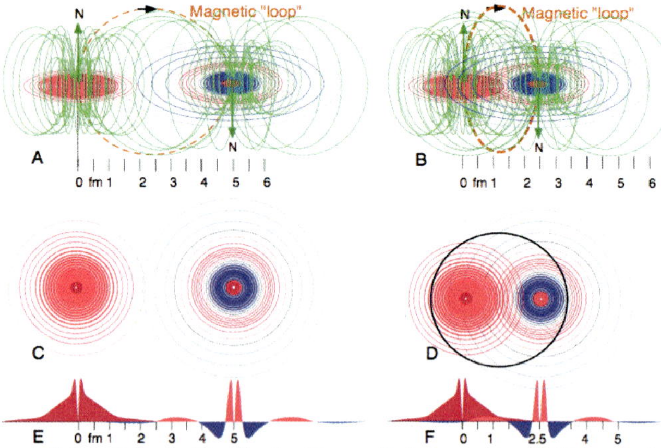

Figure 5.5: *Schematic illustrations of oblique, A and B, and plan, C and D, views of the charge distribution of a proton and neutron at 5 fm (A) and 2.5 fm (B) centre-to-centre spacing. A and B also show the magnetic fields associated with the nucleons. E and F show the normalized charge distributions.*

The proton's P/9 oscillations just overlap the neutron's negative N/27 oscillations. (The red and blue circles represent lines of equal charge intensity gradients. The denser the circles, the stronger the electric field.)

This leads to some electric attraction between them. Because both the P/9 and N/27 oscillations represent less than 10% of the charge of the nucleons, their proximity indicated in figure 5.5 A and C only shows a small overlap, making this attractive force quite weak.

The proton's magnetic moment is positive while the neutron's is negative. This means that when the two nucleons have the same spin, their magnetic moments will point in opposite directions. That results in a magnetic attraction between the two nucleons, illustrated by the brown dashed line in figure 5.5. At 5 fm centre-to-centre separation, their combined force is too weak for it to form a stable bond.

The situation Is different when their centre-to-centre spacing is 2.5 fm, as shown in figures 5.5 B and D. The proton's positive P/3 oscillations, which contain approximately $+0.5e$, begin to overlap the neutron's negative N/3 oscillations, which have a total negative charge of approximately $0.4e$. These high charge fractions mean there is always an overlap when opposing oscillations cross. Being in the same plane, the small distance between the opposing charges gives strong electrostatic binding between them.

Added to that is a stronger magnetic binding because of the much greater overlap of their magnetic fields. The magnetic "loop," combined with the proton-neutron opposite charge overlap, is sufficient to hold the nucleons together against the repulsion of their inner positive charges.

It is accepted that magnetic field strength varies as the inverse cube of distance $\left(\frac{1}{r^3}\right)$. Magnetic binding at 2.5 fm would be 8 times stronger than at 5 fm. It rapidly increases as the nucleon centre-to-centre distance diminishes. Magnetic binding is a major component of nuclear binding. Nucleons orient their spin axes to achieve the maximum amount of magnetic binding with their neighbours.

Figure 5.5 E and F show the charge distribution intensity profiles for a proton to neutron centre-to-centre separation distance of 5 fm and 2.5 fm respectively. Red represents positive and blue represents negative. Diagram E shows that only a small amount of the neutron's N/27 oscillation overlaps with the proton's P/9 oscillation. There is no opportunity for binding to overcome the positive repulsion between the proton and neutron's positive sectors. Diagram F shows there is some overlap of the proton's P/9 oscillation with the neutron's N/3 oscillation when their centres are separated by 2.5 fm.

Figure 5.5 shows that there is a repulsion of the proton's $+1e$ charge against the neutron's $+0.5e$ charge. Is a small overlap of part of the proton's positive charge with part of the neutron's negative charge enough to compete with that? Even with the loop of magnetic field binding between the same spin nucleons, it may appear to be insufficient to hold the two nucleons together. It is here that the nature of the charge generation plays an important role.

The black circle In figure 5.5 D represents a diameter of 4.45 fm, double the deuteron's measured root mean square (rms) radius. Less than 15% of the total positive charge of the proton and neutron is outside that radius. That occurs at a proton to neutron centre-to-centre separation distance of about 2.5 fm.

Figure 5.2 shows that the charge is generated by harmonic oscillation "loops" in the plane of the nucleon's rotation. Where the proton's positive loops overlap the neutron's positive loops, they would repel each other and interleave, as shown in figure 5.6 A. The repulsion of the overlapped positive charges would push them further apart in the vertical plane. They are held in their vertical position by the structure of the nucleons. They do not exert much horizontal opposition on each other.

Figure 5.6 B shows a positive charge harmonic loop overlapping a negative harmonic loop. The two will attract each other with no separation. That means the positive proton loop overlapping a neutron's negative charge loop will give a strong binding while a similar positive – positive overlap will not give a strong repulsion. In both cases there will be a distortion of shape of the charged harmonic loops.

***Figure 5.6:** Schematic illustrations of a proton's P/9 oscillation overlapping a neutron's positive P/3 oscillation, A, and negative oscillation, B.*

It is suggested a positive–positive overlap would compress the loops. That would reduce the effective radius of the loops. From equation 2.10, that would diminish the strength of the magnetic field of those loops. In the same way, a positive–negative overlap would stretch the oscillations, increasing the effective radius and hence the magnetic field strength of those loops.

The transient changing nature of these oscillations means the charge planes are not in a stable plane. The charge planes of individual nucleons vibrate at high frequencies, suggested to be typically 10^{23} Hz. The result will be that the magnetic moment of the nucleus will be different from the sum of the magnetic moment of both nucleons.

The ^{2}H deuterium nucleus is the simplest compound nucleus. Figure 5.5 shows various representations of it. The neutral neutron binding to the positive proton is a combination of two factors, the overlap of portions of the proton's positive oscillations with portions of the neutron's negative oscillations, and the magnetic binding when the two nucleons have their spins in the same direction. Their spins add because that causes their magnetic poles to point in the opposite directions, increasing magnetic binding between the two. With each nucleon having a spin of ½($\hbar$), the nucleus has a total spin of 1($\hbar$). It gives the nucleus greater stability because of increased magnetic binding.

Experimental measurements suggest deuterium nuclei have a proton to neutron centre-to-centre separation of 2.5 fm. Figures 5.5 B, D and F give three representations of it. Figure 5.5 shows different ways to easily visualise what happens at different proton to neutron centre-to-centre spacings.

Figure 5.7 illustrates some details of the charge overlap at different proton to neutron centre to centre distances from 5 fm to 0.5 fm. It shows only the electric charge effect. Figures 5.5 A and B show there is also a magnetic component to the binding of nucleons within nuclei. That needs to be added to the indicated electric binding.

Between 5 fm and 3.5 fm, the only opposing charge overlap is the neutron's negative N/27 oscillations with the proton. At 3 fm there is a slight overlap between the neutron's negative N/3 oscillations with the proton's N/9 oscillations. By 2.5 fm, there is sufficient overlap for a single proton to bind to a single neutron and produce the ^{2}H nucleus.

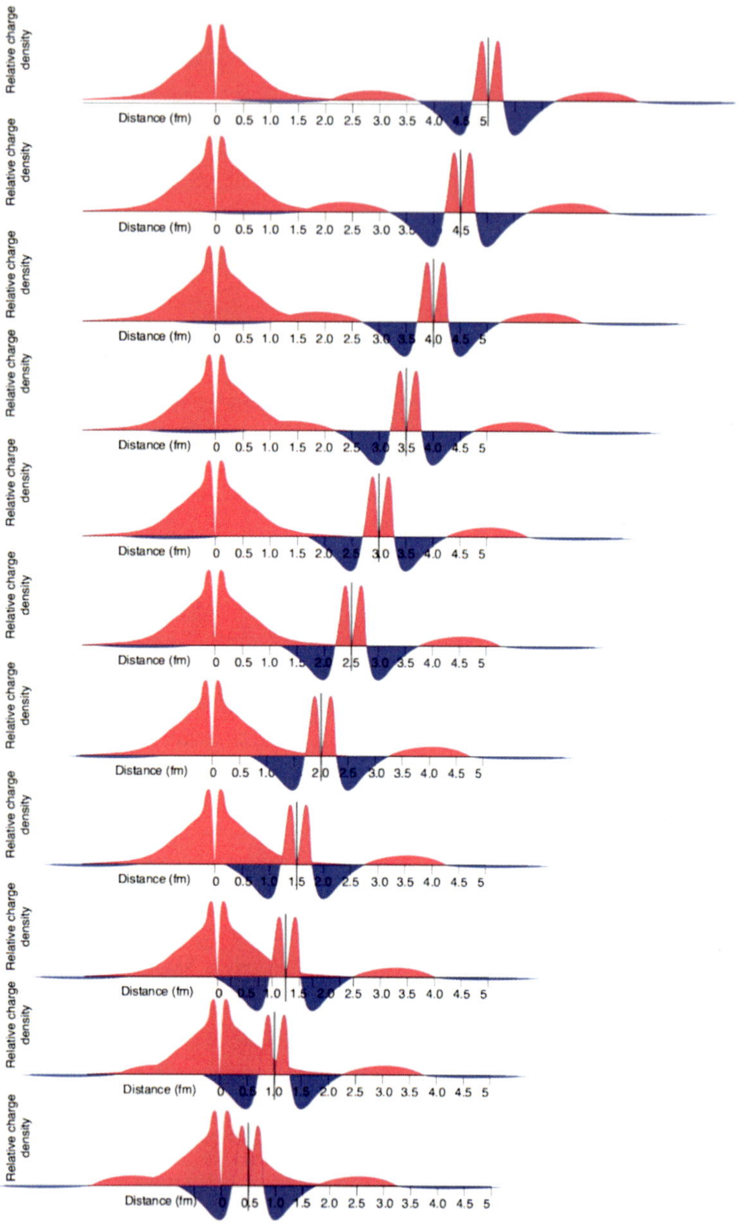

Figure 5.7: *Illustrations of side-by-side proton to neutron charge overlap.*

At 2 fm, the neutron's negative N/3 oscillations are starting to overlap the proton's N/3 oscillations. By 1.5 fm they are starting to overlap the proton's P/1 oscillation. At 1.2 fm there is large overlap between the proton's P/1 and P/3 oscillations and the neutron's N/3 oscillations. There is also significant overlap between the neutron's N/1 oscillation and the proton's P/9 oscillations. 1.2 fm is the separation distance between proton and neutron centres in alpha particles.

At 1 fm, almost half of the N/3 oscillations are within the proton's P/1 and P/3 oscillations. At the same time there is significant overlap between the neutron's positive central N/1 oscillation with the proton's P/3 oscillation. The neutron's N/1 oscillation positive field will also exert a significant repulsive force on the proton's P/1 oscillation, with which it is not interleaving.

At 0.5 fm, both nucleons' central oscillations will be close to each other without interleaving, and the neutrons' positive N/1 oscillation will be surrounded by the proton's positive P/3 oscillation. The repulsive positive force will exceed any opposing charge attraction, giving the two nucleons a strong repulsive force between them.

Figure 5.7 allows a semi-quantitative evaluation of the electric binding interactions between protons and neutrons at various centre to centre separation distances. It is suggested that their separation distance of 5.5 fm would have so little overlap between the neutron's negative N/27 oscillations with a proton's P/9 oscillations that it marks the limit of nucleon attraction. By 5 fm, there is a slight attraction. As the nucleons get closer together, their opposing charge overlap causes the attraction between them to increase faster than the distance between their centres decreases.

As they get closer, the positive charge repulsion between their centres increases as their separation distance decreases. Somewhere between 1.5 fm and 1 fm the binding force between the neutron's N/3 negative oscillations overlapping the proton's P/3 positive oscillations will reach a maximum. At closer distances, their positive centres will exert a strong and rapidly increasing repulsive force between them. That trend is not dissimilar to the measured results shown in figure 5.8 (Cook, 2010).

Although neutrons have no external charge, they do have an internal charge distribution. Figure 5.9 shows how the electric charge distributions change when neutrons have different centre to centre separation distances. It does not show the effects of the magnetic fields.

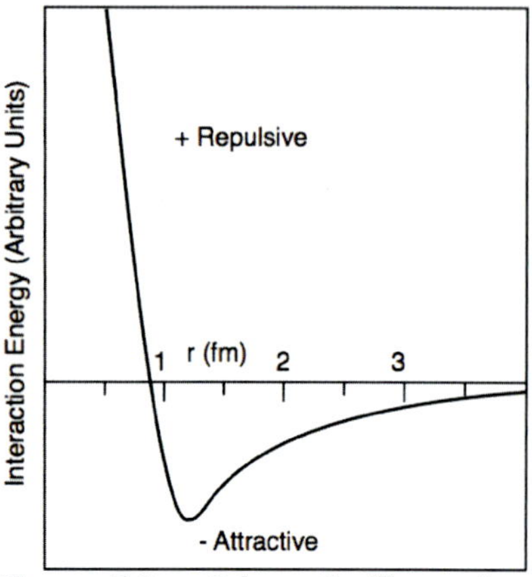

Figure 5.8: *Schematic illustration of the expected interaction energies between a proton and neutron versus their centre-to-centre separation distance.*

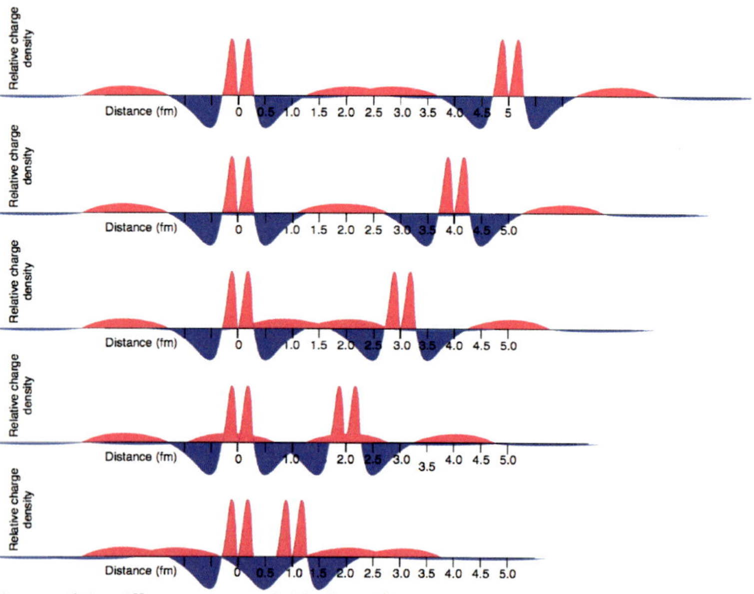

Figure 5.9: *Illustrations of side-by-side neutron to neutron charge overlap.*

At 5 fm and 4 fm, there is almost no overlap of opposing charges and just a little overlap of positive charges. At 3 fm there is possibly a little more overlap of opposing charges than of like charges. That situation continues at the 2 fm separation. By 1 fm, there is a large overlap between the N/3 negative oscillations. The positive cores are getting closer and will add to the repulsion between nucleons.

Within about 2 fm a repulsive force between the like charges will start to become noticeable. By 1 fm there will be a strong repulsive force between the two neutrons. There is never sufficient attraction between neutrons that they could bind to each other. It is considered unlikely that two neutrons could get much closer to each other than about 1.2 fm in a stable nucleus.

Figure 5.5 showed the situation for ^2H, the simplest compound nucleus. Figure 5.10 shows schematic illustrations of the tritium, ^3H nucleus, A and B, and the helium 3, ^3He nucleus, C and D. A and C show plan representations of the charge planes.

5.10 B and D show the intensities of the relative charge intensities. B and D are drawn as straight-line representations of the charge overlap. The distances are the same as used in A and C respectively. They show that the centre-to centre distances between opposing nucleons is ≈ 1.7 fm for ^3H and 1.8 fm for ^3He.

Figures 5.10 B and D show a significant overlap of negative segments of the neutrons with positive segments of the protons, as expected. Figures A and C show they are not in a straight line. That is an indication of the strong influence of the magnetic fields.

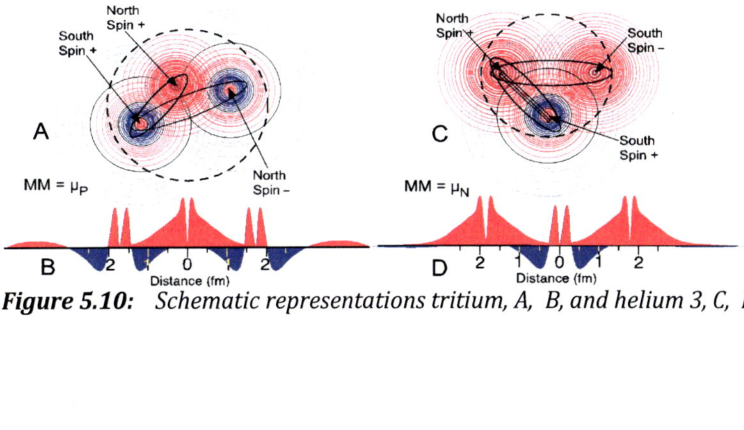

Figure 5.10: *Schematic representations tritium, A, B, and helium 3, C, D.*

The spins of both nuclei are ½$\hbar$. That requires two of the nucleons to be spin up and the third to be spin down. Figure 5.3 showed that, for the same spin, the proton's magnetic field is in the opposite direction to the neutron's magnetic field. When the two form ^2H, their same spins give them opposing magnetic fields when adjacent and in the same plane. That enables magnetic binding to contribute to their nucleon binding, as shown in figure 5.5.

The ovals around the nucleon centres in 5.10 A and C show the magnetic binding between north and south poles of the nucleons. Figure 5.10 C shows the two protons with opposing poles are sufficiently close to each other that significant magnetic binding overcomes their positive repulsion. That way, there are two spin + nucleons, as in ^2H. The second proton is spin – because the magnetic binding is stronger than the like charge repulsion. It is suggested that is due to the interleaving effect shown in figure 5.6.

In both ^3H and ^3He, the like nuclei have opposing spins. They form a closed loop magnetic field. That cancels each-others magnetic moments, leaving the nucleus to have the residual magnetic moment of the remaining nucleon. That would give the ^3H nucleus a theoretical magnetic moment of 2.793 μ_{BP}. The experimentally measured value is 2.973 μ_{BP}. That suggests the proton's rotating photons have been stretched a little by their interaction with the neutron. ^3He's residual nucleon is a neutron, magnetic moment –1.913 μ_{BP}. Its measured magnetic moment is –2.128 μ_{BP}.

The dashed lines represent the measured root mean square values of the charge distribution diameters of $\approx$ 4.2 fm for ^3H and $\approx$ 3.4 fm for ^3He. There is no clear defining diameter for any nucleus because of the large and diminishing distribution of nucleon charge with distance.

The structures displayed in figure 5.10 give a reasonable match with the observed properties of those nuclei. The angles between the nucleons are estimates. The magnetic binding between the two protons in the ^3He nucleus is a special case where protons outnumber neutrons 2:1. It is not repeated elsewhere. Throughout this study, unless otherwise referenced, nuclear data on binding energies, spin, magnetic moment, isotope abundances, and half-lives were taken from Howard (1966).

Figure 5.11 A shows a plan view of the charge distribution of the ^4He nucleus, also known as an alpha particle.

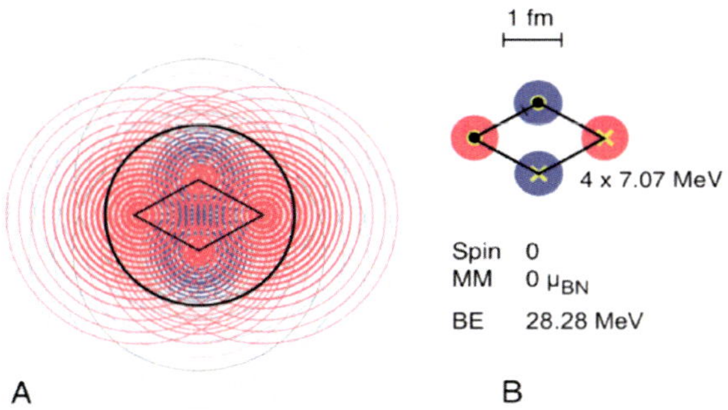

Figure 5.11: *Schematic representations of a ^{4}He nucleus.*

The protons are as far away from each other as possible. The two protons have captured two neutrons. As they try to get as far away from each other as possible, they pull the neutrons in until their mutual repulsion prevents them from moving any closer. That and the other representations of the nucleons in figures 5.5 and 5.10 are complex. 5.11 B shows a more simplified representation. The circles represent the central positions of the nucleons.

Protons are red, neutrons are blue. The dots in the centre represent the north magnetic pole pointing out of the page. The crosses in the centre represent the north magnetic poles pointing into the page. Using the convention from chapter 2, a dot on a proton represents spin up, with the oscillations rotating anti-clockwise. A dot on a neutron represents spin down, with the oscillations rotating clockwise. The black lines represent nearest neighbour bonds between proton and neutron.

The circle in figure 5.11 A represents the rms charge radius of $\approx$ 1.68 fm (Sick, 2015). It contains about 80% of the positive charge in the nucleus. It gives the proton to neutron nearest neighbour bond a length of $\approx$ 1.2 fm. For maximum magnetic binding, like nucleons orient their spins in opposite directions. 4He nuclei have no external magnetic moment.

Table 5.1 lists some of the properties of the four lightest compound nuclei. The spins are theoretical predictions of this model. They are also the experimentally measured values.

Nucleus	Spin ($\hbar$) (T)	Magnetic Moment μ_{BP} (T)	(E)	Δ† (%)	Binding Energy (MeV)	N-P dist fm (T)	Radius rms fm (E)
^{2}H	1	0.88	0.86	1	2.2	2.5	2.13
^{3}H	½	2.793	2.979	3	8.48	1.7	2.0
^{3}He	½	−1.913	2.127	4	7.71	1.8	1.7
^{4}He	0	0	0	0	28.28	1.2	1.7

Table 5.1: *Table of theoretical and experimentally determined (Angeli, 2004; Sick, 2015) values for the lightest nuclei. E indicates the results were measured while T indicates results were determined under this model.*
†Per proton μ_{BP}.

The theoretical magnetic moments are the sum of the magnetic moments with their appropriate spins.

The experimentally measured values indicate there has been a slight alteration to the shape of the oscillations of the nucleon harmonics. The differences, Δ†, are expressed per nucleon as a percent of a proton's magnetic moment.

The binding energies are as experimentally measured. The theoretical N-P distance is the centre-to-centre distance of nearest neighbour proton to neutron bonds that enable about 85% or more of the positive charge to be contained within the experimentally measured root mean square (rms) radius.

Table 5.2 gives values for the proton to neutron centre-to-centre distances and the binding energy of each nucleus. Note that the indicated binding energies are a combination of the electric and magnetic attraction.

No attempt has been made to determine the value of each. The electric attraction is limited to the overlap of positive proton oscillations with negative neutron N/3 oscillations. No oscillation overlap means only weak magnetic attraction.

Figure 5.12 shows the plot of the proton to neutron nearest neighbour binding energy per nucleon against the proton to neutron centre-to-centre separation distance. Without other information, a Bezier curve was fitted through the points.

Nucleus	Total Energy (MeV)	Proton to Neutron Centre Spacing (fm)	Binding Energy/ P to N Bond (MeV)
P to N dist		> ≈5.5	≈ 0
$^2H_1$	2.2	2.5	2.2
$^3H_1$	8.5	1.7	4.25
$^3He_2$	7.7	1.8	3.85
$^4He_2$	28.28	1.2	7.07

Table 5.2: *Table of binding energy per proton to neutron bond for different nucleon centre to centre distances determined from the nuclei.*

Magnetic field strengths diminish with the inverse cube of the distance. They combine to give the binding force between protons and neutrons the very limited range shown in figure 5.12. It also serves as a way of determining the relationship between binding energy and distance between proton to neutron centres.

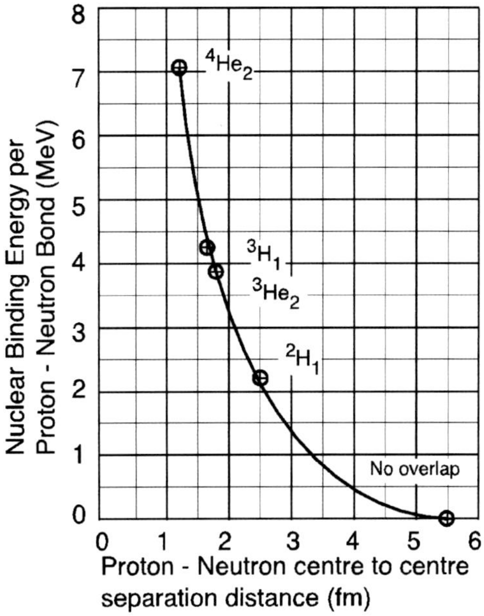

Figure 5.12 *A plot of the nuclear binding energy in MeV per proton-neutron bond versus nucleon centre-to-centre separation distance, data taken from table 5.1.*

It can be used to convert the binding energy of a proton to neutron bond into its centre to centre spacing. Conversely, knowledge of their centre-to-centre spacing can be converted to their binding energy.

The above-described binding is between nucleons in a single layer. Only a few nuclei form in one layer. Most form in multiple layers. Figure 5.4 B showed that protons of the same spin placed over each other will be magnetically attracted to each other. Figure 5.4 C showed that there will be attraction between same spin protons, even if they are off to one side.

Magnetic field strength decreases according to the inverse cube of the distance. That means there will still be a detectable magnetic field outside the rms radius distribution, ≈ 0.8 fm, attributed to a proton. Both situations also apply to neutrons.

It will be weaker than the magnetic field associated with adjacent nucleons having the correct orientation, as shown with the ^{2}H nucleus in figure 5.5 B. Nevertheless, it is present and will permeate the nucleus. Determination of the energy per nucleon pair in different layers requires knowledge of their orientation and proximity.

The objective of this presentation is to show how nuclear binding occurs. To do that, an approximation of 0.8 MeV per nucleon pair will be used in future calculations as the binding energy between nucleons in adjacent layers. It gives a reasonable fit with binding energy between the nucleons. The number of such vertical nucleon pairs will be the smaller of the number of nucleons in the two layers. A nucleon layer with 8 nucleons over a layer with 13 nucleons will have 8 nucleon pairs where layer binding occurs. The binding energy between those two layers will therefore be 6.4 MeV.

Together, the above indicate some of the rules of neutrons binding protons into a nucleus. They include:

i. Nucleons will form into planar layers.
ii. Protons and neutrons in a layer can only bind if they are in the same plane.
iii. Within a layer, protons will always repel each other because of their like charge repulsion. They cannot bind to other protons (with the exception of ^{3}He).

iv. Except for nuclei containing only two or three nucleons, all nucleons in a stable nucleus must bind with at least two opposing nucleons.

v. An external nucleon binding to a single nucleon in a planar nucleon array will be unstable.

vi. A single nucleon cannot magnetically bind to other nucleons because their magnetic attraction will cause them to move into the same plane, wherein they will be either bound by the strong force or be repelled.

vii. Nucleons within a nucleus will arrange their spins to maximize magnetic binding. That will minimize nuclear magnetic moments.

The following section indicates how these rules are applied to determine the structure and properties of some low atomic number nuclei. The relationship between binding energies and proton to neutron separation distances can be determined from the curve in figure 5.12.

5.4. Low Atomic Number Nuclei

The first nucleus to look at is an alpha particle, ^4He. Figure 5.13 shows an illustration of an alpha particle, A, overlaid on the electric field of the helium nucleus. It is extracted from figure 5.11. Attaching another nucleon is a problem.

An extra neutron, N, in the same plane could only attempt to bind to a proton. Its negative N/27 oscillation could overlap a proton and get some short-term bonding. An additional proton, P, could be

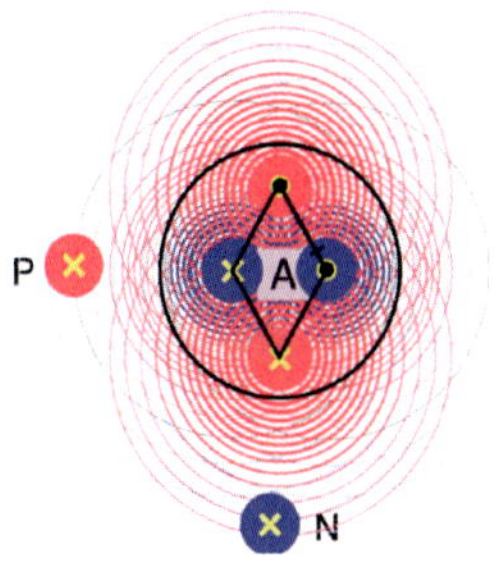

Figure 5.13: *Schematic representations of the $^4He_2$ nucleus with additional nucleons trying to attach to it.*

captured by a neutron's weak negative N/27 oscillations. It would be repelled by the strong positive charge of protons on either side of the neutron. It would only have a short half-life.

A single proton or neutron above or below the charge plane of the alpha particle would be either attracted to or repelled by the magnetic moment of one of the alpha particle's nucleons. If repelled, it would not bond with it. If attracted in, it would move in. On reaching the charge plane it would be electrically repelled. There is no way for a single additional nucleon to bind onto an alpha particle for more than a short transient time. A nucleus with atomic number A = 5 cannot exist in a stable state.

Although single nucleons cannot be added to an alpha particle, a deuterium atom can be added to form a ^{6}Li nucleus. Figure 5.14 A shows an arrangement in which a ^{2}H nucleus is superimposed on an alpha particle. For ease of presentation the layers are shown on separate lines.

The alpha particle has no external magnetic moment. When the ^{2}H nucleus is placed above it, the same spin protons will experience magnetic attraction to each other. The opposing spin proton and neutron would shift the ^{2}H nucleus slightly to one side.

The binding energies of the alpha particle, 28.28 MeV, and deuterium nucleus, 2.22 MeV add up to 30.5 MeV.

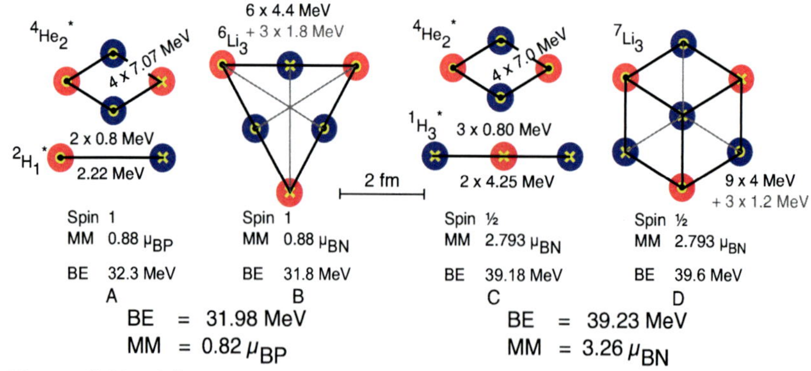

***Figure 5.14**: Schematic representations of the $^6Li_3$ and $^7Li_3$ nuclei.*

Adding 0.8 MeV for the inter layer magnetic binding energy per nucleon for the deuterium nucleus takes the total binding energy for that structure to 32.3 MeV. Both nucleons in the deuterium atom are spin up, with the alpha particle having spin 0. That gives the structure a spin of 1. Without distortion of the oscillations, the predicted magnetic moment would be the difference between the proton and neutron magnetic moments, of $0.88\mu_{BP}$. The observed magnetic moment of 0.82 μ_{BP} suggests there has been a slight compressive distortion of the oscillations.

Figure 5.14 B shows an alternative structure for a ^{6}Li nucleus. It is a single layered equilateral triangular shape as the protons try to get as far away from each other as possible. Each has a neutron separating it from another proton and two neutrons holding it in place. The spin and magnetic moment of this structure are the same as in the two-layered structure.

The dimensions of this structure can be calculated from the binding energies. There are six nearest neighbour proton-to-neutron (p-n/1) bonds, represented by dark lines, and three second nearest neighbour (p-n/2) bonds, represented by light lines. The distances between these can be calculated from figure 5.12.

The first step is to estimate the distance between nearest neighbour protons and neutrons. It is simple geometry to then determine the distance between second nearest neighbour bonds. Using these distances, determine the binding energy per bond. They are listed in figure 5.14 B. Add 6 x p-n/1 and 3 x p-n/2. Compare the answer with the measured binding energy. If there is not a close match, change the distances accordingly and you will soon get that close match. In the example given in figure 5.14 B, the p-n/1 centre-to-centre separation distance is approximately 1.7 fm. The p-n/2 distance is approximately 2.2 fm. These distances give the total binding energy of the triangular ^{6}Li nucleus as 31.8 MeV.

The two models, A and B, can be distinguished by measuring the charge density or radius of the ^{6}Li nucleus. The ^{6}Li nucleus is known to have a lower charge density and somewhat larger radius than the ^{4}He nucleus. This suggests the ^{6}Li nucleus has the single layered structure equilateral triangle shape given in figure 5.14 B.

In the same manner, two possible configurations for the ^{7}Li nucleus are shown in figures 5.14 C and D. In the structure of 5.14 C, a

$^3H^*$ nucleus is magnetically bound to a $^4He_2^*$ nucleus. The asterisk (*) indicates the structures have different spin arrangements than those of the isolated nuclei. With the binding energies somewhat like those of the original nuclei, a magnetic binding energy of 0.8 MeV per nucleon, gives 39.4 MeV, a reasonable match with the experimentally measured binding energy of 39.23 MeV.

Figure 5.14 D shows all seven nucleons in one plane. Note that the additional neutron is positioned in the centre of the figure 5.14 B structure. That forces the neutrons further apart. It also gives three proton-to-neutron nearest neighbour bonds per proton, a more stable structure than the two bonds per proton of figure 5.14 C. In both cases the spin ½ means that there is one proton spinning up that is not countered by its opposite spin, giving a residual magnetic moment of 2.793… μ_{BN}.

The experimentally measured magnetic moment is 3.256... μ_{BN}, a difference of less than $\approx$ 0.08 μ_{BN} per nucleon. As in the other situations, this variation can be explained by distortions in the rotating photons' trajectories.

The spin of ^7Li nuclei is often recorded as 1½. That requires all protons to be spin up and the neutrons to have the spins indicated. It would give them a total magnetic moment of 5.586… μ_{BN}. That requires a difference of $\approx$ 0.245 μ_{BN} per nucleon.

With either of the structures in figure 5.14 C and D, it is difficult to add a single nucleon. The addition of a single proton creates an ^8Be nucleus. Its half-life is $\approx 10^{-15}$ sec before it forms 2 x ^4He nuclei. A single added neutron to form ^8Li can only bind to one proton. It is outside the binding rules and has a short half-life of 0.85 sec.

Under this model, the ^7Li nucleus is the largest single layer nucleus. Its average binding energy per nucleon is higher than that of ^6Li. This suggests the ^7Li isotope would be more abundant than the ^6Li isotope. This should hold true under any nuclear model.

These examples show that a magnetic binding energy of about 0.8 MeV per nucleon is present between the nucleon layers. This is significantly less than the binding energy per proton to neutron bond in the strongly bound layers. The nucleon types, spins, and relative positions can affect the magnetic binding energy per nucleon between nucleon layers. A value of 0.8 MeV will be used as a first approximation for the magnetic binding energy per nucleon pair between layers. It will be illustrated later that the magnetic binding can be somewhat stronger.

Figures 5.15 A and B show possible representations of the ^{9}Be nucleus, with C showing the ^{10}Be nucleus. It should be noted that the binding energies for the $^{6}Li^{*}$ and $^{7}Li^{*}$ nuclei are larger than those for the ^{6}Li and ^{7}Li nuclei in figure 5.14.

This is attributed to the magnetic binding in different layers bringing the nuclei closer together. The distances involved are ≈ 0.2 fm, which means that different structures overlaying those layers could slightly alter the nucleon separation distances, giving slightly varying binding energies per nucleon.

The experimentally measured $^{9}Be_4$ magnetic moment is −1.177... μ_{BN}. If it had the indicated spin of ½$\hbar$, it requires a distortion of the magnetic moment of about 0.08 μ_{BN} per nucleon. The spin of a $^{9}Be_4$ nucleus is sometimes indicated as 1½$\hbar$.

That requires a single nucleon to change its orientation. Changing the spin of a proton changes the nucleus' magnetic moment by 2 μ_{BP}, or ≈ 5.5 μ_{BN}. Changing the spin of a neutron changes the nucleus's magnetic moment by ≈3.8... μ_{BN}.

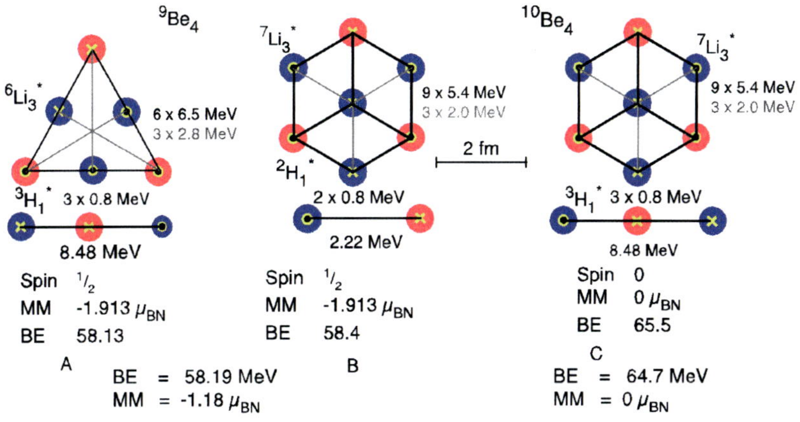

***Figure 5.15:** Schematic representations of the $^{9}Be_4$ and $^{10}Be_4$ nuclei consisting of two layers of nucleons.*

This requires significantly more distortion of the nucleon's rotating photons to match ^{9}Be's measured magnetic moment.

Not shown here is another possibility for the ^{10}Be isotope. It could have a deuteron on either side of a ^{6}Li structure. Distinguishing between the different structures would require experimental measurement of either their charge density or their dimensions. It will be shown later that many nuclei have different structures for the same isotope.

Figure 5.16 indicates possible structures for the ^{12}C, ^{13}C, and ^{14}C nuclei. The three alpha particle structures in figure 5.16 A gives the closest value to the observed binding energy of the ^{12}C nucleus. The spins of the even numbers of protons and neutrons would align to cancel each other. That occurs in all situations where there are even numbers of both protons and neutrons within a nucleus. That gives those ^{12}C nuclei no spin and no magnetic moment. It also gives them no other bonding positions to which other nucleons can attach to form additional isotopes, and it is therefore discarded as a possible structure.

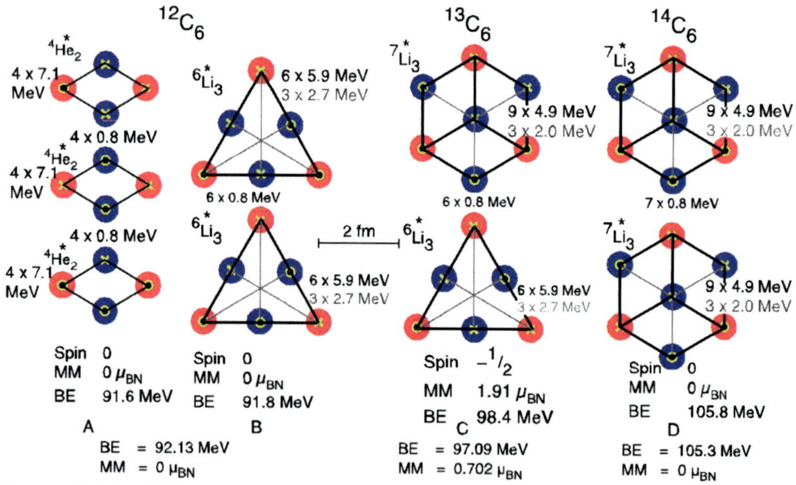

Figure 5.16: *Schematic representations of the $^{12}C_6$, $^{13}C_6$ and $^{14}C_6$ nuclei. A and B show possible variants of the $^{12}C_6$ nucleus, while C and D display structures for the $^{13}C_6$ and $^{14}C_6$ nuclei respectively. The full lines represent the bond to nearest neighbours. The grey lines extend the proton's bond to the second nearest neutron.*

The other possible structure is formed when two $^6Li^*$ layers superimpose, one on top of the other. Figure 5.16 B shows that a higher binding energy between the nucleons is required than in the ^6Li structure of figure 5.14 B. That can only occur by magnetic field attraction. This is a further indication of the strong role magnetic binding plays in nuclear binding.

The equilateral triangle shape of the ^{12}C nuclei has also been detected. This indicates that figure 5.16 B presents the correct structure of ^{12}C nuclei.

The addition of a single neutron to form the ^{13}C nucleus would result in one of the $^6Li^*$ layers becoming a $^7Li^*$ layer, as shown in figure 5.16 C. It would have a spin of $-½(\hbar)$ and a magnetic moment close to that of the added neutron, namely 1.913 μ_{BN}. This represents a difference of $\approx$ 0.09 μ_{BN} per nucleon from the observed magnetic moment of 0.702 μ_{BN}.

Adding a spin up neutron to the structure in figure 5.16 C would give the structure shown in figure 5.16 D, where the equal and opposite spins would mean no net spin and no external magnetic moment on the ^{14}C nucleus, as is observed.

As can be implied from figure 5.12, it only requires the average separation distance between the nucleons to change by $\approx$ 0.25 fm for the nearest neighbour proton to neutron binding energy to increase from 4.4 MeV to 5.7 MeV. This would give the second nearest neighbours a bond strength of 2.6 MeV. It is suggested the like nucleons magnetic binding between them would be larger than the average binding between nucleon layers when the individual nucleons are not necessarily directly above one another or of the same orientation. This allows the magnetic binding energy for each nucleon to be above the average energy, and it could also pull the nucleons in each layer slightly closer together. It is further suggested this would account for the binding energy of $^{12}C_6$ nuclei being somewhat higher than expected by comparison to the binding energy of two $^6Li_3$ nuclei.

It can be seen from the above examples that it is easy to arrange the nucleons so that a nucleus has maximum internal magnetic binding. It is known that nuclei with even atomic number (Z) and atomic mass (A) will all have spin 0 and no magnetic moment. This corresponds to all like nucleons having pairs of opposite spins. Nuclei with odd Z and even A, or even Z and odd A, have fractional spins, while those with odd Z and A have integral spins. No other combinations are possible because the spins of the nuclei are the sum of the spins of the individual nucleons. When the spins of the nucleons are arranged to match the observed nucleus spin, adding their magnetic moments produces a nuclear magnetic moment that is within a few percent per nucleon of that predicted from this addition. This difference is explicable by the distortion of the rotating photons caused by their binding to the other nucleons. Alternatively determining the layout for minimum magnetic moment of the nucleus will enable its spin to be determined.

As the spins of the largest nuclei also obey the odd and even Z and A rules of spin outlined above, it appears that this trend continues for all nuclei. This has at least two implications in this study. There is no need to extend the study of spins to larger nuclei because it is simple addition. Also, the fact that the spins always add up as mentioned indicates that the nucleons themselves are not rotating about their common centre. The spin of an individual nucleon is equal to its mass traveling at c in each of its harmonic radii.

This is as expected under the rotating photon model. It is not possible for nucleons with this structure to move sideways because photons cannot move sideways. Rotating nuclei are not possible if the nucleons are always arranged in a planar array. Such rotation could conceivably occur by causing the nucleons respective planes of rotation to vibrate and present a slightly different angle. This would then enable each nucleon to vibrate back and forth and the whole nucleus to rotate.

Even then, however, a nucleus containing one hundred nucleons would require it to have a rotational frequency of $\approx 10^{20}$ Hz to contribute significantly to its spin. Lower A nuclei would need to rotate even faster. The variation in magnetic moment per nucleon under this model would also diminish as the number of nucleons increased. For these reasons, no further calculation of the spin or magnetic moments of the nuclei will be done in this presentation.

It may be argued that the above is a simplistic presentation, which will always be able to match observed nuclear properties. That has indeed been the purpose of the above presentation. Nucleons are individual entities that are added according to the common-sense rules mentioned. Determining the structure and properties is nothing more than adding nucleons and their properties to match observation, and perhaps also predicting some unknowns in the process. It is not (and need not be) complex.

5.5. Nuclear Binding at High Atomic Number

The above illustrated that it was possible to unite protons and neutrons to form the nucleus of an atom using just the electromagnetic force. This dealt only with a few low Z, mostly stable nuclei. There are over 2,800 known nuclei. Less than 240 have an infinite half-life. Some 15 are either the decay products of heavier radioactive nuclei or have such long half-lives that they still occur naturally. For this model to have validity it must show that a mechanism exists to match the structure and properties of all the detected nuclei, low and high Z, stable and highly unstable alike. The following provides a mechanism by which the structure and some of the properties of all nuclei can be determined.

As was seen in the above examples, the larger nuclei, $Z > 3$, consisted of layers of tightly bound nucleons held together by a weaker magnetic binding. A binding energy of 0.8 MeV per nucleon between the layers is significantly less than the binding energy between protons and neutrons in the same layer.

Figures 5.14 to 5.16 illustrate that the same number of protons can accommodate different numbers of neutrons in a lattice structure. The ^{6}Li structure has a neutron vacancy in its centre, which when added, changes to the ^{7}Li structure. This ^{7}Li structure suggests that protons and neutrons form a two-dimensional hexagonal close packed lattice structure, to which neutrons can be added or removed to give different isotopes for a given number of protons. Those figures also indicate that there are multiple layers of nucleons, with each layer having a hexagonal close pack lattice structure when all neutron lattice positions are filled. The existence of a lattice structure implies fixed positions for the nucleons.

It is suggested that the rules for forming nucleon layers for higher Z nuclei are:

i. Only proton to neutron planar electric charge bonding is possible. There is no electric bonding between like nucleons.

ii. The positive electric charge on the protons means they will be as far away from other protons as they can get while still binding to the neutrons. The like charge overlaps of the neutrons mean they cannot be forced together. Protons and neutrons within a given layer will form diamond and hexagonal close packed shapes.

iii. All protons will be bound to a minimum of 3 neutrons.

iv. All neutrons will be bound to a minimum of 2 protons.

v. A proton can bind to just two neutrons in any layer in which there are no further protons outside it and no more than 3 protons in the "loop". An example would be a hexagonal layer structure in which there is one neutron removed from the centre. This only applies for layers with $Z < 8$.

vi. Most layer will have a width no more than two nucleons wider than its narrowest perpendicular dimension. This allows layers to maintain an approximately circular shape. It is not essential and will not apply when a nucleus has an oblong or other structure.

vii. In any given layer, the maximum A for a given Z is determined by filling all neutron lattice positions. When this occurs, each inner proton will be bound to 6 neutrons and each inner neutron will be bound to 3 protons.

viii. Each outer proton can only be bound to 3 neutrons and each outer neutron can only be bound to two protons. (An outer nucleon is one that would be close to or touched by an elastic band placed around a model of the layer with the minimum number of neutrons.)

ix. Neutrons can be removed from inside the lattice provided the above rules are not violated. Some of the inner protons will no longer be bound to 6 neutrons. Neutron lattice vacancy positions must not merge to form larger vacancies than the single vacancy in a $^{6}Li_{3}$ nucleus.

x. For nuclei with $Z < 83$, no layer can have more than 4 neutron lattice positions vacant. For $Z > 83$, this maximum number is 3.

xi. Nucleons will arrange their spins to give maximum magnetic binding within layers. In that way, like nucleons of opposite spin have magnetic binding between them.

Examples of layers that fulfill these requirements, for Z between 1 and 20 inclusive, are shown in figure 5.17. Other structures are possible, but only one of each is given in this display. The nucleons in figures 5.14 to 5.16 were displayed as filled circles and the structures were drawn as a plan view. That representation was adequate when only a small number of nucleons in a small number of layers were considered.

Larger nuclei have over 200 nucleons spread over many layers. Such a presentation would occupy considerable space. It is for this

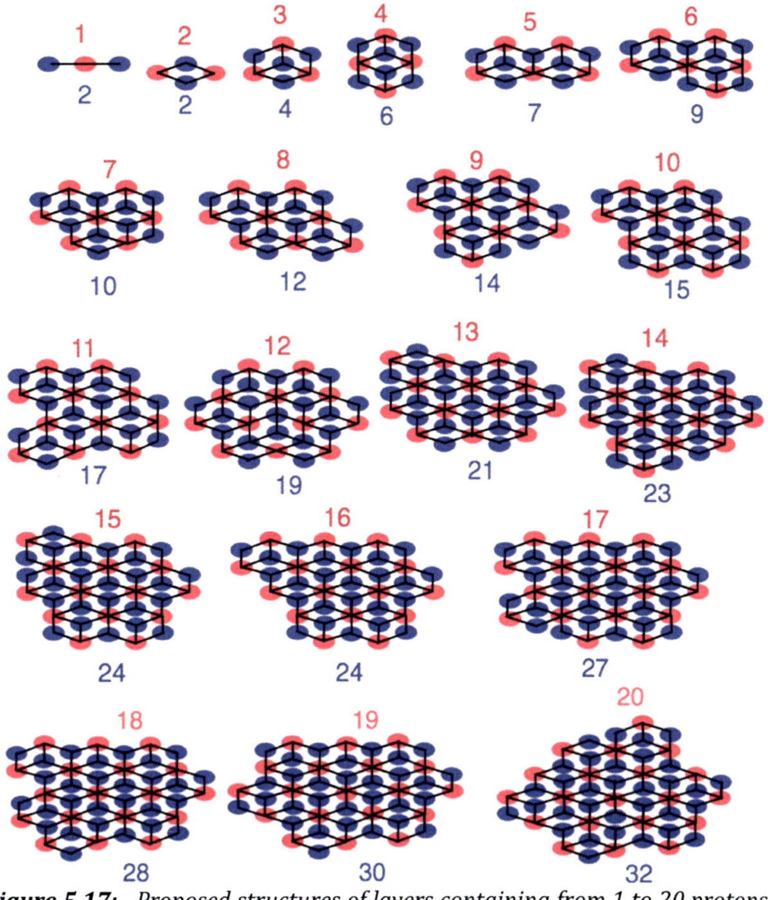

Figure 5.17: *Proposed structures of layers containing from 1 to 20 protons. All neutron lattice positions are filled. Red represents protons blue represents neutrons. The numbers above each layer refer to the number of protons in the layer. The numbers below refer to the number of neutrons in the layer.*

reason that individual nucleons are represented as filled ovals in figure 5.17. The regular hexagons are shown in oblique form. For simplicity, only nearest neighbour proton to neutron bonds are shown. They are the black lines joining neighbouring nuclei.

The object is to present the nucleon layers as if they were a planar structure viewed obliquely. It is believed this arrangement will allow all the nucleons in all the layers to be presented in the one image, making comprehension easier. Although the layers may look "squat", when stretched vertically they will appear to be circular.

Note that the Z and N values for each layer are displayed with each layer, Z above and N below, where N is the number of neutrons in the layer. This presentation shows each layer with its full complement of neutrons to fill every position in the HCP lattice. It is easy to see that, as the number of protons in a layer increase, the number of neutrons required to form the lattice increases at a faster rate. Protons and neutrons arrange in a lattice structure and otherwise spread uniformly throughout the layers.

Within the rules of binding, other arrangements are possible. No attempt is made here to show those arrangements. The important features include that the nucleons remain intact and that they do not form perfect circles with a known radius. Calculations of their radius is compounded by their charge distribution extending much further than the nucleon's centres, as shown in figure 5.11.

Under the layer binding rules given above it is possible to remove some of the inner neutrons in a manner that does not violate these rules. The same layers shown in figure 5.17 are presented again in figure 5.18 with the maximum number of neutrons removed to give the maximum number of lattice vacancies.

The results from figures 5.17 and 5.18 are summarized in table 5.3. The A Max – 2Z column shows an important trend. The number of neutrons per layer increases faster than the number of protons in that layer. A layer with 16 protons can accommodate up to 24 neutrons, a 50% Increase in neutrons over protons. A layer with 20 protons requires 60% more neutrons to fill all its neutron lattice positions. It will be shown later how that increased number of neutrons over protons matches observation.

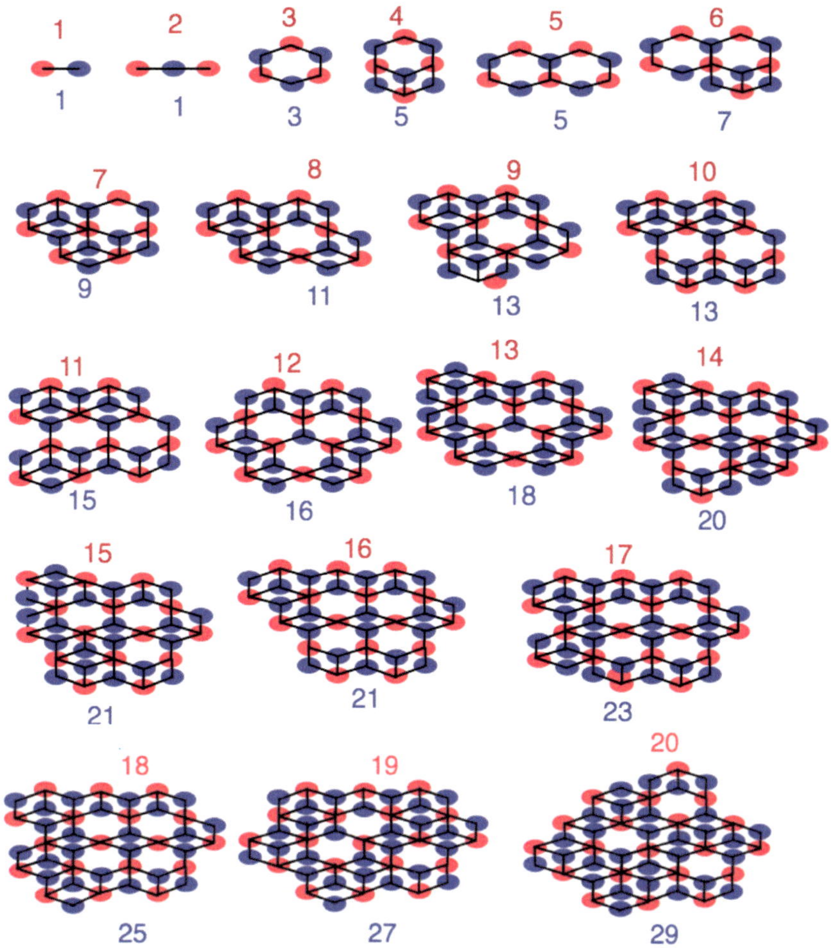

Figure 5.18: *Proposed structures of layers containing from 1 to 20 protons, showing the maximum allowed number of neutron lattice positions vacant.*

Another feature to note is that the illustrated diagrams do not show an exactly circular structure for each layer. This is illustrated in figure 5.19. It shows the maximum, A, and minimum, B, number of neutrons in a layer containing 18 protons. These layers are taken from figures 5.17 and 5.18 respectively but are shown here in top view.

Although the shapes are generally circular, there are irregularities within that circular shape. They could give the appearance of bumps and dimples on the surface of an otherwise smooth nuclear surface. The widths mentioned in table 5.3 refer to the number of nucleons at its widest part, but not necessarily across a diameter.

Two features emerge from figures 5.17 and 5.18. This model indicates that nucleons remain completely intact, whether it be a single proton or a multiple assembly of protons and neutrons. They neither exchange particles nor merge into a "soup mixture."

At most, the rotating photons will become slightly distorted, resulting in the magnetic fields not necessarily totalling the sum of the fields of the individual nucleons. Nuclei will also not be perfectly circular. As figure 5.19 suggests, many will have bumps and dimples within an approximately circular structure.

Z/layer	A Max	A Min	Width	A Max – 2Z
1	3	2	3	1
2	4	3	3	0
3	7	6	3	1
4	10	9	4	2
5	12	10	5	2
6	15	13	5	3
7	17	16	5	3
8	20	19	6	4
9	23	22	6	5
10	25	23	6	5
11	28	26	6	6
12	31	28	6	7
13	34	31	7	8
14	37	34	7	9
15	39	36	7	9
16	40	37	8	8
17	44	40	8	10
18	47	43	8	10
19	49	46	9	11
20	52	49	9	12

Table 5.3: *A list of some properties of different atomic number layers, showing the maximum and minimum number of neutrons per each Z layer.*

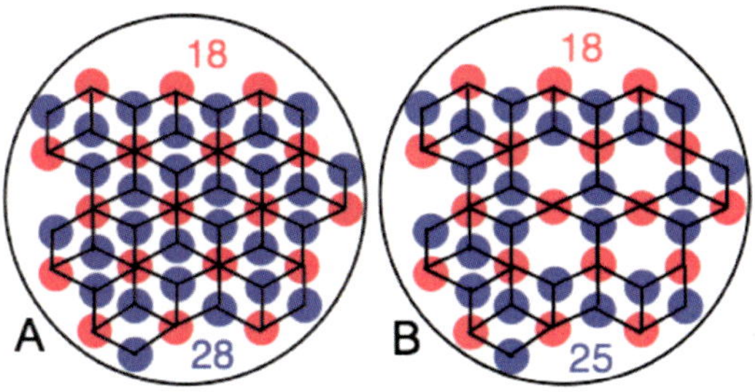

Figure 5.19 *A plan view of the structures of layers containing 18 protons.*

To determine the structure of any nucleus, the following procedure can be used:

1. Select the nucleus by the number of protons, Z, in it. For example, aluminium has 13 protons. Figure 5.20 shows how the different Al isotopes occur under these rules.

   Divide Z into layer segments that add up to Z. For example, 3+5+3+2=13. That division should be based upon the number of layers being within 1 of the number of nucleons in the width of the widest layer, low Z excluded. In our example, this means the fact that there are 4 layers is within 1 of the fact that the widest layer is 5. The layers on either side of the central layer (two layers if there is an even number of layers) should then be progressively and somewhat uniformly reduced to the lowest number.

(It should be noted that there can be different choices for Z in each layer. As long as the choice follows this guideline, it will be close to the correct structure. The selection of the number of layers is the critical step. The layer selection determines the range of isotopes theoretically available for a given Z. There will always be a layer selection that will match the spread of isotopes. For reference, the Z=5 layer has dimensions 5 nucleons x 3 nucleons; Z=8 has 6 nucleons x 5 nucleons; Z=12 has 7 nucleons by 6 nucleons, and so forth.)

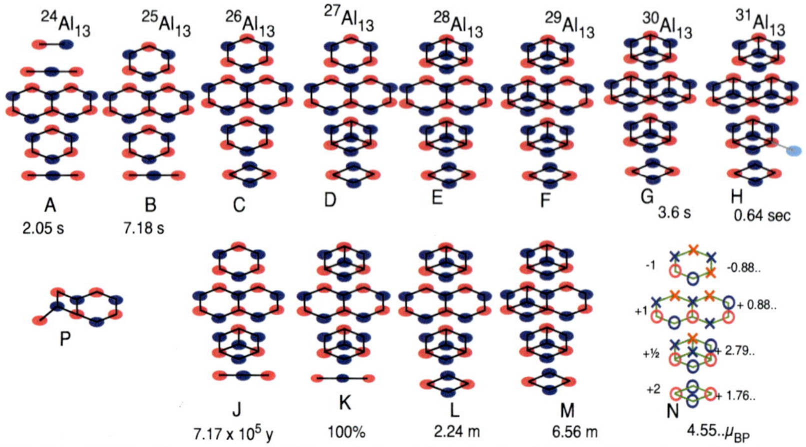

***Figure 5.20:** Structures of Al nuclei, as determined from the rules of nuclear binding.*

2. Select the layers determined in Step 2 from figure 5.17 to make up the highest A structure for the element of your choice.
3. Take the images of the layer structures chosen and arrange them one on top of the other, with the widest layer or layers in the middle. This gives the structure with the greatest number of neutrons for the number of protons in the element of interest.

(In the case of aluminium (Al), this will be a total of 30 nucleons, to form the isotope $^{30}Al_{13}$, as shown in figure 5.20 G. For most nuclei this defines the upper range for isotopes that are either stable or have half-lives greater than a few seconds. In the diagram, the number below each structure is either its half-life or its abundance.)

4. In order to determine the isotope of choice, remove the desired number of the neutrons from the structure, taking them from the largest layers first. Neutrons should be removed until the desired isotope is reached. Figure 5.20 shows that different structures are possible for the same isotope, for example C versus J, or D versus K. For most nuclei, their structure under these rules defines the range of isotopes that are either stable or have half-lives above a few seconds.
5. If insufficient neutrons can be subtracted from a particular combination of layers to match the required A for the chosen

Z, another combination should be used. The lattice structure of isotopes can change with A, even though Z remains constant. This is seen in figure 5.20 P, where a further neutron was removed from the central layer of the ^{24}Al isotope. That proton is held to only one neutron, making the whole structure quite unstable. ^{23}Al decays to ^{23}Mg within a few milliseconds.

6. If more neutrons need to be added to get the chosen isotope, for example in ^{31}Al, H, the additional neutron, pale blue, will only be attached to one proton, making it quite unstable. It has a half-life of 640 milliseconds.
7. Whenever either a proton is bound to a single neutron (low A) or a neutron is bound to a single proton (high A), the half-life of the isotope will be measured in units or fractions of seconds.
8. As far as possible, neighbouring nuclei within a layer will orient themselves such that the north magnetic pole of one nucleon is oriented in the same direction as the south magnetic pole of its neighbour. That maximizes magnetic binding energy between nucleons within the layers. Even A and even Z nuclei have all like nucleons arranged in opposite pairs, giving the nucleus no net spin or magnetic moment.

Note that two different sets of layers are displayed for some isotopes in figure 5.20. The structures A through H show the greatest array of possible Al isotopes, varying from 11 to 18 neutrons. Structure A has five layers, that being one way all 13 protons could be bound to the 11 neutrons within the rules of binding. Figure 5.20 P shows how the largest layer in the ^{25}Al structure in B could lose one neutron to form ^{24}Al. It leaves one proton bound to just one neutron, giving it a short half-life. The ^{31}Al nucleus has one neutron bound to just one proton and will also have a short half-life.

Because the neutron is a neutral particle, it will not be repelled as rapidly as a single proton. Figures 5.20 J to M show alternative filled neutron vacancy positions to those in C to F respectively. They show that it is possible for the same isotope to have different structures. Note that the most stable isotope, $^{27}Al_{13}$, has one extra neutron over the number of protons. That reflects the preference for even numbers of neutrons to give maximum binding.

Figure 5.20 N shows an arrangement for the spins of the protons and neutrons in the ^{27}Al nucleus. The colours represent the nucleons, red for protons, blue for neutrons. The circles represent spin up. The crosses represent spin down. Adding the magnetic moments of these nucleons gives ^{27}Al a theoretical magnetic moment of 4.55… μ_{BP}. The measured nuclear magnetic moment of $^{27}Al_{13}$ is 3.64… μ_{BP}. That difference of about 0.03 μ_{BP} or ≈ 1% per nucleon reflects the distortion of the nucleons' harmonic oscillations resulting from their overlap.

It should be noted here that the removal of the central neutron from the $^{7}Li^{*}$ structure could result in a slight distortion of its hexagonal shape. They are shown here in their undistorted state for ease of interpretation.

Figure 5.21 illustrates how neutron numbers outside those under the binding rules can be handled. A and G represent the limits, under the rules of binding, for a four layered structure. Five layers are possible. They appear to be needed to explain the lower A isotopes. One hexagonal ring becomes an alpha particle with a proton weakly attached in the isotope ^{23}Al, as shown in figure 5.21 N.

One proton is bound to one neutron, caught by the neutron's negative N/27 outer oscillation. It is like a satellite nucleon, and it is very unstable, with a half-life of ≈470 milliseconds. Removing another neutron from another hexagonal ring results in another alpha particle with a proton weakly attached to one neutron.

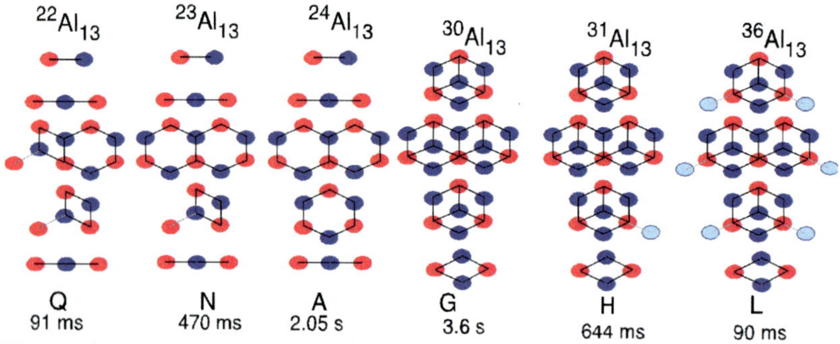

Figure 5.21: *Representations of the various Al isotopes with neutron numbers outside the rules of binding.*

That structure is shown in figure 5.21 Q, and it has a half-life of 92 milliseconds. Removing yet another neutron is unlikely. One reason is that excessive positive charge may repel at least one proton, and the structure would no longer be an aluminium isotope.

Another is that it is difficult to add a proton to a ^{20}Mg nucleus, which only has a half-life of 90 milliseconds. In any event, such an isotope has not been detected.

On the heavier isotope side, the only way for an additional neutron to bind is for its negative N/27 harmonic to overlap a proton's positive charge. There are 12 easily accessible protons on the outside of the aluminium nucleus. A neutron can easily access any one of these 12. There are no sites available, however, where a neutron can access two protons and still be within the binding rules. Figure 5.21 K shows the situation of one neutron binding to one of the easily accessible protons, but it could attach to any one of them. The resulting half-life is less than 1 second. Figure 5.21 L shows the situation where six of the available protons have each captured a neutron to form ^{36}Al. Its half-life is less than 100 milliseconds. Those neutrons are shown in lighter blue.

In the presence of excess neutrons, this capture process continues until all available protons have captured a neutron. The half-lives of the isotopes will progressively diminish, with slight deviations between even and odd number of neutrons. If they exist, the lifetimes of isotopes 40 and above will be much shorter, typically less than 1 millisecond. After that, there are no more proton positions to which a neutron can attach. It will be shown later that isotopes where neutrons can access to more than 10 exposed outer protons are rarely, if ever, detected. Some more recent results suggest that number could be as high as 12 (Holden, 2019).

Figure 5.12 can be used to work out the distances between the nuclei. That enables the dimensions of the different isotopes to be determined. Note that the nuclei will not necessarily be circular with a defined radius. Accurate measurements of nuclei should show dimples and protrusions. This method can give the structure, dimensions, and magnetic moment from the measured binding energy and spin of the various isotopes. It does not give any indication of those isotopes that are likely to be stable and those that are not stable. That is discussed later in this section. Information about the half-life of isotopes came from

Firestone et al. (1999) and The Berkeley Laboratory Isotope Project. Some more recent updates come from the work of Holden et al., (2019).

This process can be repeated for all nuclei. Figure 5.22 shows the structures of different isotopes of oxygen (O). It has 8 protons shown in a 2-3-3 layer configuration to fit the binding rule requirements. The isotope ^{13}O has the two protons in the alpha particle layers outside the rules of binding. They are shown in pink. Normally that would suggest the lifetime of such a structure to be less than a second. With the same spin on both protons and neutrons, they can get additional magnetic binding and have a slightly longer half-life. The most abundant isotope, ^{16}O, has both its hexagon rings with a vacant neutron lattice position.

The ^{18}O nucleus is stable and has all its neutron lattice positions filled. Its 8 protons all occur on the outside of the structure. This allows up to 8 neutrons to be positioned onto the ^{18}O isotope. The added neutrons are illustrated as lighter blue ovals with a light grey line joining them to the proton. The ^{20}O isotope is in the same situation as the ^{13}O isotope. With both protons and the neutrons to which they have attached having the same spin there will be additional magnetic binding that helps it have a half-life of over ten seconds.

The ^{24}O isotope has a short half-life of about 60 milliseconds. There are so many additional neutrons that the same spins are not going to help produce a longer half-life. The ^{26}O isotope is the end of the line for additional neutrons because there are no more positions to which neutrons can attach. It has been detected, but its half-life was not known when this study was undertaken.

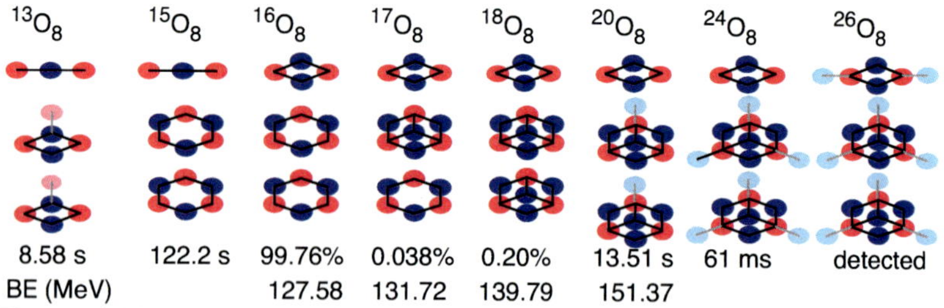

***Figure 5.22:** Structures of isotopes of oxygen, Z = 8, as determined from the rules of nuclear binding.*

Note that all the layers in the oxygen isotopes have their protons on the outside of the structure.

A measurement of the charge distribution of oxygen nuclei should therefore show a lesser charge in their centres, rising to a maximum at a radius of just over 1 fm. Figure 5.23 A shows measurements made by Hofstadter (1956). This diagram was copied from Cook (2010).

To understand how these are measured it is necessary to refer to the origins of the special relativity corrections. When particles move, they orient their rotation so that their charge plane is perpendicular to their travel direction. That exposes a combination of the charge distributions shown in figures 3.14 and 5.1 to the incident electrons.

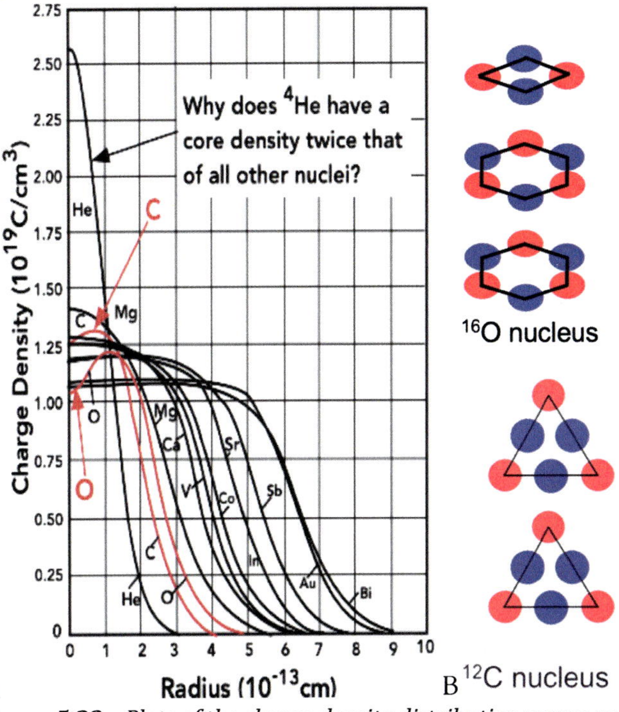

A B

Figure 5.23: *Plots of the charge density distribution versus radius of different nuclei, A, and the nucleon arrangement of the ^{12}C and ^{16}O nuclei, B.*

The red "O" and its red arrow point to a red curve that represents oxygen's nuclear charge density versus radius. It increases from about 1 unit of charge density, measured in terms of 10^{19} Coulomb per cm^3, to about 1.25 at a distance of about 1.1 fm. The ^{16}O nucleus in 5.23 B shows it has three layers, each with the positive charge on the perimeter. The random orientation of those shapes results in more scattering from the outside charge than the inner space.

The red curve that represents carbon's nuclear charge rises from just under 1.25 x 10^{19} Coulomb per cm^3, to a slightly higher maximum at about 0.8 fm. The two layers of ^{12}C nucleus are shown in figure 5.23 B. It will also give more scattering from the external charge than the internal void.

The ^{16}O nucleus has three layers. The ^{12}C nucleus has two layers. There will be more electron scattering from the centre of the ^{12}C nucleus than the ^{16}O nucleus. That is as observed.

Figure 5.23 A also shows the skin effect associated with nuclei. It is well illustrated by the curves for Au and Bi. They are at 90% of their maximum value at $\approx$ 5.5 fm. They drop to 10% of their maximum at about 8 fm. That gives them their skin effect spread over $\approx$ 2.5 fm. That distance is the same 2.5 fm charge tail shown in figures 3.14 and 5.1. The skin effect is produced by the nuclei orienting their rotation plane to perpendicular to their travel direction. The skin effect is further verification of this model. It also justifies the special relativity corrections described in chapter 2.

Figure 5.24 shows probable nucleon arrangements for different isotopes of neon (Ne), Z=10. The feature of interest is that the most abundant isotope, ^{20}Ne, like ^{16}O, has all its internal neutron lattice positions empty.

The low abundance of the ^{21}Ne isotope is yet another example of how odd numbers of nucleons are not good for nuclear binding. The ^{22}Ne isotope has a much greater abundance. The ^{23}Ne's short half-life again indicates the lower stability of odd numbers of nucleons. The additional neutron in the ^{24}Ne isotope, indicated by the pale blue oval, is outside the binding rules because it is attached to outer protons that already have three neutrons binding to them.

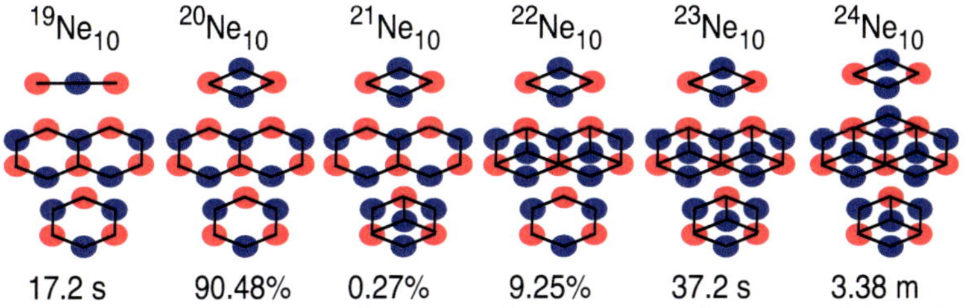

***Figure 5.24:** Structures of isotopes of Ne, Z = 10, as determined from the rules of nuclear binding.*

This is one of the few cases where neutrons in that position give nuclei a half-life of more than a couple of seconds. When it decays by beta emission, the structure rearranges itself, transiting through ^{24}Na, and ending up as ^{24}Mg.

The ^{24}Ne isotope still leaves all 10 protons external and available for neutrons to temporarily attach to it. ^{34}Ne is the highest atomic number neon isotope detected. Beyond that, there are no more sites to which neutrons can attach.

This process can be repeated for as many of the over 2,800 known nuclei as required. It is too great a task for this publication and can be easily done by other interested parties. Nevertheless, it is worth looking at the nuclei of a few specific elements.

The next nucleus of interest is calcium (Ca), Z = 20. Some of its isotopes are shown in figure 5.25. It represents the transition from low Z nuclei to high Z nuclei. It is the heaviest nucleus that has equal numbers of protons and neutrons in its most abundant structure, and for which all the internal neutron lattice positions are vacant.

Calcium, like the multi-layered even Z>4 nuclei before it, has its most abundant isotope with all its internal neutron lattice positions vacant. Odd Z nuclei with Z < 20 have one internal neutron lattice position filled and the rest vacant. Calcium also has a large number of stable isotopes – six in total. There is some suggestion that ^{48}Ca is unstable, though with a half-life of over 10 billion years. That is close enough to stability to be considered stable.

The calcium nuclei from A = 38 to A = 46 are shown with a symmetric 2-3-5-5-3-2 layer arrangement. In order to maintain the rules of binding required for the ^{48}Ca isotope to be stable, it is necessary to re-

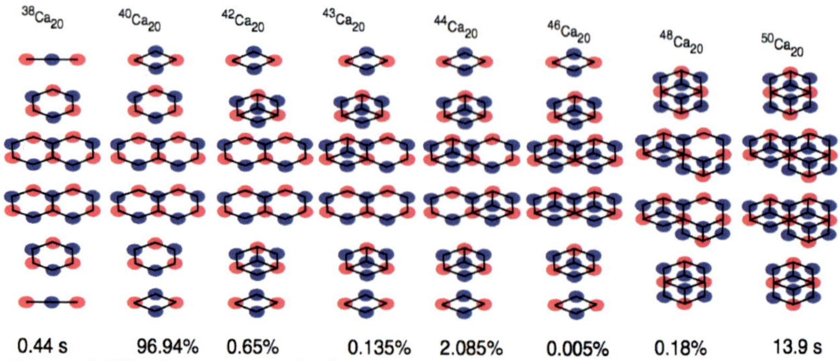

Figure 5.25: *Structures of calcium isotopes, Z = 20, as determined from the rules of nuclear binding.*

arrange the number of layers and the nucleon arrangements within them to a 4-6-6-4 layer arrangement That is not unusual among the higher Z nuclei.

The ^{38}Ca isotope is within the binding rules. Its half-life is a little lower than the 1 – 2 seconds. On the other hand, eight neutrons cannot be removed from the ^{48}Ca nucleus and have it remain within the rules of binding for ^{40}Ca. That is why ^{40}Ca nuclei and some of the higher Z nuclei have the 2-3-5-5-3-2 layer structure indicated. The highest A calcium nuclei have the 4-6-6-4 layer structure. This shows that different isotopes of the same element can have different layer structures.

The ^{50}Ca isotope has all neutron lattice positions filled. It has 18 external protons to which a neutron could temporarily attach. As mentioned earlier, an additional 10 neutrons are the maximum number of excess neutrons per nucleus that can temporarily attach, in practice. It appears that after that many additional neutrons, the rate of decay of the nucleus exceeds the rate at which extra neutrons can be added. The ^{60}Ca isotope is the highest mass isotope for calcium listed in isotope tables.

The next nucleus of interest is iron (Fe), Z = 26. It is interesting for two main reasons. Commercially, iron is the most widely used metal in industry. Scientifically, the ^{56}Fe isotope has the highest binding energy, per nucleon, of any nucleus. Figure 5.26 shows some selected iron isotopes. Those from 52 to 62 obey the rules of binding. The isotope ^{51}Fe has a half-life of about 300 milliseconds. That is as expected because

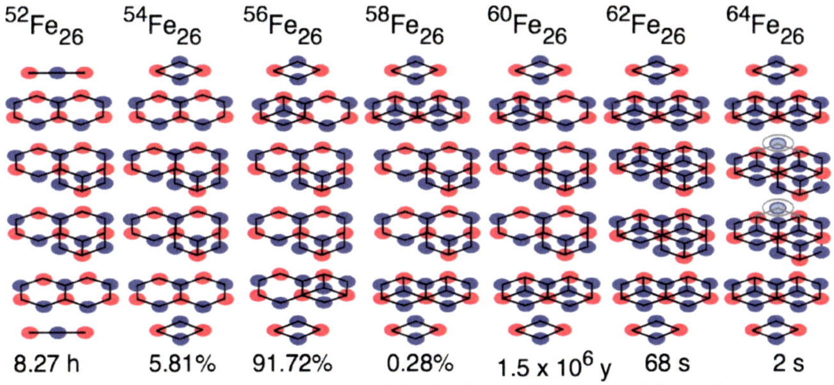

Figure 5.26: *Structures of isotopes of Fe, Z = 26, as determined from the rules of nuclear binding.*

it is outside the binding rules. The isotope ^{64}Fe has two neutrons more than the binding rules.

The isotope ^{54}Fe has all internal neutron lattice positions vacant. It is stable, but not the most abundant isotope. That role is held by ^{56}Fe, which comprises over 90% of the iron. It has two of its eight internal lattice positions filled. The isotope ^{64}Fe has two neutrons outside the rules of binding. They are indicated by the pale blue circles.

These neutrons are attached to outer protons that are already bound to three other neutrons. The “strong” nuclear force is not so strong that it can accommodate an extra neutron pulling protons away from their stable bound states. As a result, the half-life of this isotope is only 2 seconds.

The instability causes at least one of the neutrons to emit an electron and anti-electron neutrino, becoming a proton. It is suggested that, if the spins of the two neutrons are in the same direction, their magnetic binding would give them an extra degree of stability, which gives them a half-life of more than 1 second. The isotope ^{58}Fe has half its vacant neutron lattice positions filled. Its abundance of 0.28% is the lowest of the stable iron isotopes.

The ^{62}Fe isotope has all neutron lattice vacancies filled. Extra neutrons can only attach loosely to 10 of those protons before their rate of decay exceeds the rate at which new neutrons can be added. Isotope

^{72}Fe is the highest mass iron isotope detected. Its half-life is measured in nanoseconds.

The iron isotopes illustrate two important features of this model. Iron has four stable isotopes, each with a different number of unfilled neutron lattice vacancy positions. As the number of protons per nucleus increases, the number of neutrons required to form stable nuclei increase at a faster rate. The larger the number of nucleons in a layer, the more vacant neutron lattice positions there are available. That increases the number of possible isotopes, both stable and "quasi-stable." A "quasi stable" nucleus is one that has a half-life greater than ≈ 2 seconds.

There is another feature, which will be demonstrated later, that is worth pointing out now. As the number of protons in a nucleus increases from 20 to 82, the most abundant nuclei go from being those with all internal neutron lattice positions vacant (at 20) to all filled (at 82). One element that illustrates this well is palladium (Pd). It has 46 protons, and is therefore halfway towards 92 protons per nucleus, the highest even number of protons in a non-radioactive element.

Figure 5.27 shows stable palladium isotopes, all of which obey the rules of binding. In this presentation, the nucleons are shown in layers containing 2-5-10-12-10-5-2 protons respectively. Palladium has six stable isotopes with varying degrees of unfilled internal neutron lattice positions.

Their natural abundances are indicated below them. Because Z is even, most of the stable isotopes also have even A. The exception is ^{105}Pd. ^{107}Pd, which is next to the most abundant ^{106}Pd, has a half-life of over six million years. Further neutrons can be removed.

The lowest stable isotope, ^{102}Pd, has 11 vacant neutron lattice positions. The most abundant isotope, ^{106}Pd, has four of those vacancies filled. The highest atomic number stable palladium isotope, ^{110}Pd, has three neutron lattice positions vacant.

Some of the effects of changing the number of neutrons in a palladium nucleus are shown in figure 5.28. The isotopes ^{96}Pd and ^{100}Pd are still within the binding rules. ^{96}Pd is the lowest mass isotope that complies with them for this structure.

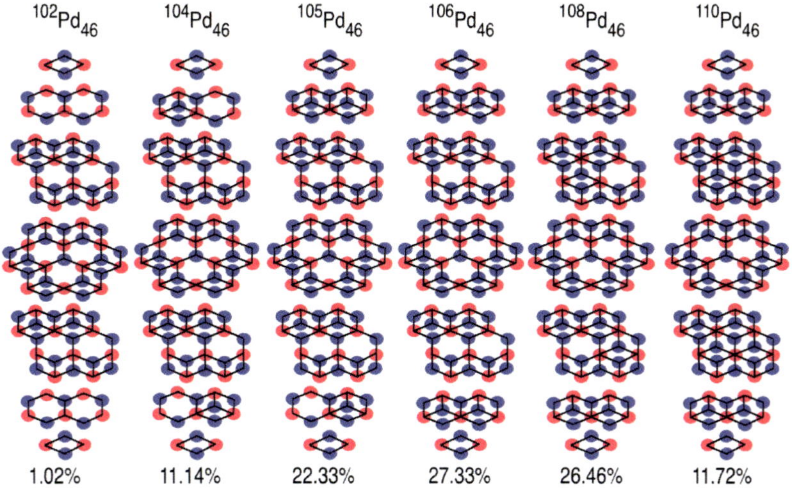

Figure 5.27: *Structures of stable isotopes of Pd, Z = 46, nuclei as determined from the rules for forming nuclei.*

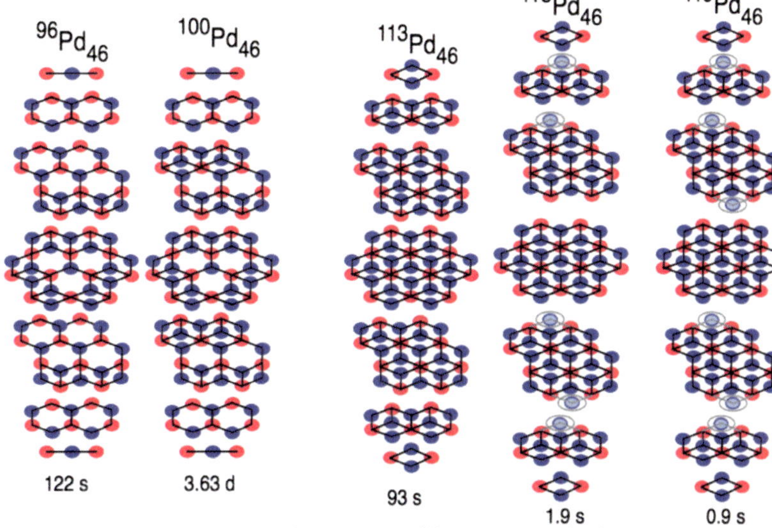

Figure 5.28: *Structures of some unstable isotopes of Pd, Z = 46, nuclei as determined from the rules for forming nuclei.*

Variants of the ^{95}Pd, ^{94}Pd, and ^{93}Pd isotopes have half-lives of up to 10 seconds. Those isotopes are referred to as variants because they have different half-lives. It is suggested that, under this model, different possible layer arrangements give different stabilities, and this, in turn, leads to different half-lives.

The isotope ^{113}Pd is the highest mass number that fulfils the binding rules for that chosen geometry. Its half-life is over 90 seconds. It has many positions where additional neutrons can find temporary binding.

Isotopes ^{118}Pd and ^{119}Pd have a few sites occupied. Their half-lives are around 1 and 2 seconds. That may also be due to a different layer arrangement. Either way, it would not be surprising if the addition of 10 further neutrons beyond the highest A quasi stable nucleus formed the highest mass number palladium isotope. The isotope ^{129}Pd is the highest mass number isotope found in lists of palladium isotopes.

Continuing this trend, tin (Sn), Z=50, is the even Z nucleus that is nearest to halfway between Z=20 and Z=82. It has ten stable isotopes. They are on either side of its most abundant isotope. Only three of those ten have odd numbers of neutrons. Odd Z nuclei have very few stable isotopes and those that are stable almost always have even numbers of neutrons. This arrangement can be continued up to any elements with 92 protons and beyond.

Figure 5.17 showed that the number of neutrons required to bind protons in a layer increases faster than the number of protons in that layer. Note also that the number of vacant neutron lattice positions for the lowest A of a given Z layer increases with the number of protons in that layer. This adds to the number of stable and quasi-stable isotopes that an element can have.

That trend reaches its maximum at tin, Z = 50. It is the nearest even Z element that is halfway between calcium, the last of the "light element nuclei" and lead, the last of the stable heavy elements. Tin has the highest number of stable isotopes. It also has among the highest number of detected isotopes, see for example Holden et al., (2019). Tin's most abundant isotope, at almost one third of its composition, is ^{120}Sn. It is also in the middle of the isotope range, which extends from ^{99}Sn to ^{139}Sn. It is also where about half of the vacant inner neutron lattice positions are filled.

Tin, Z=50, represents a significant change between the lower Z and higher Z nuclei. To clarify why, the next element to be discussed will be lead (Pb), Z=82.

Figure 5.29 shows an arrangement with 8 layers of nucleons, containing 3-8-12-18-18-12-8-3 protons respectively. An important feature is that its most stable isotope, ^{208}Pb, has all its neutron lattice positions full. For that reason, it is the end of the line for stable nuclei.

It is variously claimed that ^{209}Bi has that distinction. There is a suggestion that it has a half-life more than 10^{17} years, which is essentially stable. The other feature of ^{208}Pb is that it is the main end product of the decay of higher Z nuclei. When there are a lot of vacant neutron lattice positions, such as in the ^{190}Pb isotope, it is unstable because there are not enough neutrons to keep the structure stable.

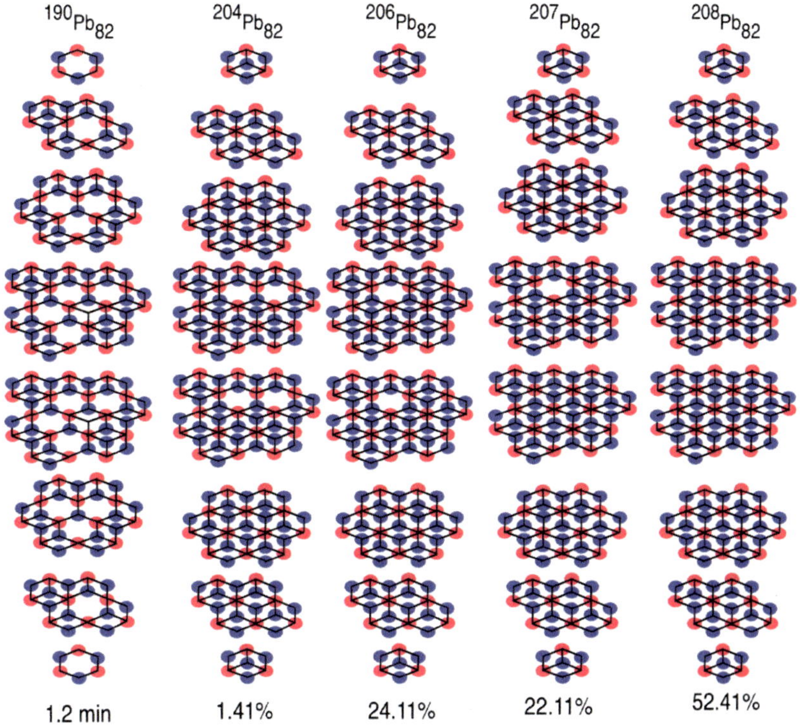

Figure 5.29: *Structures of isotopes of Pb, Z = 82, as determined from the rules for forming nuclei.*

There are a couple of features to note about the lattice arrays. Every nucleon in all the stable isotopes is arrayed in an alpha particle structure. That is also a feature of all the filled nuclear layers shown in figure 5.17 that have more than 1 proton. This model helps explain why it is so easy to knock alpha particles out of nuclei. They are made of them.

The hexagonal close packed array depends upon the neutrons' repulsion from each other, giving their centre-to-centre spacing the same dimension as the attraction between a proton and a neutron. If these are not the same, the array may differ slightly from hexagonal close packing.

Isotope ^{190}Pb has a half-life of about 1 minute. It is unstable because there are not enough neutrons to keep it stable. It is not until 14 neutrons later, with isotope $^{204}Pb_{82}$, that there are sufficient neutrons for the nucleus to be stable. There are then three stable isotopes with some vacant neutron lattice positions, and a fourth, the most abundant isotope ^{208}Pb, whose most significant feature is that it has no vacant neutron lattice positions.

Removing further neutrons below ^{190}Pb goes outside the layer structures in figure 5.18, and results in isotopes with half-lives measured in seconds or fractions thereof. Adding further neutrons above ^{208}Pb requires the neutrons to violate the binding rules, making for quite unstable isotopes. Adding the statutory 10 (or 12) neutrons takes the mass number to isotope ^{218}Pb (or ^{220}Pb). They are the heaviest lead isotopes registered.

Another significant feature of the lead nucleus is that it is the heaviest stable nucleus. There are five naturally occurring heavier nuclei, ^{209}Bi, ^{232}Th, ^{234}U, ^{235}U, and ^{238}U. Of those, ^{209}Bi is said to have a half-life of more than 10^{17} years and is therefore stable for all practical purposes. Its structure requires only seven layers. The rest decay at various rates. The end product of their decay is ^{208}Pb. Other elements such as Po and Rn are transient nuclei produced by the radioactive decay of the five aforementioned nuclei.

Between them figures 5.20 to 5.29 illustrate important characteristics of nuclei. Nuclei with an even $Z \leq 20$ (excluding beryllium) have their most abundant isotopes with all internal neutron lattice positions vacant. Nuclei with odd $Z < 20$ have their most abundant isotope with one filled internal lattice position. Some have just one stable isotope. As the number of protons increases beyond 20, the most abundant even Z nuclei have some of their internal neutron lattice positions filled. As the number of protons goes from 50 to 82, there is a decrease in the number of unfilled neutron lattice positions for the most abundant isotopes.

At 82 protons, the heaviest stable nucleus, all neutron lattice positions are filled in the most abundant isotope. All other stable Pb isotopes have lower atomic masses (A) than the most abundant isotope. That trend continues down the periodic table. The closer nuclei are to $Z = 82$, the more likely they are to have a greater number of stable nuclei below their must abundant isotope. That continues until the mid-range elements, $46 < Z < 56$. Their most abundant isotopes are when approximately half the vacant neutron lattice positions are filled. They have stable isotopes with A above and below that of their most abundant isotope. Those even Z nuclei tend to have several isotopes of similar abundance. The half-life range of their radioactive isotopes are spread approximately equally on either side of those most abundant isotopes. Even Z elements in that range tend to have several stable isotopes, with tin, $Z=50$, having 10. Odd Z nuclei have so few stable isotopes. This trend does not translate across to them in any significant manner.

It Is suggested that this Indicates the Importance of magnetic binding. Protons have a higher magnetic moment than neutrons and even numbers of protons can align their spins to give greater binding. Odd Z nuclei have one additional proton whose magnetic field cannot be easily captured. Odd Z nuclei with even numbers of neutrons can orient their spins so that they give maximum magnetic binding to the protons by all the neutrons. Internally closed loop magnetic fluxes in which there are no external magnetic moments give stronger magnetic binding than do aligned magnetic moments that have an external magnetic moment. This appears to be associated with the bi-polar nature of the neutron's magnetic field.

The importance of the variation of isotope abundances with neutron lattice vacancies is illustrated in figure 5.30. The mass number, A, is plotted against the atomic number, Z. The $A = 2Z$ line represents equal numbers of protons and neutrons in a nucleus.

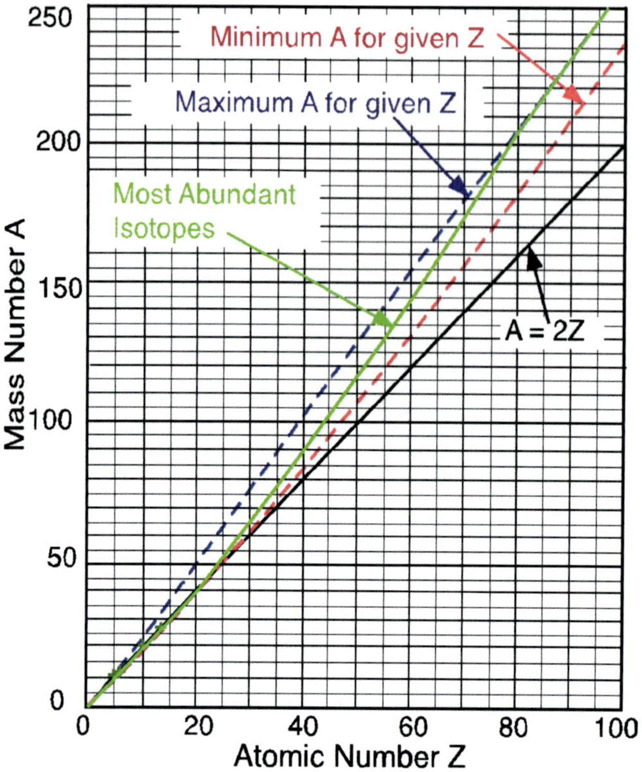

Figure 5.30: *Plots of mass number A versus atomic number Z.*

The dashed blue curve is theoretically generated from the layers shown in figures 5.17. It shows the relationship between A and Z for all nuclei where all the neutron lattice positions are filled. The dashed red curve is theoretically generated from figures 5.18. are typically ≈ 1 sec.

Figure 5.30 shows the relationship between A and Z for all nuclei where all the available neutron lattice positions are vacant. Most nuclei between the red and blue lines will be stable or quasi stable. Those outside have short half-lives. For $Z < \approx 70$, the half-lives above the maximum A for a given Z.

At Z = 82 the most stable isotope has an infinite half-life and is also the most abundant Pb isotope. With plenty of protons available to accommodate additional neutrons, it can accommodate several more isotopes before the half-life drops below 1 sec. It is suggested the longer stability is due to the availability of additional neutron binding sites, as mentioned with ^{64}Fe.

The curved green line shows the A and Z of the most abundant isotopes in a smooth curve. For Z = 82 and lower (with the exception of technetium, Z = 43, and promethium, Z=61), the most abundant elements are all stable. At Z < 21, the most abundant isotopes have all neutron lattice vacancies free for even Z and just one filled for odd Z. As Z increases from 20 to 82, the most abundant isotopes slowly change to having all lattice vacancy positions occupied. In the mid-range, 46 < Z < 56, there are a few stable isotopes with the most abundant about midway between the two curves. Isotopes between the minimum and maximum curves are either stable or "quasi-stable" with half-lives measured in seconds. Most of those outside that region have half-lives measured in low seconds to fractions of seconds.

There is a lot more to the structure of nuclei than just the arrangement of protons and neutrons in the nucleus. For completeness, we would also need to match properties of spin, magnetic moment, dimension, charge density, the skin effect, and radioactive decay, to name a few. In this model all nucleons retain their original properties, except for their magnetic moments. These are slightly distorted due to minor changes in their dimensions resulting from their binding. The trajectories of their rotating photons are slightly stretched or compressed as they cross over the other rotating photon trajectories.

What has not been attempted here is to place one layer of nucleons over another and determine the orientations of the nuclei that give maximum magnetic binding between the layers and minimum magnetic moment to the nucleus. As far as possible nucleons between layers will position themselves to get maximum binding to a nucleon in an adjacent layer. That involves like nucleons with the same spin and differing nucleons with opposing spin in the different layers getting as close to each other as possible.

The nucleon relationships between layers are opposite to those between adjacent nucleons in the same layer, where differing nuclei have the same spin and like nuclei have opposite spins. When nuclei are being formed, individual nucleons will both position themselves in the same plane as other nucleons and also align their spins to give maximum binding energy within the nucleus. That can, and indeed does, result in nuclei of the same isotope of the same element having different possible configurations. This is observed by the same isotope having different half-lives.

In the studies of low Z nuclei, it was shown that the same structure requires slightly different proton to neutron binding energies to explain the measured binding energies. That was best illustrated by comparing the binding energy of the ^{6}Li nucleus to that of the ^{12}C nucleus, the latter consisting of two $^{6}Li^{*}$ nuclei laid on top of each other. It suggests that when the same spins of the same nuclei are overlaid, there is considerable magnetic attraction between the individual nucleons. That appears to be sufficient to pull the nucleons in each layer closer, increasing their proton to neutron binding energy.

This suggests the orientation of the nucleons in a layer positioned above another layer appears to be important in low Z nuclei. At high Z, the differences average out. When there are an even number of protons and neutrons in a nucleus, the spin and magnetic moment are both almost always zero. Half the neutrons spin in one direction and half spin in the other direction. The same occurs for the protons. The result should therefore be no residual spin and no residual magnetic moment, as is observed.

Determining the dimensions of the nuclei can be done in the manner shown in figure 5.31, which shows several views of the $^{208}Pb_{82}$ nucleus.

Figure 5.31 A shows a stylized presentation of the positions of the protons and neutrons in one of the central layers containing 18 protons. They are the middle layers of the ^{208}Pb nucleus shown in figure 5.29. The lines joining them show the nearest neighbour bonds. Their length is approximately calculated from the binding energy of the nucleus. The bold circle represents the positions of the root mean square (rms) charge radius. The lighter circles represent the radii at which the positive charge is 90% and 10% of its plateau values, as indicated.

Modelling the dimensions of a nucleus requires knowledge of the total binding energy of the nucleus. The first step is to determine the layer structure. From that, the number of inter layer nucleon pair bonds can be determined. Multiply that number by 0.8 MeV and subtract that from the total binding energy. Then determine the number of nearest neighbour proton to neutron bonds.

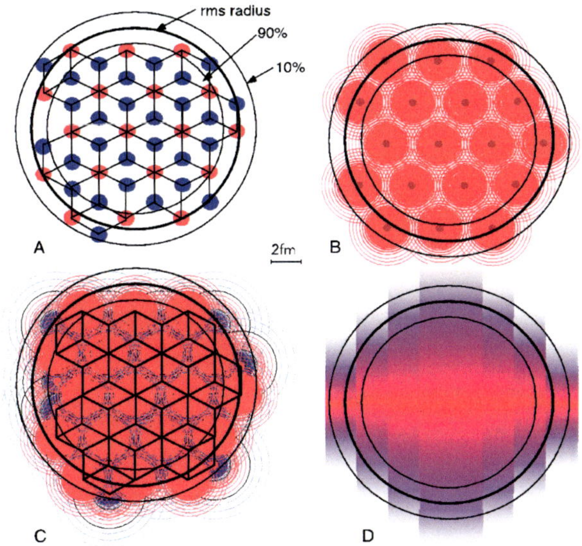

Figure 5.31: *Schematic views of the shape of lead (Pb) nuclei. A, B and C show the one of the central layers. D shows a side view of the eight-layered Pb nucleus.*

Follow that with the number of second nearest neighbour bonds. Their length is double that of nearest neighbours. Their strength is typically about one-third that of nearest neighbour bonds. Divide the second nearest neighbour bonds by 3 and add that number to the number of nearest neighbour bonds. Divide the remaining binding energy by that total.

Use figure 5.12 to determine the length of the nearest neighbour bonds. Apply that length to the central layer structure to determine its radius. When it is all done, you will get an answer that is not much different from (Cook, 2010):

$$r = 1.02(2Z)^{\frac{1}{3}} + 0.01(A - 2Z)^{\frac{1}{3}} \qquad (5.3$$

Figure 5.31 and many of the previous nuclei studied, show that they are approximate spheres but are not perfectly spherical. They have dimples and bumps. Any calculation of the radius of a nucleus will therefore be in error.

Note that, under this model, isotopes should hardly change their nuclear radius when further neutrons are added. They simply fit into vacant neutron lattice positions and do not push the other protons and neutrons much further apart. Equation 5.3, representing the rms radius of a nucleus, has a term that relates to what is called the neutron excess. It is the excess of neutrons over protons required to form stable nuclei at higher Z. It is so small that there is very little detectable change in the radius of a nucleus when neutrons fit into those vacant lattice positions. Under this model, the reason for such a small increase is the existence of neutron lattice vacancies within the nucleus. Neutrons can slot into them without much further distortion to the nucleus.

The rules for selecting the number of nucleons in the separate layers are such that the nuclei will appear approximately circular, whether viewed end on, as in figure 5.31 C, or side on, as in figure 5.31 D. While this accords with a preconceived concept, based upon a "surface tension" like effect, spherical shapes are not possible for small Z nuclei. It may well be that higher Z nuclei appear like oblong shaped cubic structures with rounded corners and indents. The origin of the indents can be seen from figures 5.17 and 5.18. Such changes do not have a noticeable impact upon the curves in figure 5.30.

The above explains the close relationship between the predicted range of isotopes and their observed range. It enables the structure, probability of abundance and stability, spin and approximate magnetic moment, and dimensions of any nucleus to be determined. It explains the skin effect and suggests that nuclei will not be spherical, having both "dimples" and projections at their surface. It strongly suggests that this one model can be used to explain all nuclear observations. This will not be possible in this short presentation, but there is sufficient information provided for those who wish to explore it further.

5.6. Uranium and Fission

Among the nuclei of importance in modern industry are some of the uranium isotopes. Figure 5.32 illustrates some of the isotope structures that this model generates using layers with Z of 3-9-13-19-19-13-9-3. Other layer arrangements are possible under this model. They may result in a change of a neutron or two for the same number of protons when eight layers are used.

Using seven layers does not meet the required rules when the central layer has nine nucleons in its widest dimension. In this layout ^{238}U, the most stable uranium isotope, has all its lattice positions filled. That is expected for the most stable nuclei with $Z \geq 82$.

The structure of the ^{235}U nucleus is also shown. In this situation, the three neutron lattice position vacancies are in one of the central layers. Proton positive charge repulsion is so great they need as many neutrons as possible. The isotope ^{226}U shows the lowest A isotope that will be predicted under this model. It has a half-life of ≈ 0.35 sec. Figure 5.32 also illustrates that vacant neutron lattice positions in the layer's nearest to the surface are filled first.

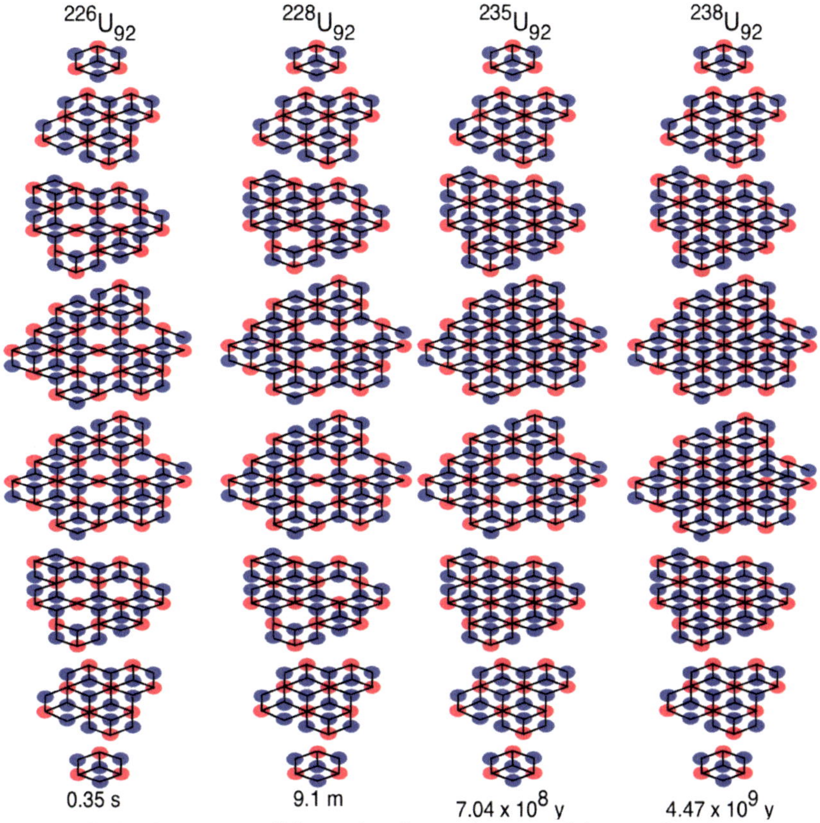

Figure 5.32: *Structures of the various layers in some of the uranium isotopes.*

With all nuclei above ^{208}Pb being radioactive, it should not be surprising that it is not possible to add 10 neutrons to ^{238}U nuclei. The highest A uranium isotope made in a laboratory is about ^{243}U.

The uranium isotope of greatest interest is ^{235}U. Figure 5.33 illustrates possible ways in which it can fission when struck by an additional neutron. A is the same representation of ^{235}U found in figure 5.32. Figure 5.33 B and C show representations of how ^{235}U nuclei can fission in different ways.

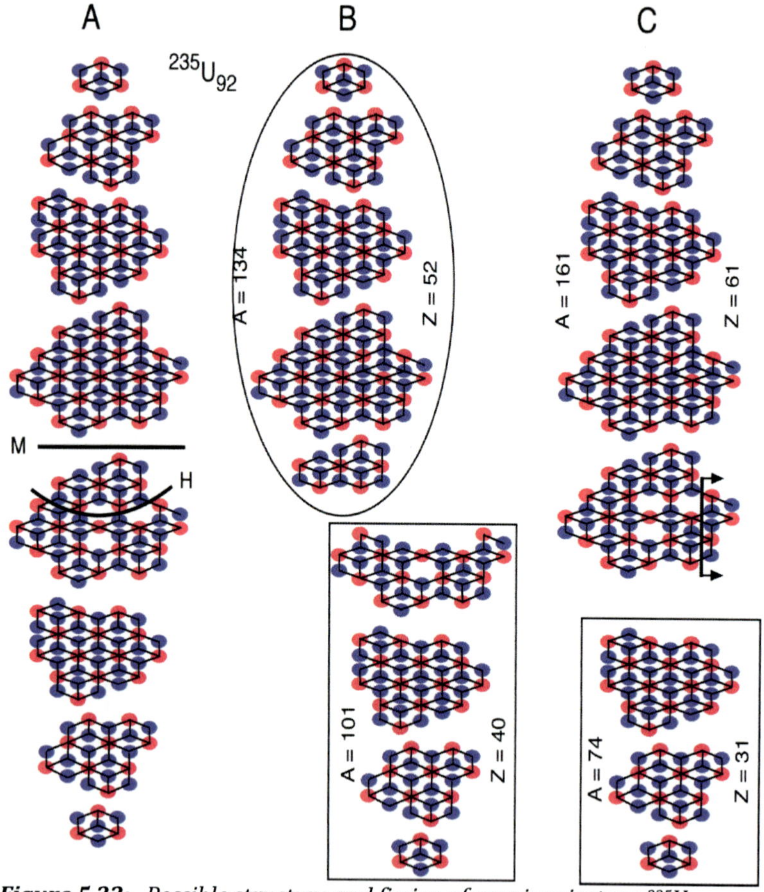

Figure 5.33: *Possible structure and fission of uranium isotope* ^{235}U.

As mentioned in the binding rules, neutron vacancies are filled from the outside in. That is the reason the last three neutron lattice vacancies are shown on one of the inner layers. The line M separates the nucleus in half between its two middle layers. The curved line H represents a partition of the layer with the lattice vacancies through a "passage" that involves the breaking of only five proton-to-neutron bonds. An incoming neutron initiates the fission.

The fission process is an intra-layer phenomenon as well as an inter-layer one. That is because the layers are weakly bound and can split when there is a disturbance. With neutron lattice vacancies in the middle layer, that is where the disturbance (involving a neutron absorption) will occur.

In this presentation, the split along line H in figure 5.33 A, leaves the layer fragment above line H connected to the upper layers, as shown in the elliptically enclosed array in diagram B. That yields isotope $^{134}Te_{52}$. The residual layers below line H constitute the isotope ^{101}Zr.

Both are before the emission of two neutrons, which, together with the original neutron, means the fission event releases three neutrons.

Alternatively, the larger segment could stay with the upper layer instead, which would then generate the isotopes ^{155}Nd and ^{80}Ge, as shown in figure 5.33 C. Again, two emitted neutrons must be subtracted from these structures.

Figure 5.33 C suggests a scenario in which three protons and four neutrons separate from the lower central layer that contains the three vacancies. In the configuration shown here, this requires the breaking of six proton to neutron nearest neighbour bonds. If that fragment remains with the other three lower layers, it yields the isotopes ^{73}Cu and ^{162}Eu (less two neutrons).

Apart from splitting between whole layers, this represents the extremes of the fractions that can be generated in the fission of this 8-layer arrangement. A different set of vacant lattice positions would allow other possible structures, and therefore, different fractions to be generated.

The instability arising from the absorption of the excess neutron is one that is specific to the layers and not to the magnetic binding between the layers. As such, splitting between the two central layers, about the line M in figure 5.33 A to create two equal fragments, would not be the most probable mode of splitting. It is more probable that the splitting would occur through the weakest point of one of the middle layers.

When a ^{235}U nucleus fissions, it releases two neutrons in addition to the original neutron that set off the reaction. This occurs when the nucleons rearrange themselves into their new layer configurations, determined from figure 5.17. The masses of the fragments depend upon the numbers of neutrons that were released from either side of the fission line. Excluding the two neutrons, the atomic masses of the two fragments (of varying masses) must add up to A=233.

The $^{235}U_{92}$ isotope shown contains A=119 above line M (just as in isotope $^{238}U_{92}$), and A=116 below line M. The middle layer with the neutron lattice vacancies has mass of A=29. If its most probable fission is half of that 29, it would add a mass of 15 to the upper layer, giving the most probable heavy fraction a mass of A=134. That would leave the most probable light fraction a mass of A=101 atomic mass units. If two mass units are removed for the two neutrons that are emitted, the most probable light fraction has a mass of A=99.

There are other directions through which that layer splitting could occur, all of which lead to different fission products. There are also other positions in the central layer where the vacant neutron lattice positions could be located. Figure 5.33 C shows a situation in which a Z=3 fragment would be the remaining segment of the middle layer that is splitting, shown by the arrowed line. This requires the splitting of 6 proton-to-neutron bonds, making it considerably less probable than the splitting of 4 proton-to-neutron bonds. It can be seen from a study of the middle layer of the $^{235}U_{92}$ nucleus that it is difficult to isolate a Z=2 fragment. This suggests the Z=31 fragment (figure 5.33 C) would be the lowest Z fragment left after splitting. Such an arrangement would give mass fractions as high as $\approx$ 161 and as low as $\approx$ 71, after subtracting the three free neutrons.

Figure 5.34, copied from the work of Wahl (2002), displays a summary of the detected mass fragments from the splitting of ^{235}U nuclei. Again, the final mass of the fragments would depend upon the

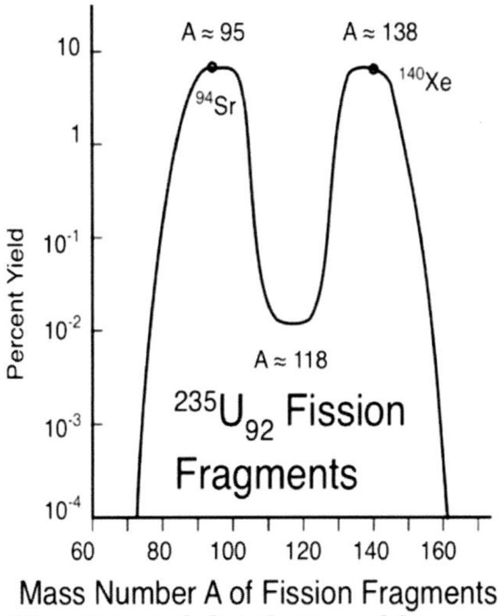

Figure 5.34: *Detected distribution of fragments from the fission of $^{235}U_{92}$ nuclei.*

side of the fission line from which they were released. This work is not inconsistent with those results.

It is believed these trends have been detected. It is suggested that the splitting of one of the central layers under this model gives a reasonable fit with the detected fission fragments. Once the nucleus has split, the layers rearrange themselves according to the rules of layer binding, resulting in neutron filled lattices for lower Z layers.

However, those lower Z layers are only stable when some of the neutron lattice positions are vacant. Excess neutrons emit beta particles, converting to protons.

The structure rearranges itself to accommodate the additional proton. The process continues until a stable layer structure is reached, that is, when the radioactive decay is complete.

When fission occurs, the nucleons do not stay in the configuration they were in. They re-arrange themselves into layers befitting their new A and Z status. It is that re-arrangement that releases the energy available from nuclear reactions. The remnants are left with more neutrons than required for their new atomic number. That is why they are radioactive.

5.7. Radioactive Decay

A well-known property of unstable nuclei is that they decay, giving off radiation. By definition, an unstable nucleus is one that emits radiation. There are three basic reasons why nuclei give off radiation. There are either too many or too few neutrons to stabilise the structure for the number of protons, or some additional energy has been given to the nucleus, bringing it into an excited state. In its excited state the nucleus gets rid of excess energy by emitting a gamma ray. It remains otherwise unchanged. Nuclei are usually only excited when they have been given excess energy in the first place, either by cosmic rays or in nuclear reactors.

Earth's age means that the natural radioactivity of all short-lived nuclei has long since decayed. The only natural radioactivity occurring is that of elements with $Z > 82$. As the number of protons in a nucleus increases, their mutual repulsion pushes them further apart. That increased distance weakens the average proton to neutron bond. Many nuclei with $Z > 83$ decay by alpha particle emission.

Viewing any of the nuclear structures shown makes clear that alpha particles pervade the structure of high Z nuclei. At 28.28 MeV, their binding energy as an isolated particle is far higher than it is when they are bound into an unstable nucleus. Alpha particle emission therefore takes a lot of energy with it.

All other forms of radioactivity are made in laboratories. Elements whose nuclei contain more neutrons than reflected in the binding rules cannot simply eject an extra neutron. There is almost no repulsive force between a neutron and a nucleus. Instead, an excess neutron decays to a proton by emitting an electron. It is suggested that the neutron experiences beta particle decay because there are insufficient protons to keep it stable over a long time period. The neutron decay mechanism was discussed in chapter 3.

It occurs because the alternating rotating photon structures in a neutron-rich nucleus result in insufficient interlocking to keep a neutron from decaying. After emission, the lattice structure of the nucleus rearranges itself to accommodate the additional proton.

The physical cause of nuclear decay comes from the structures of the nucleons and their binding mechanism. Part of the strong force comes from direct overlap of a proton's positive P/3 and P/9 oscillations with a neutron's negative N/3 oscillations. These oscillations are not permanent. They appear and disappear at different times. An opposing charge overlap disappearance weakens the binding between the nucleons.

In a neutron-rich nucleus, two result are possible. The most common is insufficient interlocking to keep a neutron from decaying. After emission, the lattice structure of the nucleus rearranges itself to accommodate the additional proton. The other is the neutron is only weakly bound by its negative N/27 oscillation. When it disappears, there is nothing binding it to the nucleus. It decays by neutron emission.

A good illustration of this is the isotopes of aluminium. Isotopes ^{28}Al, ^{29}Al, and ^{30}Al are neutron-rich, where the excess neutrons are within the rules of binding. This can be seen in figures 5.20 and 5.21. They decay by β^- emission. Aluminium isotopes with $A > 30$ can also decay by the emission of a neutron because they are only weakly bound. This applies to many nuclei with excess neutrons beyond the number of neutron lattice positions available. Those with neutrons bound within the lattice structure almost invariably decay by electron emission only.

Nuclei with an excess of protons within the binding rules decay by electron capture or the emission of a positron. Outside those binding rules, the strong repulsion between like positive charges causes protons, or even alpha particles, to be emitted.

Another possibility is a weakened bond allows the nucleons to re–arrange to a more stable state, emitting the extra energy as a gamma ray.

As mentioned earlier, the same isotopes of the same element can have different nucleon layers and lattice structures. It is suggested that is the reason the same isotope can have different decay modes with different half-lives.

Figure 5.30 shows the easiest way of getting an indication of the approximate lifetime of unstable nuclei. Nuclei of isotopes that sit on the green line are either stable or close to it. Even Z and A nuclei are more stable than others. That is because they have maximum magnetic binding. Odd Z nuclei with odd A have an even number of neutrons. They are more stable than odd Z nuclei with even A. They have odd numbers of protons and neutrons and have the lowest magnetic binding.

Between Z = 1 and Z = 20, the most abundant even Z isotopes are on the A = 2Z line. The most abundant odd Z nuclei are on the A = 2Z+1 line, ^{1}H excluded. Away from that line, the half-lives of the unstable nuclei drop rapidly. By the time they reach the blue line of maximum A for a given Z, half-lives are measured in minutes near the low Z end and seconds near the high Z end. Neutron deficient nuclei in the same range have half-lives measured in seconds.

At Z = 20, the most abundant isotope is on the minimum A for a given Z line. At Z = 82 and above, the most abundant isotopes are on the maximum A for a given Z line. The most abundant isotope is almost invariably the one with the maximum binding energy per nucleon. In between those two, the closer an isotope is to the green line, the longer its half-life. As the isotope gets closer to either the maximum or minimum A for given Z lines, its half-life shortens. Most isotopes on those lines have half-lives of the order of seconds, typically 1 to 2 seconds.

Figure 5.30 is a good guide to the stability of nuclei. It is suggested that it provides a reasonably accurate framework into which the great majority of the over 2,800 known nuclei will fit for the reasons given in the rules of binding under this model. Nuclei that are on the minimum A for a given Z lines in figure 5.30 have half-lives measured in low seconds down to high fractions of seconds. Those on the maximum A for a given Z line have similar half-lives up to Z ≈ 70.

Beyond that, they increase. At Z = 82, that line represents the most abundant element and is stable. The model needs a little adjustment at the high Z nuclei. Outside those lines, the further away are the nuclei, the shorter is their half-life. Inside those lines, the closer they are to the most abundant isotope line, the longer is their half-life. The closer they are to Z=50, the more stable isotopes they will have, Ca excluded. Odd Z nuclei have far fewer stable isotopes than even Z nuclei. Odd Z and even A nuclei are the least stable because they lack equal numbers of both nucleons, considerably reducing their magnetic binding capability.

These trends are not "hard and fast." They hold reasonably well for most nuclei and have been known for a long time. The above presentation on nuclear binding and the structure of nuclei gives a reasonably easy explanation as to why. It suggests why some numbers may appear to have additional stability. These have become known as "magic numbers."

Doubly even A and Z nuclei are always more stable than the others because they have maximum magnetic binding. This model still needs further evaluation of the structure of the layers in order to find out why some are more stable than others. In any event, the possibility of stable nuclei existing above Z=100 appears to be highly unlikely.

5.8. The Neutron Drip Line

Everything above has dealt with nuclei bound by both electric and magnetic fields. Attempts to determine the nature of nuclear binding and possible nuclear structures have included forming high A nuclei of some low Z elements. The recent discovery of ^{39}Na (Ahn et al., 2022) provided some evidence in support of nuclear shell models.

Figure 5.35 shows how this model supports high A Na nuclei. 5.35 A shows a proposed 4 layer structure for ^{23}Na, its most abundant isotope. 5.35 B shows the highest A isotope under this 4 layer model. It has a half-life of ≈ 1 minute. A and B both show that all of the protons are on the outside of the nucleus. Each one could act as a source for a single neutron to align its weak N/27 negative oscillation to proton. That would give an extra 11 neutrons for ^{36}Na.

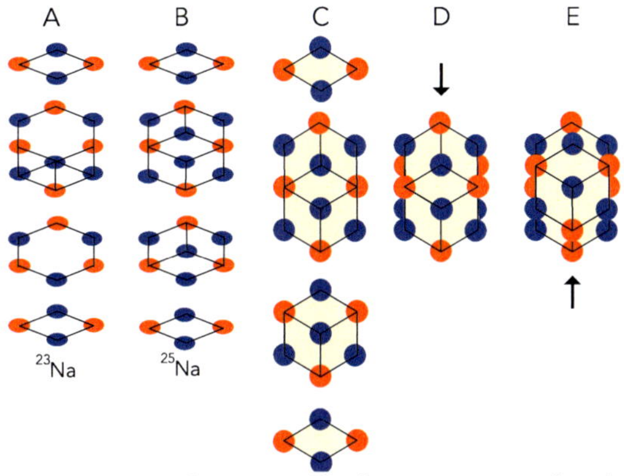

Figure 5.35: *Suggested origins of high A Na isotopes heading towards the neutron drip line.*

Figure 5.35 C shows B with the nucleons drawn as circles. Each layer was covered with the pale colour to hide the underneath layers when the four layers were stacked centrally upon each other, D shows the C stack looking down from the top. E shows the stack viewed from the bottom. A feature of D is that there are sets of protons stacked slightly to one side of each other. With the appropriate alignments, neutrons could be held magnetically between them in the different layers.

Each additional neutron would make the nucleus less stable. Their yields would be lower and their lifetimes shorter. This model gives no reason to suggest it could not sustain ^{39}Na over a short time. Nor does it suggest that ^{39}Na is necessarily its highest A isotope.

The neutron drip line was studied in the hope it could provide more information about the nature of nuclear binding. It is suggested much better results will be achieved by checking this model out in the same way that the structure of the ^{12}C nucleus was determined.

5.9. Discussion

It should be remembered that this rotating photon model of nuclei relies only upon the mass and magnetic moments of the proton and neutron and their known electromagnetic properties. No artificial particles or constraints have been introduced. They automatically give nucleons their dimensions and nuclei their structure and other properties. No assumptions have been made about the dimensions of the nucleons, which come from their structure. The ability of empty space to support them was presented in chapters 1 and 2.

Under this rotating photon model, electric charge originates at the circumference of the rotating photon. That places it in the two dimensions of the photon's rotation plane. The magnetic field is generated in the third dimension by the rotating photon's magnetic field and possibly its circulating charge. The proton is entirely positively charged, while the neutron consists of segments of $+½e$ and $-½e$, for a zero net charge.

An overlap between a negative segment of the neutron with a positive segment of a proton can occur at a charge separation distance of less than 0.1 fm. Allowing such electric attraction and repulsion is based upon the inverse square of distance law. An overlap of $+0.1e$ with $-0.1e$ at a distance of 0.1 fm can overcome the repulsion between two positive charges separated by a distance of 2.25 fm, which is the approximate distance of closest approach between protons in the $^4He_2$ nucleus.

Figure 5.7 shows that the overlaps between a proton's positive charge and a neutron's negative charge segments give the correct dimensions for the observed nuclear binding properties. Figure 5.9 shows that the neutrons' charge distribution prevents them from being forced too close together.

The strong nuclear force that holds the positively charged protons together in a nucleus is the direct overlap between segments of the proton's planar distributed positive charge with segments of the neutron's planar distributed negative charge. It is reinforced by the nucleons orienting their spins to give maximum magnetic binding. In this situation it is easy to see that this strong nuclear force has a strictly limited range of up to ≈5.5 fm for proton to neutron centre-to-centre distance. A slight overlap of the neutron's outer negative ring leads to the weak binding that gives rise to nucleons forming a short term "satellite" around a nucleus.

Beyond ≈5.5 fm there is no direct overlap between the circumferences of opposing electric charge, and hence no possibility of any binding. Between ≈5.5 fm and 3 fm, the overlap produces a relatively weak attraction. It can temporarily support satellite nucleons. They are held in place by the weak negative field associated with the neutron's N/27 oscillation and its associated magnetic field.

Between ≈ 3 fm and ≈ 1 fm, the overlap increases the attractive force between a proton and neutron. It is the interaction between the proton's positive charge and the neutron's inner negative ring for separation distances between 1 fm and 3 fm, together with the magnetic field, that yields the much stronger bonds that hold the protons in the nucleus. In alpha particles it yields 7.07 MeV per proton-to-neutron bond. With increasing numbers of protons in a layer, their mutual repulsion forces them further apart. The greater distance means a weaker proton to neutron bond.

At distances less than ≈1 fm, the overlap of the neutron's positive core with the proton's positive charges causes a strong repulsion between the nucleons. The results in tables 5.1 and 5.2 and figure 5.12

were obtained from experimentally measured binding energies of nuclei with A < 5. The trend has been schematically illustrated in figure 5.8.

All of that is supported by the magnetic moments of the nucleons. Within layers, individual nucleons align their spins to give maximum magnetic binding with their neighbours. The magnetic field attraction diminishes as the inverse of distance cubed ($1/r^3$), which drops off very rapidly. In that way, overlapping opposite electric charge segments combine with oriented magnetic fields to give a strong force with a limited range.

Because of their positive charge, protons repel each other and position themselves as far away from other protons as possible, ^{3}He excluded. Figure 5.10 C shows that the strong magnetic attraction between opposing spin protons overcomes some of the electric repulsion between them. The four nucleons in the ^{4}He nucleus consist of the two protons separated by a greater distance than the two neutrons, as shown in figure 5.11 B.

The neutron centres are positively charged and repel each other. That repulsion between neutrons means they will not overlap, instead producing the "diamond" like structure illustrated in figure 5.11. The angles for this structure are suggested to be 60° because that is the angle in hexagonal close packed structures. It may vary a little.

Maximum nuclear binding occurs when the magnetic fields of individual nucleons are oriented to attract the magnetic fields of surrounding nucleons. Within individual layers, magnetic attraction between adjacent like nucleons is maximised when they have opposite spins. Maximum magnetic attraction between adjacent protons and neutrons occurs when they have the same spin.

Maximum internal magnetic binding means minimum external magnetic field. Magnetic binding between layers is not totally dependent upon the superposition of opposing poles over nucleons. Magnetic lines of force extend beyond the nucleus' core and become available for nucleons in other layers to attract each other.

The spin of a nucleon determines the direction of its magnetic field. Nucleons orient themselves in the nucleus to give maximum nucleon binding and minimum external magnetic field. Because the spin and magnetic moment of nucleons are directly related, that determines the spin of the nucleus. The total spin of each nucleus is the sum of the spins of the individual nucleons. There is no rotational motion of any nucleus that significantly contributes to the total angular momentum or spin of that nucleus.

It has been shown that the sum of the magnetic fields associated with the spin of individual nucleons gives a close approximation to the measured magnetic moment of the nucleus. It is suggested that the difference between the predicted and observed magnetic moments, based upon the measured spin of the nucleus, is due to slight distortions of the rotating photons' trajectories.

In this model, nucleons within a nucleus pack together in layers that have a hexagonal close packed structure. The protons are separated as far as possible from each other and are held together by at least two neutrons, the exception being the ^{3}He nucleus. Most protons in a "quasi-stable" nuclear isotope must be held in place by at least three neutrons. The exceptions include some of the low Z nuclei. All neutrons in a "quasi-stable" isotope must be held in place by at least two protons, ^{2}H and ^{3}H excluded.

In this situation, a "quasi-stable" isotope is defined as one that holds together for longer than a couple of seconds. It should be noted that it is possible for a proton or neutron to be held to a compound nucleus, $A > 5$, by a single opposing nucleon, up to a separation distance of ≈5 fm. It will not hold together for very long because of the existence of other charges in the same plane and the weak binding.

Beyond ≈5.5 fm, planar binding is not possible under this model. It is believed that this is observed. $A=5$ nuclei are not stable because the fifth nucleon has only one opposing nucleon to which it can bind. The ^{8}Be nucleus is unstable because it will rearrange itself into two alpha particles.

There are two types of bonding within nuclei. With electric charge distributed in two dimensions and generated at the circumferences of rotating photon segments, direct overlap of opposite charge segments can give a strong bond between opposing electric fields. They essentially have no separation distance between them. Their overlap has been called "division by zero" attraction. The magnetic field is distributed in three dimensions and hence direct overlap will generate a lower strength binding. This creates a situation in which layers of nucleons bound together by a strong force are also held together by a weaker, reinforcing magnetic force. This replicates the conditions of the Ising model (MacGregor, 1976), which successfully explained aspects of nuclear fission.

Figure 5.23 A shows two problems that have concerned nuclear physicists for decades. Why is the charge density of a ^{4}He nucleus much higher than the average density of typically ≈1.2 x 10^{19} C/cm^3, while its neighbour, ^{6}Li is significantly lower? The answer emerges from this model. Part of the answer is found in the distributions of the protons in ^{6}Li versus ^{4}He. The protons in ^{6}Li nuclei are at the corners of an equilateral triangle, much further apart from those at the ends of a diamond shaped ^{4}He nucleus. The other part of the answer comes from treating nuclei as spherical. If they are treated as the planar arrays that they are, the differences would not be as great.

The other problem that has concerned nuclear physicists is the skin effect. Experimentalists have detected a large charge distribution protruding from the edge of the nucleus and tailing off with distance. It was measured by firing energetic electrons at beams of protons moving towards them. The moving protons can only move with their magnetic field leading, which means their charge plane lies perpendicular to that direction. We will return to this concept shortly, showing how this model can account for the observation.

The average charge densities in higher Z nuclei range from ≈ 1.25 x 10^{19} C/cm^3 for Z = 20, to ≈ 1.1 x 10^{19} C/cm^3 for Z = 83 (Pauling and Robinson, 1975). This model appears to explain these trends because of the increasing distance between protons due to the increasing charge per layer. Any calculation of binding energy should take this increasing distance and hence lower binding energy into consideration.

That task is complicated by the magnetic field that tends to pull the nucleons together. The result is that it tends to make the dimensions of the higher Z nucleons slightly larger than would be determined if all proton to neutron bonds were treated as being the same length. This model also suggests that, with the added neutrons for different isotopes fitting into lattice vacancy positions, there will only be a minor increase in the dimensions of nuclei between low A and high A isotopes of the same Z.

The binding energy per nucleon reaches a maximum of 8.79.. MeV in ^{56}Fe. That gives a total binding energy of 492.2 MeV. Under this model there are 43 magnetic bonds of 0.8 MeV each, giving a binding energy of 34.4 MeV. There are 66 nearest-neighbour proton-to-neutron bonds and 24 second-nearest-neighbour bonds of approximately half the nearest-neighbour strength. Figure 5.11 can be used to match the binding energy, and hence, calculate the dimensions of the ^{56}Fe bonds and therefore also the nucleus. Numerical iteration gives the nearest-neighbour bond strength a value of 6 MeV and a length of $\approx$1.25 fm and the second-nearest-neighbour bond a length of $\approx$ 2.5 fm and a strength of 2.3 MeV. Adding these gives a ^{56}Fe nucleus a binding energy of 490.4 MeV. These dimensions are determined using only this model and the masses of the proton and neutron. At 7.07 MeV, the ^{4}He nearest-neighbour proton-to-neutron bond is by far the strongest bond, even if it does not have the strongest binding energy per nucleon. This is reflected in the ^{4}He nucleus having the highest charge density of all nuclei.

This model strongly suggests that nuclei are made with nucleons arranged in the lattice patterns shown in figures 5.17 and 5.18. Even with neutron vacancies within the lattice, the space will still be filled by the electric and magnetic fields from neighbouring nucleons. In occupied lattice positions, whether it is a proton or a neutron, there will be a region of intense positive charge, containing $\approx 0.35e$ distributed within a radius of $\approx$ 0.2 fm. Outside that region there will be a weaker positive charge density for each proton and a negative charge density around the neutron. With nuclei composed according to this model, there will be a tapering off of charge density associated with all nuclei. It will be like that shown in figure 5.23, with that tapering being spread over $\approx$ 2 to 2.5 fm. This has been observed and is called the "skin" effect. Its origins can be seen in figure 3.14.

It is worth noting that the highest naturally occurring isotope, $^{238}U_{92}$, has a saturation of neutrons within the lattice layers that form the nucleus. Even so, it is still not stable and will decay. $^{208}Pb_{82}$ is the highest Z and A stable nucleus, with $^{209}Bi_{83}$ being the only higher Z and A (almost) stable nucleus. The appropriate choice of layers, as outlined above, will result in A increasing with Z in the manner that fits the observed increase. The most abundant isotopes of nuclei with $Z \leq 20$ have all internal neutron lattice positions vacant for even Z and one filled position for odd Z. At Z just above 20, the proportion of unfilled internal neutron lattice vacancies for the most abundant isotope is close to those given in figure 5.18. As Z increases, that proportion increases for the most abundant nucleus. At Z = 82 and beyond, the most abundant isotopes have all internal neutron lattice positions filled.

Figure 5.30 shows that the most abundant isotope of any element is always found within the range of structures allowed under this model. At Z < 20, they are generally the lowest possible A. Fewer neutrons are required to hold the protons together when there are not many protons. At Z = 82+, all possible neutron lattice positions are filled for the most abundant isotopes. High Z nuclei require the maximum number of neutrons to hold the protons together in that nucleus. It follows that the majority of stable isotopes of low Z nuclei will have a higher A than the most abundant isotope. The stable isotopes of high Z nuclei will have a lower A than that of the most abundant isotope. Intermediate, Z ≈ 50, nuclei have stable isotopes either side of the most abundant isotopes and they also tend to have more isotopes. This has been observed (Bohr and Mottelson, 1969).

This gives a strong indication that there will not be any higher Z stable nuclei beyond those occurring naturally. Speculation about the existence of stable heavy nuclei near Z = 114 have been predicted (Wahl, 2002). Despite decades of searching no stable nuclei have been detected (Seaborg and Loveland, 1987). This would agree with the prediction herein that the increasing number of neutrons required to bind protons in the larger Z layers exceeds the number of neutron lattice positions available for the nuclear layers to remain stable.

This presentation has shown how protons and neutrons bind together to form nuclei. The following shows how this model has matched some of the known properties of nuclei.

1. Nuclei are automatically subjected to the special relativity corrections of mass, length, and time when they move.
2. The spin of a nucleus is the sum of the spins of the individual nucleons, the spin of each being the angular momentum of the rotating photon that comprises the nucleon.
3. The magnetic moment of a nucleus is the sum of the magnetic moments from the spin orientation of the individual nucleons, with a slight correction due to the distortion of the rotating photons as they overlap those of neighbouring nucleons.

4. The nucleons within a nucleus orient their spins to give maximum magnetic attraction with neighbouring nucleons and minimum external magnetic moment.
5. This strong nuclear force has a range of less than ≈5.5 fm, being very weak at that distance and increasing at shorter separation distances, particularly between 1.9 fm and 1.2 fm.
6. Layers of strongly bound nucleons are held together by a weaker force.
7. Proton-to-proton and neutron-to-neutron binding are only possible when the nucleons are in separate layers, and it is a magnetic force.
8. Short term weakly bound isotopes can be formed by either a proton or neutron being bound at distances of less than ≈3 fm from the surface of the nucleus. It occurs when the nucleon is bound to just one opposite nucleon.
9. Nuclei bombarded by energetic particles have a high probability of alpha particles being knocked out of them because those structures are present throughout almost all compound nuclei.
10. The greater the number of protons in a nucleus, the more neutrons are required to bind them together.
11. Low Z nuclei have more isotopes above their most abundant isotope, while high Z nuclei have more isotopes below their most abundant state. This is because low Z nuclei have many vacant neutron sites within the lattice array for their most abundant isotope, while high Z nuclei have almost all of their neutron lattice positions filled for their most abundant isotope.
12. When a ^{235}U nucleus fissions, it generates an array of different nuclear fragments which will have maxima near A = 90 and A = 142, with a lesser probability of getting equal mass fraction at A ≈ 116. The maximum and minimum fractions will be near A = 170 and A = 62 respectively. These values come about from the splitting of one of the two central layers of the ^{235}U nucleus shown in figure 5.34.
13. The charge density of the ^{4}He nucleus is much higher than that of all other nuclei because it has a minimum proton repulsion

and its radius includes $\approx +0.8e$ from the neutron's centre, as shown in figure 5.10.

14. The dimensions of nuclei are automatically determined by the structure of the individual nucleons, the lattice structure, and the binding energy of the nucleus.
15. The charge density of high Z nuclei is less than that of low Z nuclei because of their higher mutual proton repulsion.
16. The charge density of all nuclei diminishes from its maximum to zero over a distance of about 2 - 2.5 fm. That is the extent of the charge distribution of the proton with an allowance for the non-circular distribution of nucleons within the nucleus.
17. Measurements of the dimensions of nuclei will yield a slightly larger diameter when they are measured using neutron scattering than when using charged particle scattering.
18. Protons are held in place by neutrons and will not migrate to the outer regions despite their high positive charge exerting a significant repulsive force on neighbouring protons.
19. Protons and neutrons are uniformly distributed throughout the nucleus.
20. A = 5 nucleons cannot exist because the ^{4}He nucleus has only one nucleon of any type available for another nucleon to attach to it.
21. ^{8}Be nuclei cannot exist because they will rapidly become two ^{4}He nuclei.

In addition to the above, this model makes the following predictions about nuclei that are unknown (to the author) at the time of writing.

22. Nucleons align their spins so their rotation plane is perpendicular to their travel direction.
23. Nucleons combine within a nucleus as individual particles, largely retaining their individual particle properties.
24. Nucleons unite into a nucleus by forming into layers in a hexagonal close packed lattice structure, $A < 8$ excluded.
25. The strong nuclear force is due to the two-dimensional nature of the charge distribution of the nucleons and their magnetic

field. Parts of a proton's positive charge directly overlap parts of the neutron's negative charge segments at very small distances, giving rise to what can be called "division by zero" attraction. The nucleons orient their spins to give maximum magnetic binding.

26. Individual nucleons within a layer are arranged in a lattice structure within the nucleus, with specific sites for each of the protons and neutrons. A proton cannot switch its lattice position with a neutron.
27. With high Z nuclei, inner protons are bound by up to 6 neutrons, while outer protons in stable nuclei are bound to at least 3 neutrons.
28. Stable nuclei with Z > 3 are composed of multiple layers of nucleons, with nucleons within an individual layer held together by the strong "division by zero" electric attraction and magnetic field alignment. The layers are held to one another by the weaker magnetic attraction.
29. The ^{6}Li and ^{7}Li nuclei have a dimension somewhat larger than predicted from their Z and/or A because they form in a single layer.
30. In "quasi-stable" nuclei, all protons must be bound to at least three neutrons (H, He, ^{6}Li, and some other low Z isotopes excluded).
31. In "quasi-stable" nuclei, all neutrons must be bound to at least two protons (^{2}H excluded).
32. Nuclei in which one or more nucleons are bound to just one opposing nucleon will form isotopes with a very short half-life.
33. For low Z nuclei, the minimum number of neutrons for a given Z is determined by the minimum number for each proton to be held by two neutrons (^{2}H and ^{3}He excluded).
34. Nucleons are bound in layers within the nucleus by the electromagnetic force. Because charge is only distributed in the two dimensions of the rotating photons that constitute the nucleons, overlapping charge segments can occur at effective distances of less than 0.1 fm. This allows small overlapping

opposite charge segments to overcome the repulsion of much larger proton charges at distances of ≈ 2.5 fm.

35. Individual nucleons are bound in the lattice structure and are not free to migrate around within the nucleus.
36. The spin of a nucleus is the sum of the angular momenta or spins of the individual nucleons. There is no contribution to the spin of a nucleus from nucleons rotating about its common centre of mass.
37. Within a given layer, like nucleons cannot bind to each other.
38. The different numbers of neutrons that form different isotopes are due to neutrons filling the empty positions within the neutron lattice structure.
39. Protons and low Z nuclei can be added to existing nuclei without the need for tunnelling energy if their axis and magnetic moments are aligned and they are sufficiently close. When this occurs with a nucleus that has a magnetic moment, it attracts the incoming proton and causes the two to fuse more rapidly.
40. No stable high Z (> 92) nuclei will be synthesized.
41. Although larger Z nuclei are approximately circular, many will have bumps and dimples over their surface.

5.10. Summary

Nuclear physics is a topic that has been extensively studied for almost 100 years. This model was forwarded as one means of matching known properties of nuclei and predicting new properties. It does answer many questions about the nature of nuclear binding. It suggests that the strong nuclear force is due to the planar nature of the charge distribution of individual nucleons together with their spins orienting to give maximum magnetic binding.

The electric binding is caused by parts of a neutron's negative charge overlapping parts of a proton's positive charge at what is effectively zero distance. Its range is strictly limited. Unless there is direct charge overlap, there is no strong nuclear binding. That binding only occurs between a proton and a neutron side by side in a layer. The closer the two centres are to each other, the greater the opposing charge overlap and the stronger the binding, down to ≈ 1.2 fm.

When the nucleons get too close, their strong positive cores begin to repel each other. These properties match the known properties of the strong nuclear binding force. It can only occur because the charge at the circumference of the individual nucleons is two-dimensional. This binding mechanism enables the nucleons to be strongly bound together into layers. The magnetic field and some of the electric field occupies the third dimension. Layers of strongly bound nucleons are held together by a weaker magnetic field. This forms the basis of the Ising model of nuclei that was successful in explaining nuclear fission.

The rules for the formation of nucleon layers and nuclei, given in section 5.5, were designed to give nuclei that were approximately spherical in shape. It is apparent from some of the low Z structures that the nuclei will not be spherical. There is also no reason that such spherical shape should be imparted to the higher Z nuclei. Indeed, it is suggested that many higher Z nuclei could be more like a "rounded rectangular box with bumps and dimples" than a sphere. There is ample opportunity for that to be further explored.

This presentation is not sufficient to describe all the properties of all nuclei. It does provide a means by which a structure and some properties of the nucleus of any isotope of any element can be determined. It shows that the nucleons remain intact inside the nucleus. There is no orbital motion of any nucleon within the nucleus or any rotation of the nucleus that could contribute any spin or magnetic moment to the nucleus. For a spinning nucleus to contribute a significant fraction of the ½$\hbar$ spin of a single nucleus, the whole nucleus would have to rotate on its axis at some 10^{20} Hz.

The spin and magnetic moment of a nucleus is the sums of the spins and magnetic moments of the individual nucleons within it. In nuclei, the spins of the nucleons are oriented to give maximum magnetic field binding strength That results in minimum external magnetic field. Even Z and A nuclei have no external magnetic field.

The planes of the rotating photons for all nucleons in a nucleus are parallel to each other. This means there is no or minimal charge in the direction of the axis of the nucleons' rotation, and the electric field in that direction is not as strong. That is also the direction in which nuclei travel, magnetic field leading with the electric field spread laterally.

This model suggests there is no special significance to "magic number" nuclei. The so-called "magic numbers" of 82 and 126 are nothing more than the number of protons and neutrons that form the lattice structure required to make ^{208}Pb, the heaviest stable nucleus, ^{239}Bi excluded. This lattice structure shows there is no reason why there should be an "island" of heavier stable nuclei

Hofstadter showed that the charge density of the central cores of nuclei varied from ≈ 1.3 x 10^{19} C/cm^3 at low Z (Mg_{12}) to ≈ 1.1 x 10^{19} C/cm^3 at high Z (Bi_{83}), see figure 5.23. It is suggested that this is due to the repulsive force between protons pushing them further apart. Figure 5.12 shows that small changes in proton to neutron separation distances make significant differences to their binding energy.

Even if the "magic number" 126 had special significance for protons, the nucleons would be pushed too far apart to overcome proton repulsion at such high Z. Together, this explains why higher Z stable nuclei have not been detected.

Nuclei with even numbers of protons and neutrons have zero external magnetic moment because the protons and neutrons adopt opposing spins in order to strengthen their magnetic binding. They form more stable nuclei than those with odd Z or odd (A-Z). The reason why lead ($^{208}Pb_{82}$) is the last stable nucleus can be seen in figure 5.29. It is a lowest even Z nucleus with all its neutron lattice positions filled in its most abundant isotope.

The lattice structures given in figures 5.17 and 5.18 give a very good indication that these structures do indeed represent the real world of nuclei. The plot in figure 5.30 shows that the most abundant low Z nuclei have all internal lattice positions vacant, with other stable isotopes having some of the vacant neutron lattice positions filled. The most abundant high Z nuclei have their entire neutron lattice positions filled, with all other stable isotopes having some vacant neutron lattice positions.

Figure 5.30 illustrates the smooth transition between the two limits. That gives significant credence to this proposed means of nuclear binding. It could be argued that other structures such as trilobal photons could give the same result. It appears difficult to overcome the observation that this lattice structure explains a large number of known properties of nuclei.

It could be argued that quark exchange could generate similar lattice positions. That requires some mechanism by which quarks are oriented in the correct manner for all nucleons within the nucleus. It also requires the strength between quarks to decrease as distances increase to account for the low binding energy of a single nucleon to the outside of a nucleus, such as the single neutron in ^{31}Al, as shown in figure 5.20 H.

This one model is capable of explaining more observed nuclear properties than the over 30 other models that have been forwarded over the past few decades of study into the nature of the nucleus and nuclear binding. There is significant scope for expanding this study. It is not suggested that the model presented herein is complete. The topic is a complex one and there is some need and considerable scope to refine it. However, it is strongly suggested that this model offers the best platform upon which to expand our understanding of nuclear physics. It is able to match observations across a wide range of A and Z, and to offer simple explanations of what is occurring without the need to introduce constituents that have not been detected.

We now know how protons and neutrons bind together to form nuclei. But we still cannot start to build our universe until we have the nuclei forming in the ratios in which they are observed. The next chapter describes how the ratio of nuclei or elements seen in the sun and other stars occurs. Only when we have the correct ratio of light to heavy nuclei around us, we can add the electrons to form the ratio of atoms that make up our world. They can then be let loose to interact according to their nature and their structure.

5.11. List of References

Ahn DS, et al., (2022); "Discovery of ^{39}Na, Physical Review Letters, **129**, 215502

Angeli A, (2004); "A consistent set of nuclear rms charge radii: properties of the radius surface R(N, Z)", Atomic Data and Nuclear Data Tables, **87**, pp 185 – 206

http://indico.vecc.gov.in/indico/getFile.py/access?resId=0&materialId=14&confId=19

The Berkeley Laboratory Isotopes Project,

http://ie.lbl.gov/education/parent/Al_iso.htm : See also:-

https://periodictable.com/Properties/A/KnownIsotopes.html

Bleuler K, (1984) ; "lecture Notes in Physics", **197**, p 1

Braibant SA, Giacomelli G and Spurio M, (2009); "Particles and Fundamental Interactions, An Introduction to Particle Physics", Springer, Berlin, p 313

Bohr N, (1936); "Neutron Capture and Nuclear Constitution, Nature **137**, pp 344 – 348

Bohr N and Wheeler J A, (1939); "The Mechanisms of Nuclear Fission", Phys. Rev., **56**, pp 426 – 450

Bohr A and Mottelson B, (1969); "Nuclear Structure", Benjamin, Reading MA, p 253

Brown GE and Jackson AD, (1976); "The Nucleon-Nucleon Interaction", North Holland Publishing, Amsterdam, NL

Cook ND, (2010); "Models of the Atomic Nucleus, Unification Through a Lattice of Nucleons", Springer, Berlin

Chadwick J (1932); "Possible Existence of a Neutron", Nature, **129**, 312

Elliot JP, (1990); "The Shell Model Today", Nuclear Physics **A507**, pp 15c-24c

Firestone RB et al., (1999); "The 8th Edition of the Table of Isotopes", John Wiley and Sons, NY

Geiger H and Marsden E, (1913); "The Laws of Deflexion of a Particles through Large Angles"; Philosophical Magazine, **25**, pp 604 – 623

Glashow S, (1961); "Partial-symmetries of weak interactions", Nuclear Physics **22**, pp 579-588

Goldman T, Maltman KR and Stephenson GJ, (1988); "The Importance of Nuclear Substructure in Nuclear Ground States", Nuclear Physics **A481**, pp 621 – 667

Greiner W and Maruhn JA, (1996); "Nuclear Physics", Springer, Berlin: http://webéducation.com/wp-content/uploads/2019/02/Walter-Greiner-Joachim-A.-Maruhn-D.A.-Bromley-Nuclear-Models-Springer-Verlag-1996.pdf

Hasse RW and Myers WD, (1988); "Geometrical Relationships of Macroscopic Nuclear Physics", Springer, Berlin

Hodgson PE, (1982); "Alpha-clustering in nuclei", in The Uncertainty Principle and Foundations of Quantum Mechanics, Price WC, Ed, Wiley, New York, NY, p 485

Hofstadter R, (1956); Electron Scattering and Nuclear Structure, Reviews of Modern Physics, **28**, pp 214 – 252

Holden NE et al., (2019); "IUPAC Periodic Table of the Elements and Isotopes (IPTEI)", https://iupac.org/wp-content/uploads/2015/02/IPTEI_postprint_20190301.pdf

Howard RA, (1966); "Nuclear Physics", Wadsworth, Belmont CA, Table of Nuclidic Data

Mayer MG and Jensen JHD, (1955); "Elementary Theory of Nuclear Shell Structure", Wiley, New York, NY

Miller GA, (1984); Lecture Notes in Physics, **197**, p 234

MacGregor MH, (1976): Il Nuovo Cimento, **A36**, p 113

Myers WD, (1977); "Droplet Model of Atomic Nuclei", Plenum, New York, NY

Moretto LG and Wozniak GJ, (1993): "Multifragmentation in Heavy-Ion Processes", Annual Review of Nuclear and Particle Science, **43**, pp 379 – 455

Moszkowski SA, (1955); "Models of nuclear structure", Encyclopedia of Nuclear Physics, Vol. **39**, ed Flügge S, Springer, Berlin, p 411

Myers WD and Swiatecki WJ, (1966); "Nuclear Masses and Deformations", Nuclear Physics **81**, pp 1 – 60

Pauling L, (1965); "The Close-Packed-Spheron Theory and Nuclear Fission", Science, **150**, pp 297 – 305

Pauling L and Robinson AB, (1975); "Prediction of the shapes of deformed nuclei by the polyspheron theory", Canadian Journal of Physics, **53**, pp 1953 – 1964

Petry HR, (1984); "Lecture Notes in Physics" **197**, p 236
https://link.springer.com/bookseries/5304

Robson D, (1978); "Many-body interactions from quark exchanges and the tetrahedral crystal structure of nuclei", Nuclear Physics, **A308**, pp 381 – 428

Rutherford E, (1911); "The Scattering of α and β Particles by Matter and the Structure of the Atom", Philosophical Magazine, **21**, pp 669 – 688

Santiago AJ and Chung KC, (1993); "Do Percolative Simulations of Nuclear Fragmentation Depend upon the Lattice Structure?" Journal of Physics G, **19**, pp 349 – 355

Sick I, (2015); "Form factors and radii of light nuclei", https://arxiv.org/pdf/1505.06924.pdf

Seaborg GT and Loveland W, (1987); "The Trans Uranium Elements", Contemporary Physics, **28**, p 33

Segre E, (1965); "Nuclei and Particles", Benjamin, Reading MA

Townes CH, Foley HM and Low W, (1949); "Nuclear Quadrupole Moments and Nuclear Shell Structure", Physical Review, **76**, pp 1415 – 1416

Wahl AC (2002); "Systematics of Fission-Product Yields", Los Alamos National Laboratory, Report LA-13928, http://www.osti.gov/scitech/servlets/purl/809946

Weinberg S, (1967); "A model of Leptons", Phys. Rev. Lett. **19**, pp 1264 – 1266

Weizsacker WD, (1935); "Zur Theorie der Kernmassen", Zeits. fur Physik **96**, pp 431 – 458

Wigner E, (1937); "On the Consequences of the Symmetry of the Nuclear Hamiltonian on the Spectroscopy of Nuclei", Physical Review, **51**, pp 106 – 130

Yukawa H, (1935); "On the Interaction of Elementary Particles", Proc. Phys.-Math. Soc. Japan, 17, pp 48

6. The Abundance of Nuclei

Abstract:

The mechanism by which protons fuse to produce the higher atomic number nuclei is presented. A semi-quantitative analysis shows that this method closely matches the observed abundances of the elements in the sun and other stars.

6.1. Introduction

So far in our quest to understand the universe, we have shown all the building blocks that form the materials we see in our world and the rest of the universe around us. We have our photons, protons, neutrons, electrons, and neutrinos and have shown how those short lived "elementary particles" are produced. We have seen how all the protons and neutrons are put together to make up all the elements we have around us. We must now put the nuclei together in their observed abundances.

Free protons make up over 70% of all the mass of the baryonic universe. They are the fuel from which all other materials are produced. Our next task is to show how they fuse to form all the element nuclei. Wherever there are protons, there are usually equal numbers of electrons.

The measured abundances of the various elements in the sun are given in figure 6.1 (Lodders, 2010). This log linear plot stretches over some 12 orders of magnitude, from an abundance of 3×10^{10} nuclei for H_1 to 10^{-2} nuclei for U_{92}, based upon the abundance of $Si_{16} = 10^6$. It encompasses all elements in between, although the elements with Z of

43, 61, and 91 do not occur naturally. Even though they are unstable, elements with Z of 84 to 89 occur naturally as by-products of the radioactive decay of U_{92} and Th_{90}.

The main features of figure 6.1 are the high abundances of hydrogen, 73% of the sun's mass, and helium, 25%. They are followed by very low abundances of lithium, beryllium, and boron. Carbon and oxygen make up about 0.4% and 0.97% of the sun's mass respectively. The remainder make up less than 1%.

From oxygen, which has an abundance of about 1.5 x 10^7, there is a general downward curve at higher atomic numbers. Odd Z nuclei are less abundant than their even Z neighbours, beryllium excluded. There is a downward trend in abundances with increasing Z up to Z≈49, after which the average almost flattens out.

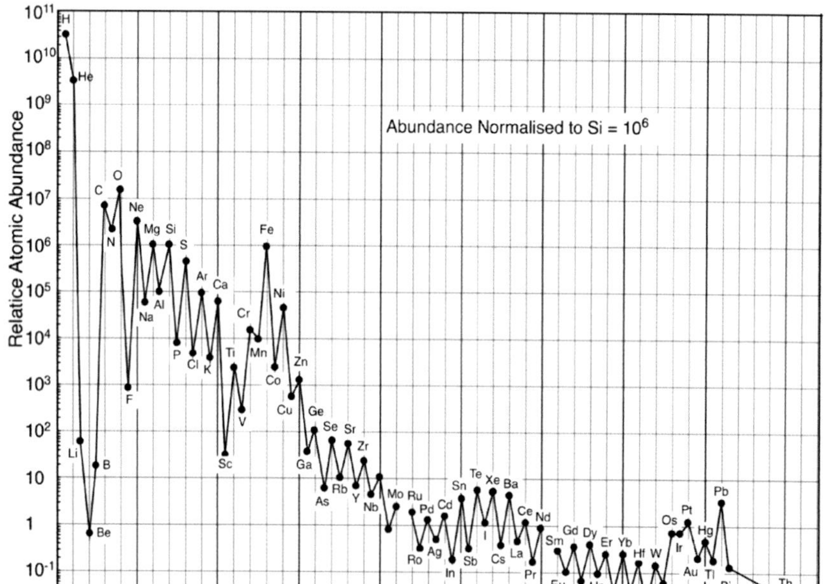

***Figure 6.1:** The measured abundances of elements in the sun (Lodders 2010).*

There are some occasional spikes of higher or lower abundances in that overall trend. Iron and the elements around it have exceptionally high abundances.

Although there is general agreement on the overall shape of that abundance curve, there is some disagreement about the abundances of some of the low Z elements. All measurements suggest that the abundances of Li_3, Be_4, and B_5 are much lower than those of He and C on either side of them. Some suggest that the Li_3 abundance is lower than the Be abundance (Asplund et al., 2009). All suggest the B abundance is higher than the Be abundance. There is general agreement that the higher Z abundances are approximately as indicated in figure 6.1. Astronomical studies suggest that these abundances are typical of other stars as well (Asplund et al., 2009).

The sun is powered by the conversion of hydrogen to helium (Hoyle, 1946). There are several possible mechanisms by which protons can fuse to form helium nuclei. These include:

$^{1}H + {}^{1}H \rightarrow {}^{2}H + e^{+} + \nu$ {6.1

$^{1}H + {}^{2}H \rightarrow {}^{3}He + \nu$ {6.2

Having formed ^{2}H and ^{3}He, there are several pathways to the formation of $^{4}He_2$ and the subsequent liberation of large amounts of energy. These include:

$^{2}H + {}^{2}H \rightarrow {}^{4}He$ {6.3

$^{3}H + {}^{2}H \rightarrow {}^{4}He + n$ {6.4

$^{3}He + {}^{2}H \rightarrow {}^{4}He + p$ {6.5

$^{3}He + {}^{4}He \rightarrow {}^{7}Be$ {6.6

$^{7}Be + {}^{1}H \rightarrow {}^{8}B \rightarrow 2 \times {}^{4}He + e^{+}$ {6.7

$^{7}Be \rightarrow {}^{7}Li + e^{+}$ {6.8

$^{7}Li + {}^{1}H \rightarrow {}^{8}Be \rightarrow 2 \times {}^{4}He$ {6.9

Note that it is only the ^{7}Be pathway in reactions {6.7 and {6.9 that allow a single proton to participate in the final step. Other pathways including the CNO cycle in which those elements fuse with protons and travel through a series of C, N, and O isotopes to eventually eject an alpha particle have also been suggested. In any case, all pathways must include reactions {6.1 and {6.2.

The conditions required for fusion to occur in the core of the sun are a stellar sized mass that will generate very high temperatures and pressures in its central core. Those extreme conditions are required to get the nucleons sufficiently close and with sufficient energy that they will be able to move together and fuse to form the products indicated in reactions {6.1 to {6.9. The result of all those pathways and conditions is that they end up liberating a lot of energy, and that they have a lot of helium as well as hydrogen. They are therefore the start of this study.

The largest bottleneck for the fusion of hydrogen to helium is the fusion of two protons to form deuterium, ^{2}H, by either positron emission or electron capture. As may have been gathered from the previous chapter, neutrons are absolutely essential for binding protons together in nuclei. The fusion of the second proton to form ^{3}He is somewhat easier. The different pathways, including the CNO cycle, all contribute to the formation of ^{4}He and the liberation of significant amounts of energy. In total when 4 protons are fused to form ^{4}He, they collectively release 28.28 MeV. Understanding that process can allow for significant advances in the generation of energy.

6.2. Background Under the Rotating Photon Model

In the rotating photon model for the structure of nucleons, the electric charge was generated in the plane of the rotating photon. The magnetic field was perpendicular to that plane. As shown in chapters 3 and 5, this model was able to successfully match many of the observed properties of both nucleons and nuclei. In them, the binding between nucleons is due to the planar nature of electric charge distribution and the existence of the magnetic field in the third dimension. Parts of the positive charge of protons overlap portions of the negative charge of neutrons at "zero" distance separation. This effectively gives "division by zero" attraction between protons and neutrons.

Neighbouring nuclei align their spins to give maximum magnetic binding between them. That forms a strong bond that can overcome the repulsion between the positively charged protons. It is the magnetic field in the third dimension that gives nucleons their thickness. Nuclear binding is a dual action event. Protons and neutrons combine into layers using a strong binding force through this "division by zero" electric attraction of opposing nucleon charge segments. That is combined with the magnetic field attraction between neighbouring nuclei. These layers are held together by a weaker residual magnetic field.

The source of energy from the sun is fusion of hydrogen to helium. Figure 6.2 illustrates how some of this would occur in a stellar interior. A deuterium nucleus is shown surrounded by protons.

Under the high pressure and temperature at the centre, the protons would approach the deuteron in a random manner, as indicated by the seven protons surrounding it. Although they can all move towards the deuteron, not all will be able to fuse with it.

Protons A, B, E, F and G will be repelled by the charge. Proton C above will be repelled because the two nucleons have like magnetic poles pointing at each other. Proton D has its north pole pointing towards the neutron's south pole. It will not be repelled and be able to fuse with the deuteron. In this diagram, the tick means it can fuse to form ^3He. The cross means it cannot fuse.

That effect occurs for all other nuclei as well. Figure 6.3 shows multiple protons approaching a compound nucleus from different directions. It is only protons B and F that have their charge planes aligned parallel to that of the nucleus and their magnetic poles are aligned with the nucleus's magnetic pole. They also need to be aligned to the neutron they are approaching.

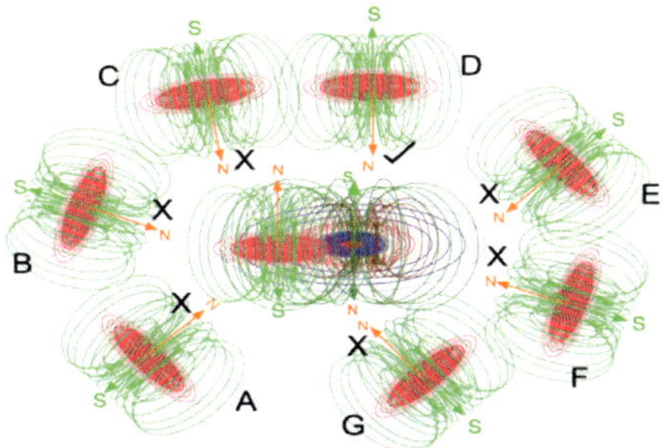

Figure 6.2: *Schematic illustration of protons approaching a deuterium nucleus from different directions.*

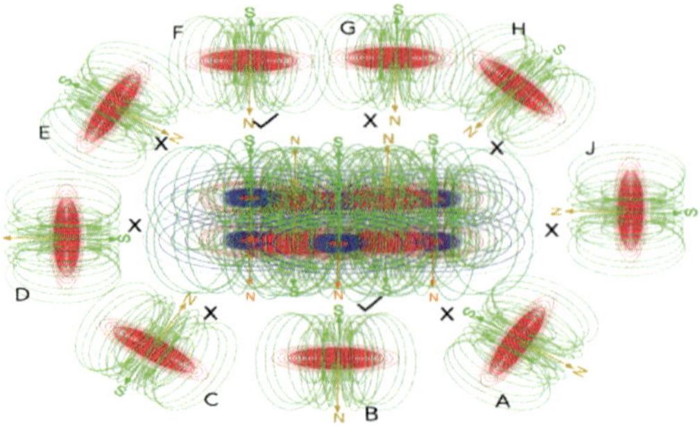

Figure 6.3: *Schematic illustration of protons approaching a larger compound nucleus from different directions.*

Proton G has its charge plane aligned with that of the nucleus. Its magnetic moment is opposite to that of the nucleus. It would require extra energy to overcome that magnetic repulsion. Protons A, C, D, E, H and J will not fuse because their charge planes are not aligned with that of the nucleus. They will be electrically repelled.

If the nucleus had even Z and A, it would not have any residual magnetic moment. Proton G would fuse with almost the same ease as protons F and B. Fusion to even Z and A nuclei without a magnetic moment is more difficult because of limited magnetic attraction.

Figures 6.2 and 6.3 suggest that, under this rotating photon model, fusion between nucleons does not require the electric repulsion between nuclei to be overcome to form higher Z nuclei. It is not a matter of having protons "tunnel" through a potential barrier surrounding each nucleus. Fusion requires close proximity and alignment.

The high pressure and temperature conditions in the centre of a star are required to get the nuclei sufficiently close for a reaction to occur. After that it is a matter of the alignment of the two nuclei so they will be magnetically attracted to each other and fuse. If they are aligned and have no magnetic field, such as ^{4}He, ^{12}C, ^{16}O, and similar, they will have to move closer with their planes of rotation parallel. Then they will still have to move together to fuse. If both the target nucleus and the incoming nucleus have a magnetic moment and they are aligned, they need only be sufficiently close that their magnetic fields can attract each other.

Provided the nuclei are arranged such that their rotation axes and hence electric charges are parallel to each other, and they move together, there will be no, or only minimal, electric field repulsion to be overcome for them to fuse. This suggests that fusion only occurs when protons have the correct alignment to a target nucleus and their trajectory overlaps its dimensions.

An example of this is illustrated in figure 6.4. Figure 6.4 A shows a proton approaching a nucleus when both have their charge planes aligned. Figure 6.4 B, suggests that, in the absence of magnetic fields, the collision cross section for fusion to occur is proportional to the square of the radius of the nucleus given, to a first approximation, by $A^{\frac{1}{3}}$, plus a portion of the diameter, d, of the incoming nucleus, in this case a proton, for a total area of $(A^{\frac{1}{3}}+\mathrm{d})^2$. The π terms cancel out and are not used.

Fusion becomes a matter of aligning the nuclei in close proximity, rather than overcoming the electric repulsion between them. If their spins and hence charge planes are parallel and sufficiently close, they can move together as a result of magnetic attraction.

This model suggests that the heavier elements are formed by the fusion of hydrogen and helium into low Z nuclei to form higher Z nuclei. The abundance of nuclei with atomic number Z+1 would depend upon the abundance of the lower Z nuclei, their collision cross section, the rate at which they were formed from lower Z nuclei, and the rate at which they fused with other nuclei to form even higher Z nuclei. The rate at which they were formed from lower Z nuclei also depends upon the number of nucleons that must be added to form the higher Z nuclei.

For example, between Z=9 and Z=20, odd Z+1 nuclei require the addition of one proton to form the most abundant even Z+2 nuclei.

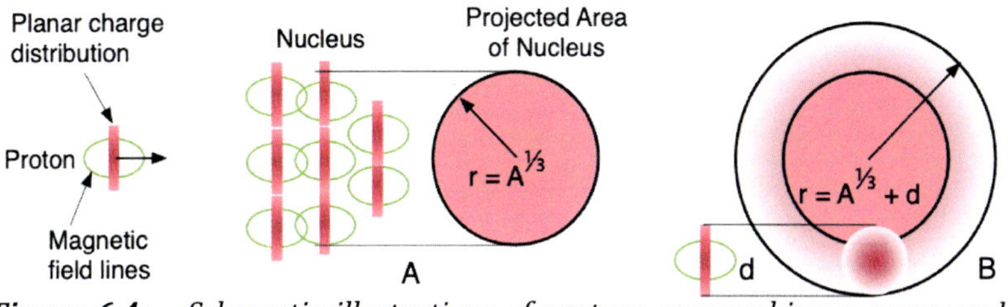

Figure 6.4: *Schematic illustrations of protons approaching a compound nucleus with spins aligned.*

Even Z nuclei need three protons, two of which must become neutrons, to form from odd Z+1 nuclei. For this reason alone, it would be expected that odd Z nuclei would be less abundant than the even Z counterparts on either side of it.

Similarly, the rate at which Z nuclei fuse to form Z+1 nuclei depends upon the presence or absence of a magnetic field to attract fusing protons. The most abundant even Z nuclei (Be excluded) have no magnetic moment. All odd Z nuclei have a magnetic moment, giving them greater opportunity to fuse to form the next highest Z nucleus. This would also contribute to the lower abundance of odd Z nuclei than their nearest even Z counterparts.

It should also be noted that the capture cross-section for a nucleus increases with $A^{\frac{2}{3}}$. This suggests that it is actually easier to fuse protons into nuclei that have a higher Z because A ≈ 2Z. If this is indeed correct and the elements were fused in the sun, it would be expected that the overall trend would follow a curve that was a function of $A^{\frac{2}{3}}$ or $Z^{\frac{2}{3}}$ multiplied by the abundance of the lower Z-1 nuclei from which they are fused. That makes it proportionally easier to fuse the higher Z nuclei than the lower Z nuclei. This is a significant difference from the predictions of the standard model. The latter suggests that charge is uniform and approaching protons must tunnel through an ever-increasing barrier to form the higher Z+1 nuclei as Z increases.

6.3. Fusion to Form the Heavy Elements

Fusion is a complex process involving many variables. Figure 6.1 shows there is a general downward trend in the abundances of the elements. The exceptions are the very low abundances of Li, Be and B, as well as their aftermath, the higher abundances of C and N. It is considered easiest to determine the nature of the variables involved by studying the fusion of heavy elements from oxygen to the highest Z nuclei. If this model cannot provide a reasonable match for the elements between oxygen and uranium there is no sense in extending the calculations. If it can provide a reasonable match, it sets the basis of the parameters that can be used in determining the abundances of the low Z elements.

In this determination of element abundance, it is assumed that the elements are all fused by a process in which the nuclei align themselves such that they can approach each other along the axis of rotation of their constituent nucleons. To simplify the calculations, only the trend to form the most stable isotope is presented.

The probability of fusion depends upon:

i. the projected area of the nucleus and that of the incident particle,
ii. the abundance of the element Z from which it is being formed,
iii. the abundance of nucleons (usually ^{1}H) to be fused to form the Z+1 element,
iv. the number of protons required to fuse the next Z+1 nucleus from the Z nucleus,
v. the correct conditions of temperature and pressure for fusion to occur,
vi. the path taken through the various isotopes.

To evaluate the overall trend, this presentation deals only with the first four. Point (v) is taken into consideration with a constant of multiplication. It would be expected that this constant would vary with the mass of the star. The greater the mass, the higher the temperature and pressure and the faster fusion will occur. Point (vi) is assumed to be constant to allow the trend to be generated. It is acknowledged that the calculated results could be improved by taking that into consideration.

Once nucleus Z + 1 had been fused from nucleus Z, it would still be in the same high pressure and temperature environment with the same availability of protons and could itself be further fused. The process of fusion from low Z nuclei to higher Z+1 nuclei by the addition of protons will continue until it is stopped by the rate of fusion of the highest Z nuclei being less than the rate at which the highest Z nuclei decay due to instability. When it has matured, the abundance of the elements will be in equilibrium. The rate at which the elements are formed matches the rate at which they are fused to form the next element. Maturity also means that enough time has passed for the nuclei formed in the central core to migrate to the surface in approximately the same ratio as they form, highly radioactive nuclei excluded.

The formation of Z + 1 nuclei from Z nuclei can be expressed by the equation

$$N_{Z+1} = N_Z \left[(A_Z)^{\frac{1}{3}} + d_F\right]^2 . Fn(F_P . D_{P1}) . fn_{(\mu Z)} \quad (6.1$$

where N_Z and N_{Z+1} are the abundances of the elements with atomic number Z and Z + 1 respectively, A_Z is the mass number of element Z, Fn is a function of D_{P1}, the number of nucleons required to form the most abundant Z + 1 nucleus, F_P is the probability that a proton will be available to fuse with the nucleus, d_F is the effective diameter of a proton to fuse to a nucleus and $fn_{(\mu Z)}$ is a function of the magnetic moment of the Z nucleus.

The term $(A_Z)^{\frac{1}{3}}$ is used for the radius of the Z nucleus only because it is more convenient than $r = 1.285A^{\frac{1}{3}} + 0.01(A - 2Z)^{\frac{1}{3}}$, which is slightly more accurate (Cook, 2010). The term d_F describes that fraction of the radius of the incident proton that will fuse with the Z nucleus with its mass number A_Z. It is not necessarily the radius of the incident nucleon, which in most cases will be a proton.

6.4. Even Z Nuclei

Equation 6.1 only determines the rate at which element Z + 1 is formed from element Z. Element Z + 1 will also be fused to form element Z + 2. Apart from ^{4}Be, all the most abundant even Z nuclei have even A and hence both their spin and magnetic moment equal zero. The odd Z nuclei have a magnetic moment that will attract protons and thus cause them to fuse faster, reducing their abundance. Irrespective of the abundance of the odd Z nuclei, the even Z nuclei will still form at the same rate. That rate will depend upon the availability of protons to fuse to the nuclei along the path between Z and Z + 2. Equation 6.1 is modified to

$$N_{Z+2} = N_Z \left[(A_Z)^{\frac{1}{3}} + d_F\right]^2 . Fn(F_P . D_{P2}) \quad (6.2$$

where D_{P2} is the number of protons required to fuse element Z + 2 from element Z. Equations 6.1 and 6.2 require the availability of protons to be constant and refer to the mid-life stars.

Of all the terms in equation 6.2, the only unknowns are d_F and $Fn(F_P D_{P2})$. To calculate the abundances of the nuclei it is necessary to set fixed values for those unknowns for the fusion of protons into larger nuclei. It should also be noted that, because there is no positive charge repulsion to overcome, other nuclei, particularly ^2H, ^3He, and ^4He, can also fuse into larger nuclei. Other nuclei are several hundred times less abundant than ^1H and will not make a significant contribution to nuclei with Z > 6. As can be seen from figure 6.2, the term d_F is a function of the proton's diameter. It has been found that setting d_F = 0.62 fm gives reasonable agreement with observed results. A proton needs to overlap the diameter of the nucleus to which it will fuse by slightly less than its rms charge radius. In a similar manner, it has been found that setting

$$Fn(F_P D_{P2}) = \frac{1}{5.5F_P} \qquad (6.3$$

gives good agreement for the predictions to match measured results. F_P in the numerator represents the number of protons to be added. The term $5.5F_P$ takes into consideration that they must be sequential (and it will be further clarified below).

By any reading of the isotopes of the various elements with Z protons, there are several pathways by which protons can fuse into nuclei Z, producing radioactive Z + 1 nuclei that either decay before they again react with another proton, or are sufficiently stable for more protons to react with them and form yet other unstable Z + 2 nuclei. Calculating all possibilities is both tedious and unnecessary. The assumption is made that protons will be available at a fixed rate for all. Even when the target nuclei have a magnetic moment and attract a proton more readily than a non-magnetic nucleus, the overall effect is averaged out in these calculations. The various pathways to the different isotopes in this averaging will be caught up in the overall averaging process.

Equations 6.2 and 6.3 are combined to give

$$N_{Z+2} = N_Z\left[(A_Z)^{\frac{1}{3}} + 0.62\right]^2 \cdot \frac{1}{5.5(A_{Z+2} - A_Z)} \qquad (6.4$$

A plot of equation 6.4 against Z for even Z going from Z = 8 to Z = 92 is given in figure 6.5. It commences with the abundance of oxygen, O, at 1.5 x 10^7. It is overlaid on the measured element abundances shown in figure 6.1. Note that F_P (= $A_{Z+2} - A_Z$) was set to 4 for the even Z nuclei between 8 and 20 because that is the value. Some of the even Z nuclei for Z ≤ 20 have several stable isotopes.

In most cases the dominant isotope makes up over 90% of the naturally occurring nuclei (Z = 12 excluded with 78%). This would not significantly alter the results. F_P was set at 8 in going from 20 to 22 because that was a reasonable average for Ti_{22}.

After Z=22 there are usually several different isotopes that make up the abundance of any element, with the most dominant isotope sometimes constituting less than 25% of the element. The average number of nucleons required to go from even Z = 20 to even Z = 92 is

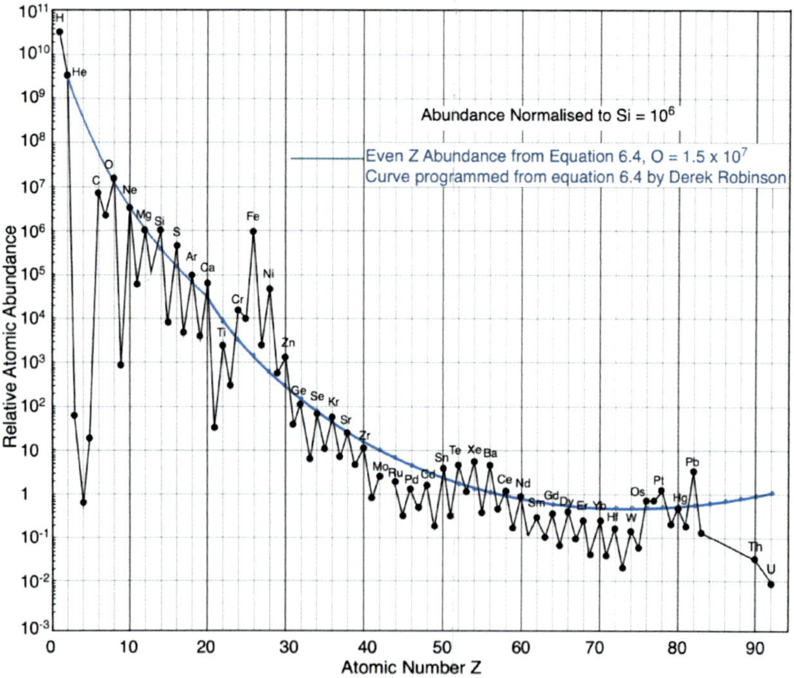

Figure 6.5: *The calculated abundances of even Z elements for Z>8, based upon the abundance of oxygen of 1.5 x 10^7, superimposed upon the measured solar abundances.*

between 5 and 6. Rather than summing them all, an average of 5.5 nucleons was used when going from Z to Z + 2, as shown in equation 6.3 above.

There is reasonable agreement of the predicted values for the even Z nuclei, Z = 8 to Z = 20 inclusive, as it gives a steady decrease in the abundances. With each of those elements, only 4 protons are required to go from Z to Z + 2. The curve shows that the eight protons required to fuse ^{40}Ca to ^{48}Ti significantly drops its abundance. From then on, it requires an average of almost 5.5 protons to increase the atomic number from Z to Z+2. While there is more than one stable isotope for most even Z elements, for Z = 20 and below, one isotope dominates.

Although many elements above Z = 20 have several isotopes, for simplicity of calculation, they have been rounded to the nearest whole number. This makes only a minor difference to the observed abundance. The curve in figure 6.5 suggests that equation 6.4 and the introduced constants give a reasonable agreement with the observed even Z abundances. That suggests that the mechanism proposed in figure 6.4 may give a reasonable representation of the actual process.

It should be noted that there is about 1 helium nucleus for every 10 protons in the sun. The larger helium nucleus diameter means there will be about 1 helium fusion event for every 5 protons fusion events. That means equation 6.4 would be responsible for the majority of the fusion. To a first approximation the helium fusion will not change the overall trend.

The largest differences between observation and prediction are seen in iron and nickel, closely followed by chromium and zinc. It is suggested these are all related to iron. Its isotope ^{56}Fe has the highest binding energy per nucleon of all nuclei. Its high abundance has always been explained by iron being the element with the highest binding energy per nucleon. It is also the last nucleus with an exothermic fusion reaction. That does not change under this model. The higher abundances of Ni and Zn could be explained as a continuation of the high Fe abundance.

The continuation between exothermic fusion before iron to endothermic fusion after iron suggests there is no difference between the endothermic and exothermic reactions as far as fusion is concerned. That would require a constant and adequate source of energy for the endothermic reactions to continue. Estimates suggest the temperature of the sun's core is about 15 million degrees C. That corresponds to each particle having an energy approaching 1,500 eV. It is suggested that the high density, about 150 gm/cm^3, and speed of a 1500 eV proton, about 2 km/sec, is sufficient to impart about 1 MeV to a fusing proton and nucleus combination in pico-seconds. The half-lives of proton rich isotopes in the fusion chain are measured in milli-seconds and above. That would give sufficient time for a nucleus to absorb the proton and sufficient energy to allow endothermic fusion to occur.

There are some significant differences between predicted and observed even Z nuclei at Z = 54, 56, 82, 90 and 92. It is suggested these are all related. The lower values of thorium and uranium are due to their radioactive decay. They and any possible trans-uranium fused elements decay. The limited half-lives of their decay products mean they would not reach the sun's surface to be detected in their fused abundances. The end product of heavy element decay is generally lead. The high lead peak can therefore be explained as reflecting the end product of the decay of the radioactive nuclei fused in the sun.

It is present with about 6 times its predicted abundance. That is about the expected abundance if the fused nuclei of atomic number greater than bismuth, Z=83, decayed to lead. That would include some trans-uranium elements that have short half-lives and would also decay back to lead. Others would continue to fuse to higher Z nuclei before themselves decaying back.

As uranium, and possibly higher Z nuclei, were fused in the sun, the fusion process would have to make its way through the isotope ^{235}U. It is suggested that during that process, more ^{235}U than ^{238}U would be generated. Under the turmoil of the fusion process, some could also fission. As shown earlier, the fission products of ^{235}U form two peaks, one around barium and xenon, and the other around strontium and krypton.

The calculated curve shows that the abundances of all isotopes of uranium and barium should be at about the same level, namely 1 on this plot. If half of the ^{235}U nuclei fissioned, that would double the Ba and Xe abundance. They would be detected. There would also be equivalent amounts of strontium and krypton being generated. Strontium's natural abundance is over twenty times higher than that of barium. Such a small increase is not detected on the log scale abundance plot.

The fusion of heavy elements in the sun and other stars leads to the possibility of significant ^{235}U accumulation near stellar cores. That could be the cause of stellar explosions, such as supernovae. Horowitz and Caplan (2021) raised such a possibility for white dwarf stars becoming type 1a supernovae.

Apart from those, the predictions of the even Z nuclei are within a factor of about 3 to 4 of the observed values. It is suggested that an approximation theory that gets within a factor of 3 to 4 of the measured values over a range of 10 million is considered a reasonable agreement. It suggests that some fine-tuning, along the lines of isotope considerations mentioned earlier, could get more accurate predictions.

That a match is obtained between a model that predicts the fusion of heavy elements in the sun and the observed abundances of the elements on the sun's surface, is a reasonable indication that the sun is indeed fusing the heavy elements. That is a possibility that can only occur under a two-dimensional planar charge distribution, such as the one predicted by this rotating photon model of matter. Its most important feature is that there is no term involving a potential barrier to be overcome before hydrogen and helium can fuse to form the heavy elements. Fusion requires alignment and close positioning. When those conditions are met, it can occur.

The above has demonstrated the formation of even Z nuclei. The actual situation is vastly more complex than that which has been indicated and calculated. That does not alter the trend. Protons are added one at a time to element ^{A}Z, forming element $^{A+1}Z+1$, which is often unstable. It decays to ^{A+1}Z, which may also be unstable. In turn another proton converts it to $^{A+2}Z+1$, which may not be stable and decays to ^{A+2}Z. In its turn ^{A+2}Z captures a third proton to form $^{A+3}Z+1$, which is often stable at least for $8 < Z < 20$. There are several pathways for forming stable isotopes of any nucleus. It is suggested these are averaged out in equation 6.4, producing the results in figure 6.5.

This agreement has been reached without the need to overcome a potential barrier between the positive charge of the proton and that of any other nucleus, including those with $Z > 80$.

6.5. Odd Z Nuclei

There are added complexities with the fusion of odd Z nuclei from even Z nuclei. One is the number of protons required to fuse with an even Z nucleus to produce the odd Z + 1 nucleus. Another is that most odd Z nuclei have a magnetic moment. A third is the abundance of the even Z isotope into which the addition of one proton can fuse the stable odd Z nucleus.

Equation 6.4 can be rewritten from N_{Z+2} as a function of N_Z, to N_{Z+1} as a function of N_Z, to give:

$$N_{Z+1} = N_Z\left[(A_Z)^{\frac{1}{3}} + 0.62\right]^2 \cdot \frac{1}{5.5(A_{Z+1}-A_Z)} \qquad (6.5$$

Applying only that correction to the odd Z nuclei, based upon the calculated abundances of the even Z nuclei, gives the results shown in figure 6.6. The blue dots show the result of applying that correction to the theoretically calculated values. The correction was calculated for each Z+1 from the previous Z. Applying the same corrections to the measured values gives the red dots. Those corrections start out at 2 for Z=9 and finish up at 7.8 for Z=83. Above Z=30 those differences are not that great. Below that, there are some significant differences. Where there is a blue dot and no red dot, it indicates the two calculations are the same.

The feature to note is that the predicted odd Z nuclei are always lower than the predicted even Z nuclei. That mimics the trend as measured experimentally. Another factor that should also be considered is the isotope abundance path between the even and odd Z nuclei. If an odd Z+1 nucleus is fused from a low abundance Z nucleus by the addition of just a single proton, it is likely to also have a low abundance. That is the situation for F, Na, P, Cl, K and Sc nuclei.

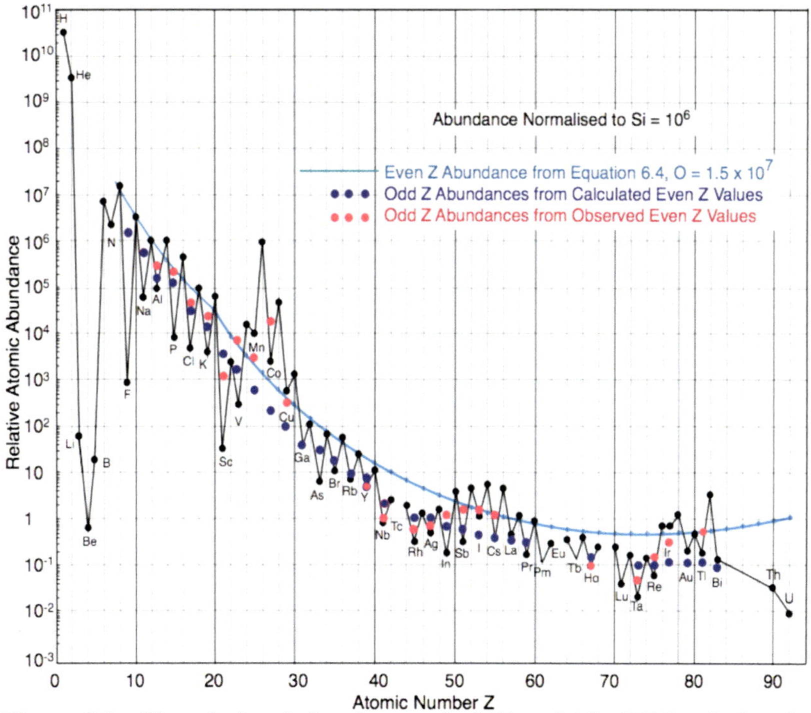

***Figure 6.6:** The calculated abundances of odd Z nuclei for Z ⍰ 9, calculated from the theoretical curve, blue, and measured, red, even Z abundances and superimposed upon the measured solar abundance curve.*

When that is taken into consideration, their expected abundances are significantly lower than the equations for their predicted abundances. As an example, ^{19}F is formed from ^{18}0, which is only 0.2% of O nuclei. On the other hand, ^{27}Al is formed directly from ^{26}Mg, which is over 11% of all Mg nuclei (Rosman and Taylor, (1999). It would therefore not be surprising to find that fluorine's abundance relative to its neighbouring even Z nuclei, is less than that of aluminium.

These results are displayed in figure 6.7. The green dots represent the abundances of the lower even Z isotope from which the odd Z+1 stable isotope is formed. It shows that the calculated odd Z+1 nucleus would be expected to have a slightly lower abundance. It is not highly accurate. That would require the total element fusion path to be

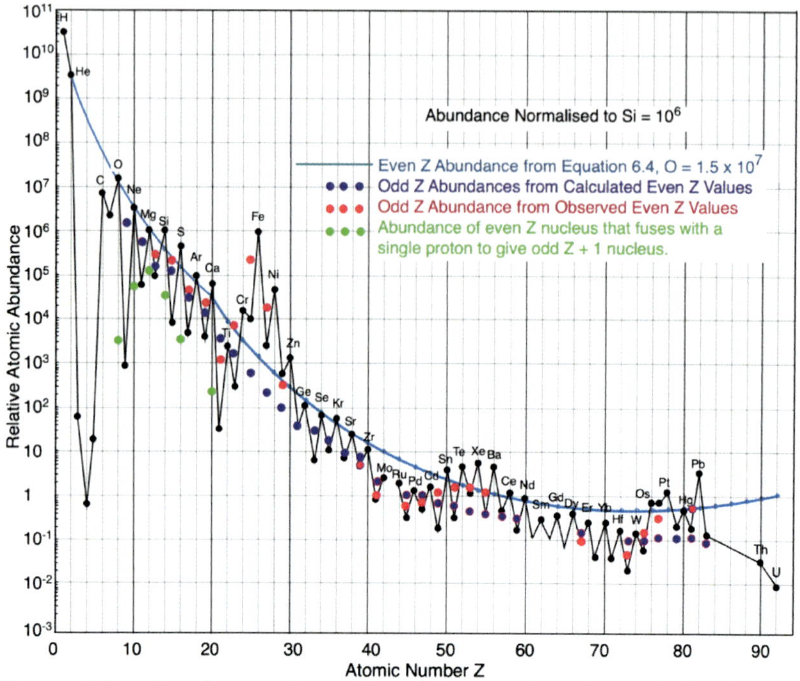

***Figure 6.7:** Abundances of even Z nuclei, green dots, from which odd Z+1 nuclei can be formed by fusion with a single proton.*

considered. It would also require the influence of the magnetic moment of the odd Z+1 nucleus on attracting protons to fuse even faster.

That task is left for others. It is suggested at this stage that the combinations of figures 6.5 to 6.7 inclusive, along with the explanations for some of the differences, give an expectation for the abundances of about half the nuclei to within a factor of about 3.

Most of the rest come within a factor of about 5. An error of 5 over a range more than 10 million is considered a reasonable fit. That is of particular importance when it does not include any term to overcome the positive repulsion of protons from the high Z nuclei.

6.6. Forming Lithium, Beryllium, and Boron

Discussions around figures 6.5 to 6.7 inclusive, have not shown why lithium, beryllium, and boron are present in such small quantities.

After all, if they, like the other elements, are fused in the sun, their low abundances would make it difficult for fusion to get past them. One of the difficulties in fusing them has been to get past the barriers at A=5 and A=8. All isotopes with those mass numbers have such short half-lives that it would not be possible for them to be formed by single proton fusion with an A-1 nucleus. Getting past A=5 can be done with

$$^{4}He + ^{2}H \rightarrow ^{6}Li \qquad \{6.10$$

That reaction is possible because both nuclei are present in the sun and there is no charge barrier to be overcome.

Equation 6.4 can be changed to:

$$N_{Li}^{6} = N_{He}^{4}\left[(A_Z)^{\frac{1}{3}})0.62\sqrt{2}\right]^2 . \frac{N_{2H}}{5.5N_{1H}} \qquad (6.6$$

N_{Li}^{6} is the abundance of fused ^{6}Li nuclei and N_{He}^{4} is the abundance of ^{4}He, which is about 10% of that of ^{1}H. A_Z, the mass number of ^{4}He is 4. The $\sqrt{2}$ is because the radius of a deuteron is ≈ $\sqrt{2}$ that of a proton. N_{2H}/N_H is the ratio of ^{2}H nuclei to ^{1}H nuclei. Putting those into equation 6.6 gives:

$$N_{Li}^{6} = 3.5x10^9[1.4 + 0.88]^2 . \frac{1.5x10^{-4}}{5.5} \qquad (6.7$$

or the abundance of ^{6}Li ≈ 5.2 x 10^5.

That is much higher than the observed Li abundance of ≈ 16. It gets past the A=5 barrier with a significant surplus. It is also possible that the reactions

$$^{6}Li + ^{1}H \longrightarrow ^{7}Be \longrightarrow ^{7}Li \qquad \{6.12$$

occur. That does not increase the Li abundance. It does increase the ^{7}Li abundance. That can also occur via the reactions:

$$^{4}He + ^{3}He \longrightarrow ^{7}Be \longrightarrow ^{7}Li \qquad \{6.13$$

The low concentration of ^3He in equation 6.7 means reaction 6.13 does not make much of a contribution to the total Li abundance.

There are two reactions that can get past the barrier at A=8.

$^6Li + {}^4He \longrightarrow {}^{10}B$ {6.14
$^7Li + {}^4He \longrightarrow {}^{11}B$ {6.15

The small difference between ^{6}Li and ^{7}Li allows these two reactions to be averaged to give

$$N_B = N_{Li}[2.5 + 1.2]^2.\frac{0.1}{5.5} \quad (6.8$$

where N_B is the boron abundance, 2.5 (fm) is the average Li radius and 1.2 (fm) is the radius of a ^{4}He nucleus. 0.1 is the ratio of ^{4}He to ^{1}H nuclei. The term 5.5 is as previously derived. Using $N_{Li} = 5.2 \times 10^5$ gives $N_B \approx 1.3 \times 10^5$.

The most probable pathway to ^{9}Be is

$^7Li + {}^2H \rightarrow {}^9Be$ {6.15

Rewriting the reaction equation 6.4 gives

$$N_{Be}^9 = N_{Li}^7[2.5 + 0.88]^2.\frac{1.5x10^{-4}}{5.5} \quad (6.9$$

That gives beryllium an abundance of ≈ 400. Those Li, Be, and B abundances are plotted as the orange points in figure 6.8. They are still much greater than the measured values. The predicted abundances of Li and B are about four orders of magnitude too high and that of Be is about 2.5 orders of magnitude too high. The only way the abundances of these elements can be reduced is by further fusion. If that occurred, it would mean that, if fusion were stopped, those abundances would be significantly higher.

An exploding supernova stops fusion at the time of explosion and send the fused nuclei out as cosmic rays. Measuring the relative abundances of those elements in cosmic rays and comparing them to those in the sun would give a good indication as to whether that

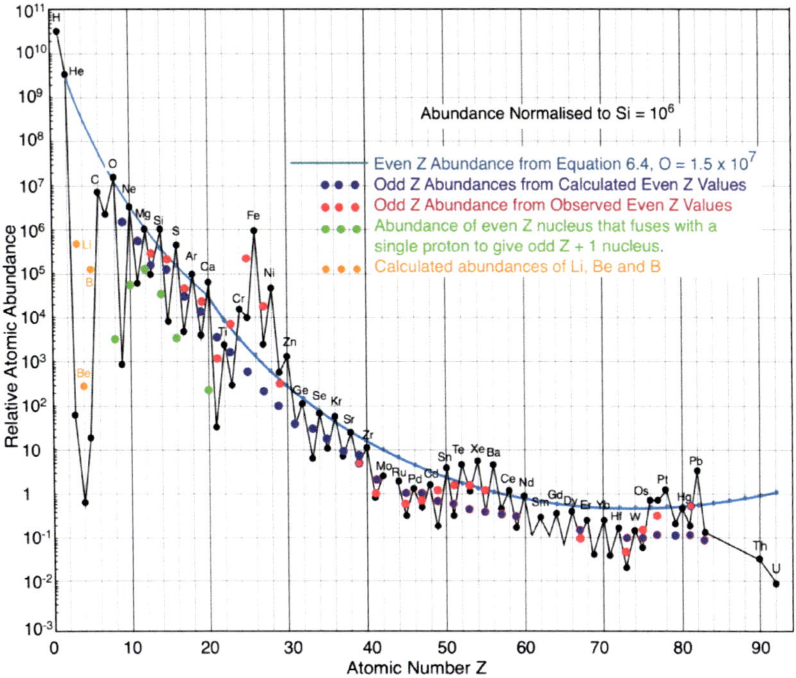

Figure 6.8: *The calculated abundances of Li, Be, and B superimposed on previous results.*

occurred. One such measurement is shown in figure 6.9 (Delahaye, 2010).

The Li abundance for galactic cosmic rays is very close to that predicted in equation 6.8. The B abundance is a little higher than predicted. The Be abundance is about 3 orders of magnitude higher. This is easily seen in figure 6.9.

The calculated values of Li and B are closer to the measured cosmic ray abundances than they are to the measured solar abundances. It is suggested that justifies the earlier comment that the low abundances of Li, Be, and B are due to their continued fusion within the sun, and hence, other stars. The higher cosmic ray abundance of Be is due to other mechanisms not mentioned here.

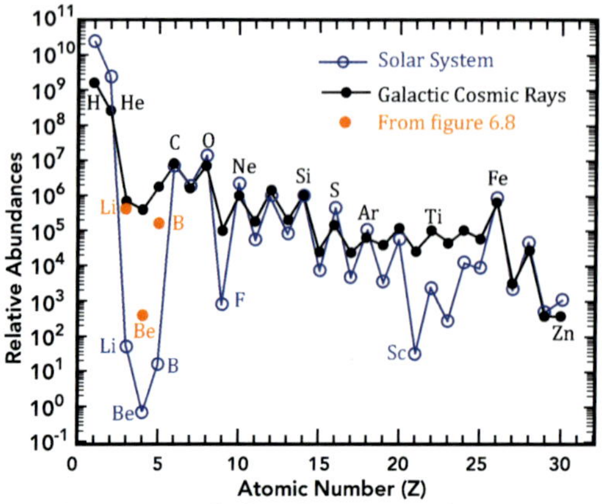

***Figure 6.9:** The measured abundances of galactic cosmic rays imposed upon the measured solar abundance curve.*

The most probable mechanisms for fusion of lithium to reduce its abundance are those involving the two most common nuclei, ^{1}H and ^{4}He. They are:

$$^{7}Li + {}^{1}H \rightarrow {}^{8}Be \rightarrow 2 \times {}^{4}He \text{ or } {}^{4}He + {}^{3}H + p \qquad \{6.16$$
$$^{7}Li + {}^{4}He \rightarrow {}^{11}B \qquad \{6.17$$
$$^{6}Li + {}^{4}He \rightarrow {}^{10}B \qquad \{6.18$$

Protons, ^{1}H nuclei, are about 10 times more abundant than alpha particles, ^{4}He nuclei. ^{4}He nuclei have a radius over double that of ^{1}H. It would therefore be expected that about 75% of the ^{7}Li nuclei will fuse with ^{1}H and 25% will fuse with ^{4}He. Reactions {6.17 and {6.18 get the fusion process past the Z = 8 barrier.

Like lithium, boron will fuse mainly with ^{1}H and ^{4}He since they are the most abundant elements. Possible reactions include:

$$^{11}B + {}^{1}H \rightarrow {}^{12}C\ (\rightarrow 3 \times {}^{4}He) \qquad \{6.19$$
$$^{11}B + {}^{4}He \rightarrow {}^{13}C \qquad \{6.20$$
$$^{10}B + {}^{1}H \rightarrow {}^{11}C \rightarrow {}^{11}B \qquad \{6.21$$
$$^{10}B + {}^{4}He \rightarrow {}^{14}N \qquad \{6.22$$

Reaction {6.19 suggests that $^{11}B + {}^{1}H$ could give three alpha particles. Although it is possible, ^{12}C has a binding energy about 6 MeV higher than that of 3 alpha particles and could be the more stable product.

Reactions {6.17 to {6.22 suggest that, once beyond the A = 8 barrier, sufficient lithium will be fused with hydrogen to form boron nuclei. In turn, most of the boron nuclei will fuse to form carbon or nitrogen. The Li and B abundances are lower than those of C, N, and O. For the gap nuclei, Li, Be, and B, to contribute to the fusion of the heavier elements, it requires that Li, Be, and B fuse more rapidly with ^{1}H and ^{4}He and that ^{12}C and/or ^{16}O are inherently stable and fuse at a slower rate.

A possible explanation for the rapid fusion of Li and B comes from their high magnetic moment to mass ratio. ^{7}Li's magnetic moment is 3.256 μ_{BP}. Its mass is 7 AMU, giving it a magnetic moment to mass ratio of 0.465 μ/A. It is suggested that its high magnetic moment makes it easy to cause protons to change their orientation and be attracted to their large area. Equally, ^{7}Li's low mass will make it responsive to an approaching proton's magnetic field. Jointly the two will have a higher probability of fusing than otherwise expected.

Together they make the fusion of ^{7}Li with protons much more rapid. ^{11}B's magnetic moment is 2.688μ_{BP}, giving it a magnetic moment to mass ratio of 0.244 μ_B/A. Although it is only about half that of ^{7}Li, it is higher than any other compound nucleus. It is significantly higher than any of carbon, nitrogen, and oxygen. That possibility has not been explored.

6.7. The Gap is Crossed

Reactions {6.17 to {6.22 show how the abundances of Li, Be, and B can be so low, while the pathway of fusing the heavy elements in the sun passes through them. They require a high fusion rate for the Li, Be, and B nuclei compared to the fusion of hydrogen to helium and the fusion of hydrogen with C, N, and O. The close match of equation 6.4 with observation over seven orders of magnitude suggests the fusion mechanisms for nuclei with $Z > 8$ are similar for all those elements. Further it occurs more slowly than for H to He, either directly, through the CNO cycle (Hoyle, 1946), or the pathways in reaction {6.17 and {6.18.

As such, the low abundances of Li, Be, and B are not a hindrance to the fusion of heavy nuclei within the sun or other stars. Their low abundance is an indication of how rapidly they fuse with protons. The best indications of that are their galactic cosmic ray abundances shown in figure 6.9.

That gap and the fusion of the heavy elements in the sun is only possible in a nucleon model in which charge is in two dimensions. All the elemental abundances for Z=3 and beyond were determined under this rotating photon model.

6.8. Comparison with the Standard Model

Under the standard model for the structure of nucleons, charge is distributed in three dimensions over protons and nuclei. To get protons into a nucleus, it is necessary to provide sufficient energy to overcome the positive charge repulsion between them. If charge is distributed in three dimensions around each nucleon, and hence uniformly around a nucleus, the energy barrier would vary according to the electrostatic potential energy, namely

$$U = \frac{q_1 q_2}{4\pi\varepsilon_0 r^2} = \frac{kZ_1Z_2}{r} \qquad (6.10$$

where k is Coulomb's constant, q_1 and q_2 are the charges on the nuclei, which equal Z_1e and Z_2e respectively, e is the unit electric charge, and r is their centre-to-centre separation distance.

Equation 6.10 shows it gets increasingly difficult for protons to fuse into heavier elements. Various suggestions have been forwarded as to how the heavier elements are formed.

1. When stars have fused all the hydrogen to helium, they collapse, raising their temperature. This provides sufficient energy for the next stages of fusing the heavier elements such as carbon, oxygen, etc. The process stops at iron. After that, the fusion reactions are endothermic. They require heat, which cools the stars. That explains the high abundance of iron and near equivalent elements.

2. Massive stars collapse, forming supernovae. The explosion provides sufficient energy that some even higher Z nuclei can be formed.
3. Neutron stars collide and explode, producing neutron rich fragments. These decay through beta radiation to produce the heaviest elements.

The standard model explanations require multiple mechanisms to produce the heavier elements.

6.9. Summary

The above presentation is but a brief introduction to a complex topic. There is vastly more to the formation of the elements than the simple presentation above. It suggests that elements seen in the sun are fused within it. That is possible under this rotating photon model of matter because electric charge is distributed in two dimensions around the centre of each nucleon. The effect of this is best illustrated in equation 6.4 and the agreement it shows in matching the observed abundances of the even Z nuclei between Z = 8 and Z = 80, first shown in figure 6.5. Equation 6.4 does not include any term required to overcome the charge potential between a proton and a nucleus of any Z.

Covering over seven orders of magnitude, equation 6.4 matches observed solar abundance results within a factor of 2 for almost half of the even Z nuclei above Z = 8. About half of the rest are matched to within a factor of 5. Reasons for the higher abundances of the remainder were forwarded. The greatest differences between prediction and observation were in the abundance of Fe and Ni. ^{56}Fe is both the nucleus with the highest binding energy per nucleon and the last of the exothermic fusion reactions. It is expected to have a higher abundance. Ni continues that higher-than-expected abundance.

This model predicts that the odd Z nuclei will have a lower abundance than their even Z neighbours for a variety of reasons. The number of protons required to fuse even Z + 2 nuclei from odd Z + 1 nuclei is usually less than the number required to fuse odd Z + 1 nuclei from even Z nuclei. Even Z nuclei have more isotopes than their odd Z neighbours. Odd Z nuclei usually have a lower binding energy per nucleon than their even Z neighbours (Weizsacker, 1935). That occurs because of the stronger magnetic binding between even Z and A nuclei than nuclei with either odd Z and/or A. Some odd Z + 1 nuclei have very low abundances due to the low abundance of the even Z nuclei that requires only 1 proton to fuse.

The two-dimensional nature of charge on the nucleons means it is no more difficult to fuse helium to other nuclei than hydrogen. The main limitation is the lower abundance of He over H. That enables the barriers of A=5 and A=8 to be crossed in solar fusion. The low solar abundances of Li, Be, and B were shown to be due to the ease with which they subsequently fuse after formation. They are not due to any difficulty in forming them. That was demonstrated by their abundances in galactic cosmic rays being much closer to that predicted under this model than observed in the sun. The gap of their very low solar concentrations does not inhibit the fusion of the higher Z nuclei within the sun.

The higher abundances of C and O over the predicted values of Li and B is due to the greater stability of those nuclei against fusion. It was suggested that this is partly due to their high magnetic moment to mass ratio making it easier for them to fuse. Once oxygen is formed, the fusion process given in equation 4 applies and the remaining nuclei are fused as described.

Fusion is endothermic for $Z > 26$. The high temperature and pressure in the solar furnace allow 1 MeV of energy to be imparted to a nucleus in a few nanoseconds. It requires 100 particles with a temperature of 100 million °K (10 keV) to strike it. The half-lives of intermediate nuclei are measured in milli-seconds and above. That is long enough to impart sufficient energy to enable endothermic fusion to occur. The abundances of those endothermic reacting elements are typically 10^{-6} times those of He from fusion. They would not make any noticeable difference to the output of solar energy.

There are fundamental differences in the fusion predictions between the standard model and the rotating photon model proposed herein. The standard model suggests that electric charge is spread throughout the protons in all three dimensions. The higher the Z of a nucleus, the greater the energy required to get another proton into it. The high temperatures in the sun are required to provide sufficient energy for the protons to tunnel through electrostatic repulsion and fuse.

This means that the heavier elements must be formed in exploding stars such as supernovae. As the matter collapses under gravitational influence, it forces the nuclei together forming the heavy elements. All this is said to happen in "an instant." The supernova forms the elements and spreads them throughout the neighbouring galactic regions in its giant explosion.

That idea is supported by the detection of elements with more than 10 protons in supernova remnants, those being considered too high to have been fused under the conventional model (Burrows et al., 2000). They are just as likely to be detected under this model because they would be fused in the star during its lifetime and be thrown out in the explosion.

There is nothing in the measured abundances of elements in other stars in our galaxy that suggest that the abundance of elements in those stars is significantly different from those in the sun. Likewise, there is nothing to suggest that stars in our galaxy are fundamentally different from those in other galaxies. This suggests that the fusion mechanism proposed herein is responsible for the formation and abundances of the elements throughout the observable universe.

There is still a lot of work required to give a good understanding of the mechanism of formation of the elements to account for their observed solar abundances. The best that can be derived from this presentation is that it is possible to consider that charge is indeed distributed in a two-dimensional manner and that fusion occurs by aligning the axis of the fusing nuclei so that there is no (or only a minimal) charge barrier to overcome. If both nuclei have magnetic moments, magnetic attraction will result in more rapid fusion.

Fusion under this model is very different than fusion under the standard model. Current fusion techniques are not using the above information. It is suggested much better results will be achieved using techniques based upon this presentation.

6.10. List of References

Asplund M et al., (2009); "The chemical composition of the sun", Annual Review of Astronomy and Astrophysics, **47**, pp 481 – 522

Burrows DN et al., (2000); "The X-Ray Remnant of SN 1987A", Astrophysical Journal, **543**, pp L149 – L152

Cook ND, (2010); "Models of the Atomic Nucleus, Unification Through a Lattice of Nucleons", Springer, Berlin, Ch 2

Delahaye T, (2010a); "Propagation of Galactic Cosmic Rays and Dark Matter indirect Detection". PhD Thesis, Mathematical Physics [math-ph] Université de Savoie, Universitá degli studi di Torino https://tel.archives-ouvertes.fr/tel-00530422v2, Ch 3

Horowitz CJ and Caplan ME, (2021); "Actinide Crystallization and Fission Reactions in Cooling White Dwarf Stars", Phys. Rev. Lett. 126

Hoyle F, (1946); "The synthesis of elements from hydrogen", MNRAS, **106**, pp 343 – 383

Lodders K, (2010); Solar system abundance of the elements, Principles and Perspectives in Cosmochemistry, Astrophysics and Space Science Proceedings, Springer Verlag, Berlin, pp 379 – 417 http://arxiv.org/pdf/1010.2746.pdf

Rosman KJR and Taylor PDP, (1999); "1997 Report of the IUPAC subcommittee for isotopic abundance measurements", Pure and Applied Chemistry, **71**, pp 1593 – 1697

Weizsacker WD, (1935); "Zur Theorie der Kernmassen", Zeits. fur Physik **96**, pp 431 – 458

7. Electrons in Atoms

Abstract:

Electrons in atoms form the basis of chemical binding and are described by quantum mechanics. This presentation suggests the apparently peculiar behaviour of electrons around nuclei, which require defined quantum laws, are the natural occurrence of the structures of the electron and nuclei.

7.1. Introduction

Chapters 2 and 3 showed the properties that individual subatomic particles have when they are composed of rotating photons of the appropriate energy. Several weird and often apparently contradictory properties were shown to be natural consequences of the model we are using to build the universe. Chapter 5 showed how protons and neutrons were placed together to make up nuclei. Chapter 6 showed how the different nuclei were formed in the abundances that are observed.

The final step before we can understand the universe at large, is to cloak the nuclei with electrons. The first step is to remove the nuclei from the star. That is done by supernova events in which stars explode and send their constituent particles everywhere as cosmic rays. They include at least the lighter nuclei in the abundances produced and the star. Li, Be, and B have much higher abundances in cosmic rays than in stars. They also include electrons that keep the star's electric charge relatively neutral. As the cosmic rays are slowed down or captured, electrons will be attracted to the nuclei. There they cluster around them, giving nuclei their chemical properties. The following is the briefest introduction to how the electrons gather around the nucleus and how their rotating photon structure shapes the behaviour of the elements. When that is done, we can get on with the task of understanding the universe at large.

It is easiest to explain what happens when electrons cluster around nuclei by referring to the work done over past generations. Much of it is already known. This presentation shows how many of the laws of quantum mechanics, introduced to explain the unique properties of electrons in atoms, are automatic consequences of the model of matter used in this presentation.

Experiments over the past couple of hundred years have measured many of the properties of electrons around nuclei. Theoreticians have studied the arrangements of electrons and come up with descriptions of how they give different properties to atoms. In turn, experimentalists have been able to use those theories to predict and test new properties of combinations of atoms that have led to the modern technological benefits we now experience around us. The collective knowledge of theoretical and experimental electron physicists forms the basis of much of our modern technological era. Although electrons only make up about 1 in 4,000 of the mass of our world, they are responsible for almost everything that happens.

Nuclei give mass to the atoms and provide the platform that holds electrons to them. Otherwise, they take very little part in most of the properties of matter, mass, nuclear and gravitational effects excluded. It is the number of electrons held together in the atom and the unique properties of the electrons that give atoms their various properties, such as making carbon very different from nitrogen, even though they only differ by one electron, proton, and neutron.

With such a good understanding of the properties of electrons in atoms, surely a chapter in putting electrons around nuclei is going to repeat the work of others. After all those others have developed concepts such as Bohr's (1913) planetary model of the atom, Pauli's (1925) exclusion principle, electrons orbit the nucleus in quantized orbits determined by the de Broglie (1925) wavelength, and in particular, the use of wave equations to solve electron properties (Schrödinger, 1926; Dirac, 1928). They appear to work quite well. So why bother?

First and foremost, the others have treated electrons as waves and used their wave properties to calculate the expected results. Along the way they have had to introduce postulates such as Pauli's exclusion principle and Bohr's quantized orbits. This presentation has shown quite clearly that electrons are particles. This chapter is a brief presentation of how its particle structure and properties explain some effects previously considered to be the exclusive domain of quantum theorists.

The derivation of the de Broglie wavelength in equation 2.14 makes it difficult to visualize how such a property could cause an electron to have quantized orbits around a nucleus. Below, we will clarify how and why particles give rise to "quantized orbits."

7.2. The Bohr Model of the Atom

Thomson (1897 a, b) discovered the electron. Rutherford (1911) and his co-workers (Geiger and Marsden, 1913) discovered that atoms had a very small positive core. It is more than coincidence that the dimensions of the small positive cores are as predicted from this rotating photon model. Electrons associate with nuclei to maintain the charge neutrality of atoms. On average, the number of electrons per atom is equal to the number of protons in its nucleus. If there should ever be an excess of positive charge, the inverse square nature of electric attraction and the low mass of the electrons will attract them to neutralize that excess. Apart from holding the proton together inside the nucleus, the number of neutrons in an atom has almost no effect upon the number of electrons or their behaviour.

Bohr (1913) developed the model of atoms with electrons orbiting the nucleus like planets orbiting a star. Under classical electromagnetic theory, a charged particle traveling in a curved path such as an electron orbiting a nucleus would emit radiation, lose energy and collapse into the central nucleus, as illustrated in figure 7.1 A. Clearly, that does not happen. Other explanations were sought. It was noticed that electrons had fixed energy levels.

That led Bohr to the idea that orbits were quantized according to their orbital angular momentum. Some indicate they are quantized according to the de Broglie wavelength.

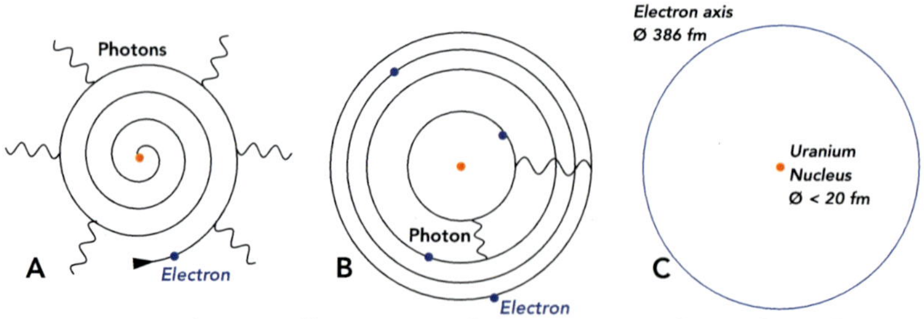

Figure 7.1: *Schematic illustrations of electrons around a nucleus, showing the classical model, A, Bohr model, B and rotating photon model C.*

Quantization prevented the electrons from losing energy as they orbited the nucleus and collapsing into it. In that manner, quantum mechanics could explain this important fact in a manner that classical electromagnetic theories could not. Fixed orbits are illustrated in figure 7.1 B.

Whenever electrons change orbit energy levels, they emit or absorb photons appropriate to the energy levels involved. Because their fixed energy levels did not include the nucleus, they would not spiral into it. Problem solved.

Quantization is an artificially induced constraint. It has no explanation beyond "it sort of works." It still requires another variable to make the electron move around the nucleus instead of orbiting in the same plane, like a planet.

An important feature of electrons under this rotating photon model is their rest radius of $\approx 1.93 \times 10^{-13}$ m. This is several hundred times larger than the radius of a proton, which is less than 10^{-15} m. It is much larger than the radius of even the largest nucleus, which is about 10^{-14} m. Figure 7.1 C shows a representation of the radius of an electron's rotating photons axis on similar scale to a uranium nucleus. Its axis is almost 200 times larger. It is simply too large to fit in the nucleus.

On top of that, all stable nuclei do not have an energy level into which they can fit. When there is such an energy level available, as in an unstable or radioactive nucleus, the nucleus will quickly capture that electron, turning a proton into a neutron. The decay process is called electron capture. If electrons were traveling in quantized orbits, one would have to tunnel through its quantized orbit to be captured by the nucleus. That is another artificial constraint.

To give an indication of dimensions, if a uranium nucleus was the size of a pea, 5 mm, an electron's axis would be about 1 m. The outer electrons would be spread over a diameter in excess of 1 km. Within an atom, the particles take up very little space. The fields they generate occupy the space. That may help to explain why field theories are so popular.

Under this model, there is no need to introduce any constraints to prevent an electron from collapsing into a nucleus. In a hydrogen atom, an electron around a proton moves within the constraints of its potential well. Rydberg (1889) derived what is now called Rydberg's constant R_H. It can be used to derive the binding energy of a single electron to a single proton. It is called the Rydberg energy, R_E, given by

$$R_E = \frac{m_e e^2}{m\varepsilon_0^2 h^2} = 13.606..\,eV \qquad (7.1$$

where m_e is the electron's mass, e is its unit charge, ε_0 is the electric permittivity of free space, and h is Planck's constant. In a single hydrogen atom, the electron will move in a series of straight lines, exchanging photons and hence angular momentum with the nucleus so that it travels around it.

That explains why electrons do not fall into a nucleus. It does not explain why electrons around a nucleus have specific energy levels. With hydrogen it is obvious. Its energy level is the Rydberg energy, 13.606 eV. An explanation of multiple energy levels is best made, however, using nuclei with $Z > 1$.

As we continue this presentation, we will refer to the energy with which an electron is bound in an atom as its ionization energy, I_E, because this is the amount of binding energy that must be overcome in order to remove an electron and make the atom an ion.

7.3. Fixed Ionization Energies

The ionization energy for the first electron around a nucleus containing Z protons is given by

$$I_E(Z) = R_E Z^2 = \frac{m_e e^2 Z^2}{8\varepsilon_0^2 h^2} \qquad (7.2$$

By convention it is labelled the $1s^1$ electron. The simplest compound nucleus is He, Z = 2. Its $1s^1$ ionization energy is given by

$$I_E(1s1)(He) = \frac{4\pi m_e^2}{8\varepsilon_0^2 h^2} = 54.42 \text{ eV} \qquad (7.3$$

The $1s^1$ electron screens out part of the He nucleus's positive charge. It requires less energy to remove the second electron because of that charge screening. That can be expressed as

$$I_E(1s2) = I_E(1s1) - fn(screening) \qquad (7.4$$

where $fn(screening)$ is a function of the $1s^1$ electron screening out some of the nucleus's positive charge. That function will be discussed later.

When the two electrons are together in the He atom they do not retain their individual binding energies. They adopt an average energy. One of the aspects of the Bohr model is that electrons are treated as probability waves for their location. Figure 7.2 shows two schematic illustrations of the electron distribution under the Bohr model, A, and the rotating photon model, B.

There is some similarity between the Bohr model of quantized orbits and this rotating photon model. Both rely upon the potential well of the captured electron.

In the Bohr model, the electron's wave function forms a radius around the nucleus. The wave function probability oscillates around the Bohr radius, allowing the electrons to be anywhere within the extent of the wave function.

The probability of finding an electron is shown by the shading intensity. It is highest near the Bohr radius and diminishes away from it. The Bohr model requires another term to make the planar orbit rotate through 180° to give complete charge shielding around the nucleus.

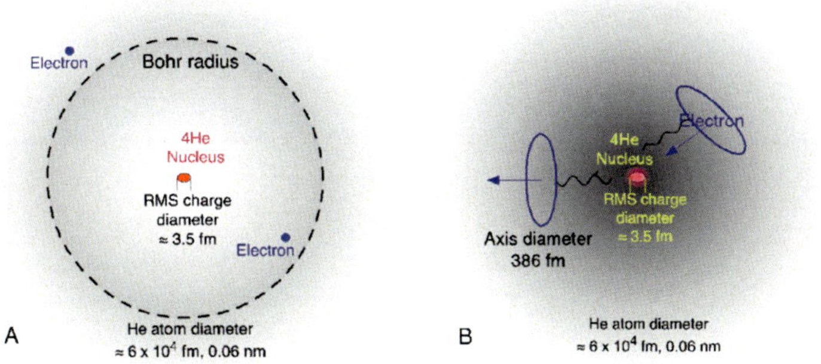

Figure 7.2: *Schematic illustrations of electrons in a helium atom under the Bohr model, A and the rotating photon model B.*

In this rotating photon model, the electrons are also constrained within the nucleus's potential well. That makes the energies of the two models the same. From then on, the electrons move back and forth within that potential well. With an average energy of about 40 eV the electrons would travel a distance of about 0.12 nm about 10^{15} times per second. Their travel is a multiple effects affair. The electrons are attracted to the positive nucleus, losing potential energy as they approach it. They gain maximum kinetic energy as they pass near or over it. Their kinetic energy takes them towards the limit of the potential well in which they are confined. There they are attracted back to the nucleus. The two electrons, being negatively charged, will repel each other as much as possible. As they travel, they exchange photons – the carriers of the electromagnetic force.

It must also be remembered that the $\frac{1}{2}\hbar$ spin of an electron is the angular momentum of the whole mass rotating at the speed of light at a radius of 193 fm. It is a very powerful gyroscope. It can only change its direction of travel by exchanging angular momentum with another particle. That is why electrons must pair whenever possible. Exchanging photons enables them to exchange angular momentum and thus change their direction of travel. With random travel at 10^{15} times per second they soon surround the nucleus with their charge. They do not need any quantization to prevent them from sliding into the nucleus. They are simply far too large to fit into any nucleus. Their trajectories take them beyond the Bohr radius, with an ever-decreasing probability that reaches zero at or near the boundary of the potential well.

In the Bohr model, paired electrons in the same orbital, e.g., $1s^1$ and $1s^2$, are said to have opposite spin. That was introduced via the Pauli exclusion principle to prevent the two fermions occupying the same place in space–time. There is no need for that postulate. Their like electric charge keeps them from occupying the same point in space at the same time. In all probability, paired electrons would orient their magnetic fields in the same direction to minimise their energy. Moving back and forth over 10^{15} times per second makes it difficult to measure the magnetic moments of individual electrons. The average will always be zero. That allows the assumption that they must have opposite spins to seem valid.

As it is, quantum mechanics specialists do not know what spin is. They often quote the Stern-Gerlach experiment as being an indication of measuring spin. The reality of that experiment is that they are measuring the magnetic field of the single unbound electron and relating that to the property they call spin.

The Stern-Gerlach experiment (Gerlach and Stern, 1922) does not measure spin as such. After all, both stable ^{47}Ag silver nuclei, isotopes 107 and 109, have equal numbers of neutrons and one unpaired proton. That gives those silver nuclei a spin of $\frac{1}{2}\hbar$, the same as that of the unpaired electron. If the experiment were indeed measuring the silver atom's spin, each atom could have one of three possible spin values: -1 when the two were aligned in the same spin down direction; 0 when they were aligned in the opposite direction, and +1 when they were aligned in the same spin up direction.

The Stern-Gerlach experiment measures the direction of the spare electron's magnetic moment. Later, Uhlenbeck and Goudsmit (1925, 1926) explained the splitting of the spectral lines from atoms in a magnetic field as being due to the electrons having an internal angular momentum. That is also what it is under this rotating photon model for the structure of matter particles.

Pauli's exclusion principle was an unnecessary restraint designed to prevent something that could not occur.

The helium atom presents a closed system to other particles. It has no need to attract another electron to either help neutralize the nucleus' positive charge or to provide photons for the electrons to exchange angular momentum to move in different directions. It is a completely inert entity.

There are only a fixed number of positive charges in a nucleus. That means there are only a fixed number of binding energies available. The addition of one electron around a nucleus screens out the effect of one proton, as suggested by equation 7.4.

Before we determine the effects of screening, it helps to consider what happens when particles are caught in a potential well. When electrons are captured by the nucleus, they travel around it at the average distance determined by the equilibrium between the potential energy of its capture and the kinetic energy of its motion.

An electron can only move in straight lines until it exchanges angular momentum through photon exchange. Then it can move in another direction dictated by its momentum and the electromagnetic field in which it finds itself. That is the reason why electrons always try to form pairs, for example the $1s^1$ and $1s^2$, the $2s^1$ and $2s^2$ etc.

Complications arise because one electron screens out the potential of one proton. The amount of screening is best determined by looking at the effect of different numbers of electrons around the same nuclei. With almost 100 different naturally occurring elements, each with the same number of electrons as protons, evaluating them all is beyond the scope of this presentation. Instead, we will look at the cases of a few electrons around the first 24 nuclei.

7.4. 1s1 and 1s2 Electrons

The equation for the energy of a single electron ($1s^1$) around a single proton in a hydrogen atom was given by equation 7.1. The equation for a single ($1s^1$) electron around a nucleus containing Z protons is given by equation 7.2. There is no screening involved with the $1s^1$ electrons. They can be calculated directly. The results are shown in table 7.1. The $1s^1$ electrons screen out the charge of 1 proton. When that occurs, the screening forces the second electron further away from the nucleus because it is attracted by a lesser force than the first electron. One such screening equation is

$$I_E Z(1s^2) = I_E Z(1s^1) - R_E[2 + 1.28(Z - 2)] \quad (7.5$$

It indicates that the ionization energy of the $1s^2$ electron is reduced by a function of the $1s^1$ electron and the atomic number difference. The results for this are also shown in table 7.1.

Z	Calculated $I_EZ(1s^1)$	**Measured $I_EZ(1s^1)$**	Calculated $I_EZ(1s^2)$	**Measured $I_EZ(1s^2)$**
1	13.606	**13.606**		
2	54.42	**54.42**	27.21	**24.59**
3	122.45	**122.45**	77.55	**75.64**
4	217.7	**217.72**	155.14	**153.9**
5	340.15	**340.22**	259.95	**259.37**
6	489.82	**490**	392.04	**392.1**
7	666.69	**667.05**	551.4	**552.07**
8	870.78	**871.41**	738.08	**739.29**
9	1102.09	**1103.12**	952.1	**953.91**
10	1360.6	**1362.2**	1193.5	**1195.8**
11	1646.33	**1648.7**	1462.31	**1465.1**
12	1959.26	**1962.7**	1758.63	**1761.8**
13	2299.41	**2304.1**	2082.34	**2086.0**
14	2666.78	**2673.2**	2434.5	**2437.6**
15	3061.35	**3069.8**	2812.7	**2816.9**
16	3483.14	**3494.2**	3219.4	**3223.8**
17	3932.13	**3946.3**	3653.8	**3658.5**
18	4408.34	**4426.2**	4116	**4120.9**
19	4911.77	**4934**	4606.1	**4610.8**
20	5442.4	**5469.9**	5124.3	**5128.8**
21	6000.25	**6033.7**	5670.4	**5674.8**
22	6585.3	**6625.8**	6244.9	**6249.0**
23	7197.57	**7246.1**	6847.5	**6851.3**
24	7837.06	**7894.81**	7478.5	**7481.7**

Table 7.1: *Theoretical and measured values for the first and second electrons in the 1s shell of a Z=1 to Z=24 atom. The binding energy units are expressed in electron volts (eV). The measured values are taken from:-*
http://www.periodictable.com/Properties/A/IonizationEnergies.html

The theoretically predicted $1s^1$ ionizations start out matching the experimentally measured values quite accurately. At Z = 24, the difference is about 0.7%, with the predictions being slightly lower than the measured values. This indicates a slight variation of unknown origin. The overall agreement is sufficient, though, to indicate that equation 7.1 is reasonably accurate.

The theoretically predicted $1s^2$ ionization energies commence predicting a value too high by 2.62 eV. By Z=6 the prediction matches measurement. At Z=24 the prediction is below the measured value by 3.2 eV. That is a strong indication that the screening effect of the $1s^1$ electron is responsible for the ionization energy of the $1s^2$ electron.

Table 7.1 is a strong indication that the ionization energies are not artificially constrained by quantum mechanical effects. The fixed ionization energies are due to the number of protons in the nucleus and the screening effect of inner electrons.

7.5. 2s and 2p Electrons

Before 2*s* electron energies can be determined, the combined effect of the $1s^1$ and $1s^2$ electrons must be considered. Together they form a closed system for two of the protons in the nucleus. As such, it would be expected that the third electron, $2s^1$, would have its ionization energy as a function of $(Z - 2)^2$, rather than as a screening factor to be subtracted. That can be expressed as

$$I_E Z(2s^1) = fn[R_E.(Z-2)^2] \qquad (7.6$$

A suitable equation is given by

$$I_E Z(2s^1) = 0.25R_E.[(Z-2)^2 + (Z-2)] \qquad (7.7$$

Suitable expressions for the next three electrons, $2s^2$, $2p^1$ and $2p^2$ will involve screening effects of the previous electron. That involves subtracting terms that are fractions of the Rydberg energy from the ionization energy of the next lowest energy electron.

When they are done, suitable equations for the $2s^2$, $2p^1$ and $2p^2$ electrons are

$$I_E Z\ (2s^2) = I_E Z(2s^1) - R_E[0.6 + 0.3(Z - 3)] \quad (7.8$$

$$I_E Z(2p^1) = I_E Z(2s^2) - R_E[0.5 + 0.5(Z - 4)] \quad (7.9$$

$$I_E Z(2p^2) = I_E Z(2p^1) - R_E[1 + 0.33(Z - 5)] \quad (7.10$$

The results of these equations are listed in table 7.2.

Z	*I_E*Z (2s¹)	***Exptl* *2s¹***	*I_E*Z (2s²)	***Exptl* *2s²***	*I_E*Z (2p¹)	***Exptl* *2p¹***	*I_E*Z (2p²)	***Exptl* *2p²***
3	6.8	**5.39**						
4	20.4	**18.2**	8.2	**9.32**				
5	40.8	**37.9**	24.4	**25.1**		**8.3**		
6	68.0	**64.5**	47.6	**47.9**	27.2	**24.4**	9.1	**11.3**
7	102.0	**97.9**	77.5	**77.5**	52.3	**47.4**	29.7	**29.6**
8	142.8	**138**	114	**113**	80.2	**77.**	53.0	**55.0**
9	190.4	**185**	158	**157**	117	**114**	85.2	**87.1**
10	244.8	**239**	208	**207**	160	**158**	128.8	**126**
11	306.0	**300**	265	**264**	211	**208**	174.8	**172**
12	374.0	**367**	329	**328**	268	**266**	222.9	**225**
13	448.8	**442**	400	**398**	332	**330**	286.9	**285**
14	523.6	**523**	470	**476.3**	396	**401**	341.3	**351**
15	618.8	**6112**	561.7	**560.8**	480.1	**479**	421.6	**424**
16	714.0	**707**	652.8	**652.2**	564.4	**564**	501.4	**505**
17	816.0	**809**	750.7	**749.8**	655.5	**657**	587.5	**592**
18	924.8	**918**	855.4	**854.8**	754.1	**756**	682.2	**686**
19	1040	**1033**	966.6	**968.0**	857.8	**861**	781.4	**787**
20	1163	**1158**	1085	**1087**	969.4	**974**	887.8	**894**
21	1292	**1288**	1218	**1213**	1096	**1094**	1011	**1009**
22	1428	**1425**	1350	**1346**	1220	**1221**	1130	**1131**
23	1571	**1570**	1481	**1486**	1345	**1355**	1250	**1260**
24	1720	**1721**	1634	**1634**	1491	**1496**	1392	**1396**

Table 7.2: *Theoretically calculated and experimentally measured (bold) values for the ionization energies of the 2s and 2p, 1 and 2 electrons around the nuclei of elements of Z < 25.*

The differences between the results theoretically predicted from the screening effect of the inner electrons and the measured values are never more than a few eV at low Z or less than 0.5% at high Z. They results again demonstrate that there is very close agreement between the experimentally measured ionization energies and those predicted from the screening effect of electrons closer to the nucleus.

It is suggested the agreement from the screening effect calculations gives a closer match with observation than do those involving quantization due to either quantization according to the de Broglie wavelength or angular momentum.

These ionization energies due to the screening effect of inner electrons are a natural consequence of the number of protons in a nucleus and the number of electrons around it.

Whether orbits are quantized according to orbital angular momentum, or their de Broglie wavelength, is of no importance. The act of quantizing electron orbits to match observation is an artificial constraint. There is no need for that constraint. Similarly, there is no need for an artificial constraint to stop an electron spiralling into a nucleus. They are simply too large to fit into a nucleus. No stable nucleus has an energy level available to accommodate an electron.

There are some other fundamental differences between treating electrons under the Bohr model and this rotating photon model. The Bohr model treats electrons as wave probability functions and has the electrons orbiting the nucleus in energies fixed by either their de Broglie wavelength or their orbital angular momentum.

This model treats electrons as individual particles with specific properties that move according to the lowest energy level available. For their mass they are such powerful gyroscopes that the only way they can change their direction is by exchanging photons with another particle. That particle is usually another electron.

It is suggested that the latter has been observed. Berry (1989), commenting on the Bohr atomic model, observed that *"... two atomic electrons kept out of the deep potential well close to the nucleus but in the same neighbouring shells behave in the most part like a collective molecular like system more than they behave like quantum mechanical analogues of the solar system."*

Berry further added: *"Energies and, in the applicable cases, angular momenta are approximate constants for collective modes of vibration and rotation - better approximate constants in most cases than the energies and angular momenta we would attribute to individual electrons in the orbitals appropriate to an independent particle model."*

The phrasing *"... two atomic electrons ... behave in the most part like a collective molecular like system"* is what would be expected from this rotating photon model of the structure of electrons. The gyroscopic property of electrons means they cannot change their direction of travel unless they have another electron with which to exchange angular momentum. If they do, they can move freely, as dictated by the potential well produced by the positively charged nucleus.

Given all the photons being emitted and absorbed by the charged particles, it is easy for one electron to associate with another electron, and for the two to move as determined by other circumstances. It does not necessarily have to be photon exchange between paired electrons. For example, photons could be exchanged between all four *p* electrons instead of with just one each. There is no rule forcing electrons to be "monogamous."

That forms the basis of Berry's second comment, that collective angular momenta more accurately reflected observation than the individual angular momenta proposed by Bohr. There is no need to introduce quantized orbit terms. Neither is there any need to introduce a term relating to electrons pairing by spin. Electrons automatically pair whenever possible to exchange photons and hence angular momentum wherever possible. That photon exchange may be no different than two charged particles affecting each other when they are close. Once paired, electrons could form an entity that could be directed by their paired wave function. That mechanism may give rise to distinct trajectories for given energy levels, or "orbitals."

An unpaired electron has difficulty exchanging photons. That causes an unpaired electron in one atom to seek another unpaired electron wherever it can find one. When it finds one in another atom, if one is close, their connection forms the basis of chemical bonding.

Figure 7.3 illustrates what happens when two hydrogen atoms approach each other. 7.3 A illustrates two independent hydrogen atoms,

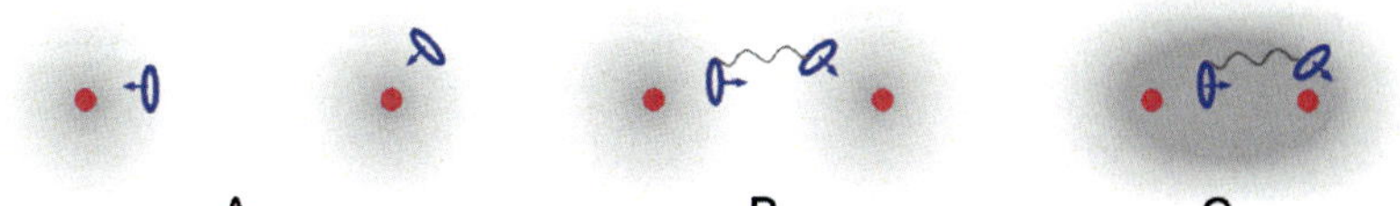

Figure 7.3: *Not to scale schematic illustrations of independent hydrogen atoms, A, forming a hydrogen molecule, C.*

each with its electron, an ellipse, moving within the potential well of its proton. Figure 7.3 B shows that when the two atoms are close enough, the electrons exchange photons, allowing them to move more freely.

Their need to exchange photons to move freely as a pair brings them together to form the molecule, shown in 7.3 C. Both electrons are free to move about both nuclei. The shading intensities are approximate representations of the probability of finding an electron within a particular region.

The dimensions and properties of the hydrogen molecule (H_2) are determined by the protons' mutual repulsion, screened by the electrons, along with the electrons' mutual repulsion, tempered by their close attraction to both protons. As they move, they exchange angular momentum and travel around the pair of nuclei. This arrangement, where electrons are shared between two atoms, is known as covalent bonding.

The situation with helium is quite different. Both electrons are already paired. Jointly, they can move freely about the nucleus. They completely screen out the nucleus's positive charge, and they have no need to interact with other electrons. Helium is therefore an inert atom that will not form any bonds with any other body.

7.6. Chemical Bonding

It has long been known that electron pairing is the basis of all chemical bonding. Under this rotating photon model of matter, the reason for that is the necessity of electrons to exchange photons, and hence angular momentum, to allow them to change their travel direction. Beyond that, the rules of chemical binding are well known and understood. They do not change because the reason is given. A couple of examples are given to indicate how bonding works under the gyroscopic effects of the large angular momentum to mass ratio of electrons. Details of chemical properties are available from such sites as https://periodictable.com/Elements/003/data.html.

Figure 7.4 schematically illustrates a situation in which lithium is covalently bonded to hydrogen, B, and ionically bonded to fluorine, D. The blue ellipses represent the electrons, the wavy lines represent photon exchange, the red centres are the nuclei. The shade intensity represents the approximate electron density within the atoms.

The two 1s electrons in Li, shown in A, are pulled in more strongly than the equivalent electrons in helium because there is an extra proton. Those two electrons keep the $2s^1$ electron away from the nucleus. Its attraction is significantly less than the single electron in a hydrogen atom. Its week binding means it can move over a larger distance. Li's atomic radius of 145 picometres (pm) is almost 5 times larger than the He atom at 32 pm. That is entirely due to the two 1s electrons screening out two protons and keeping lithium's 2s electron away from the remaining proton charge.

Both outer electrons in hydrogen and lithium are constrained in their movement. Exchanging photons with a nucleus which has almost 2,000 and 7,000 times each electron's mass does not make for easy directional change. When they come within range of each other, shown in figure 7.4 B, the two will exchange photons and form a pair.

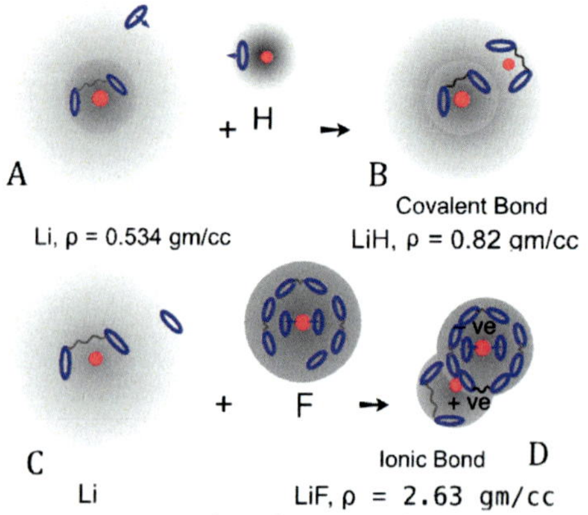

Figure 7.4: *Not to scale schematic illustrations of lithium involved in covalent, B and ionic, D bonding.*

Like all electron pairing, it is a "strained" relationship. They need each other to change direction but their like charges mean they will try to get as far away from each other as possible.

They are constrained by the positive charges of their respective nuclei. In turn, the nuclei are trying to get as far away from each other as possible and are constrained within the electrons' pairing based on their need of each other to be able to change direction.

The low binding energies of the hydrogen atoms 1s electron and the lithium's 2s electron leave the pair free to move around both nuclei. The addition of the hydrogen nucleus limits the distance the lithium's 2s electron can travel from its nucleus.

The addition of H to Li metal makes LiH denser than Li because it gets rid of the large orbit of lithium's weakly bound outer electron. LiH, density ρ=0.82 gm/cc, is over 50% more dense than Li, density ρ=0.534 gm/cc. The paired 1s and 2s electrons are not constrained around either nucleus. They move around both, making LiH an example of bonding by electron sharing, i.e., covalent bonding.

Figure 7.4 C and D illustrate ionic bonding. With 9 protons attracting the two 2s (and 7 other) electrons around a fluorine nucleus, they are pulled in quite tightly. With a radius of 50 pm, a fluorine atom is about one third the diameter of a lithium atom despite having about three times its mass. With nine electrons, fluorine is one electron short of forming a stable $1s^2\,2s^2\,2p^6$ state, which gives added stability.

The binding energy of the $2s^1$ electron to the lithium nucleus is about 5.4 eV. The binding energy of a 2p electron to a fluorine atom is about 17.4 eV. That results in the fluorine atom capturing the lithium atom's $2s^1$ electron and giving 4 pairs of electrons moving in a stable manner in the second potential well.

The fluorine atom becomes an F^- fluoride ion and the lithium atom becomes an Li^+ lithium ion. The two are held together by electric attraction of opposing charges. For that reason, it is called ionic bonding. It leaves the Li^+ ion with a central core that is smaller than a helium atom. When that is attached to the F^- ion, the assembly is very small.

Another form of bonding is metallic bonding. Electrons in an atom with a small number of free electrons in their outer energy levels can move away from their atom to be shared among, and move freely between, the adjacent atoms. That is the basis of electrical and heat conductivity. With no fixed bond energies, substances with metallic bonding can more easily change their shape.

The necessity of electrons to pair to enable them to exchange photons and move in different directions has a strong influence on atomic properties. Their strong ½$\hbar$ angular momentum represents their whole mass rotating at the speed of light with a radius hundreds of times larger than a proton's diameter. That constitutes a powerful gyroscope that cannot change its direction unless it can find a partner with which to exchange photons, and hence, angular momentum.

Figure 7.5 gives an indication of how the need for electrons to find a partner with which they can exchange photons combines with the screening effect of inner electrons to give different properties in each situation.

These illustrations are highly stylized. They are intended to show the pairing of electrons, not their positions around the nucleus. They move around the nucleus according to their attraction toward unscreened positive charge. Their motions are typically back and forth 10^{16} times per second, changing direction by picking up the photons that are continually emitted and absorbed by charged particles.

Figure 7.5 shows neon with 4 sets of outer electrons paired. It has no need to either seek or donate an electron and is an inert gas. Sodium has one unpaired outer electron. Since it is attracted by only one unscreened proton, it is loosely bound and will therefore move over a larger diameter, making it a "large" diameter and very reactive atom.

With one unbound electron it will engage in metallic bonding if placed with other atoms of similar status. Its electron needs a second electron with which to combine to move freely. This draws the two atoms together, effectively pairing their unpaired electrons.

Magnesium has two unscreened protons attracting its two outer electrons. Because they are paired, they do not screen each other from the nucleus. Mg atoms are smaller than Na atoms. Although their electrons are paired, they have a large volume over which to move. Each will readily pair with an electron from another atom if it comes sufficiently close. Mg atoms are therefore reactive and will participate in metallic bonding if combined with other metallic bonding atoms.

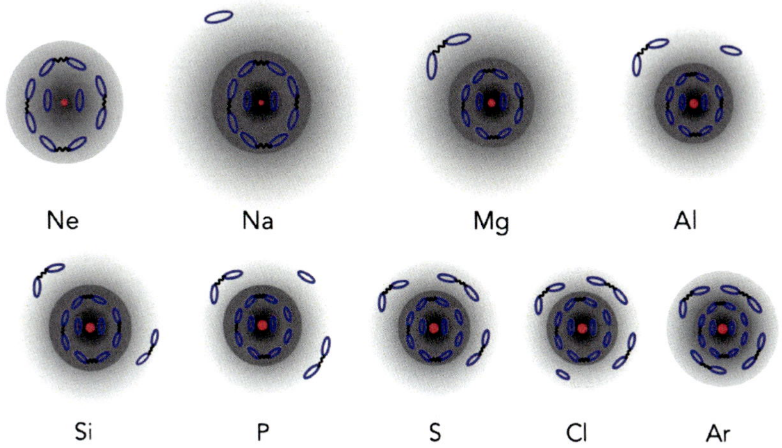

Figure 7.5: *Schematic illustrations of the electron arrangements affecting different properties of atoms.*

Continuing down the list, the higher the atomic number, the more protons are unscreened by the inner electrons. The atoms therefore become smaller under the greater nuclear attraction. Aluminium has one outer set of paired electrons and one single electron. Al will donate and share up to three electrons. Its single, loosely bound, free electron also enables it to react more easily with other materials than Mg's outer pair. Al is trivalent, a more reactive metal than Mg, and a good conductor. The three electrons still have a large volume over which to roam to screen out the positive charge.

Silicon has two sets of paired electrons. They will accept and share electrons from another source under the right conditions, such as at a high temperature. Without a free electron, they are neither good conductors nor very reactive atoms at low temperatures.

Phosphorous has two sets of paired electrons and one single electron. The number of outer electrons is too large for it to participate in metallic bonding. Its single electron makes it quite reactive as it accepts an electron from anywhere. The two sets of paired electrons will, under the right conditions, also accept electrons from elsewhere. P can therefore be a reactive pentavalent atom.

With three sets of paired electrons, sulphur is relatively stable at low temperatures under most conditions. Under high temperatures or other appropriate conditions, it can be a reactive material that displays several (usually) even valences, up to hexavalent.

The one unbound electron associated with chlorine will accept an electron from anywhere. Cl atoms are very reactive because the unbound electron is close to the unscreened proton, giving it a high binding energy.

Argon has four sets of paired electrons. It has no need to either accept or donate an electron to any other atom. It, like He, is an inert gas.

It is apparent that with Na having one loosely bound electron and Cl having a strongly bound single electron, when the two are close, they will form a strong ionic bond as the Na's electron is captured by the Cl atom.

The above is well-known chemistry, all of which has been greatly expanded on in chemistry textbooks. Mendeleev (1863) was the first to note that there was an ordered pattern associated with elements. He noted that over 40 years before Thomson's 1897 discovery of electrons and the observation of Brownian motion that firmly established the minute particle nature of matter, i.e., the atomic theory of matter.

The above was included to indicate why those atoms have their properties. They do not have their reactivity because of postulated quantum mechanical laws or binding rules. They bind in pairs for the very practical reason of being able to move freely because of their very high angular momentum to mass ratio. They have their reactive properties because of the combination of their need to pair with other electrons and screening out the positive charge in the nucleus to which they are attracted.

Electrons are not moving about the nucleus in some ordered orbits. Within those constraints, the electrons are in a "free for all," anything goes, tussle around nuclei as they seek their lowest energy level. If an electron is dislodged from an inner energy level, an outer electron immediately fills its space.

In a stable atom, electrons occupy the lowest energy level they can find. Each of these energy levels is controlled by the number of protons in the nucleus and the screening effect of the inner electrons on the positive charge available to the outer electrons. Electrons seek to bind in pairs or multiples of pairs to exchange angular momentum via photons to move more readily. Inner electrons screen out charge for outer electrons. They are not affected by the presence or absence of outer electrons.

This exercise can be repeated for all atoms. The above was just a brief introduction into how the physical structure of electrons and simple electrodynamics lead directly to what has been determined are complicated quantum mechanical effects. Those are the effects that give rise to the chemical reactivity order picked out by Mendeleev long before the atomic theory was established, and electrons were discovered.

Another feature of chemical bonding is the defined angles found in many molecules and crystals. If motion of the electrons is random, as suggested herein, how do bonds form defined angles?

That is best described using carbon's tetrahedral bonding. Its four outer electrons are available to bond. When they bond with separate atoms or molecules, they form electron pairs around the central positive core. The like charges on the electron pairs repel each other. The four bonds form as far away from each other as possible. That gives carbon its tetrahedral bond angles. They are equal in all three dimensions, as represented by methane, CH_4, in figure 7.6.

The situations with ammonia, NH_3, and water, H_2O, are similar, as shown in figure 7.7. With ammonia, the three hydrogen atoms are as far from each other as possible because of the mutual repulsion of their paired electrons.

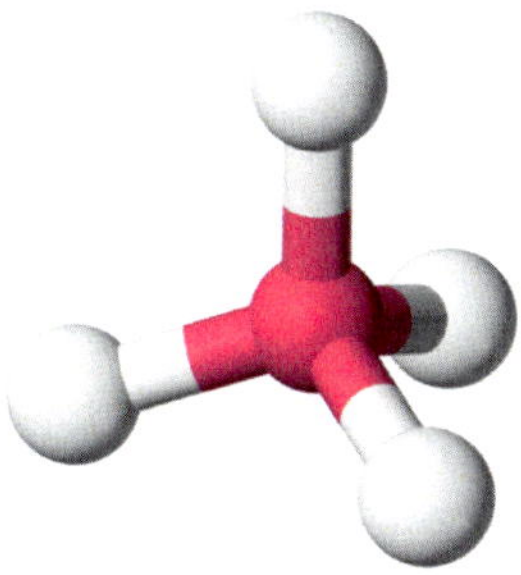

Figure 7.6: *Tetrahedral bonds associated with carbon atoms, e.g., methane.* (Wikipedia)

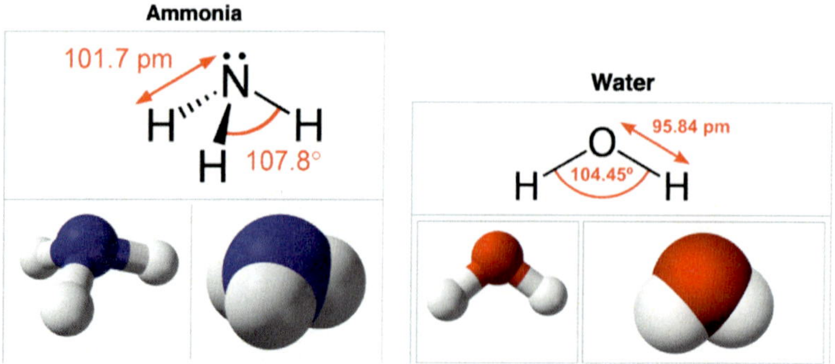

Figure 7.7: *Bond angles associated with ammonia and water molecules.* (Wikipedia).

The two residual electrons form a pair. Their presence on one side of the N atom pushes the three bonds away from the horizontal by about 17.8°.

Water has only two bonds with hydrogen. Their paired electron repulsion makes the bonds 180° apart. The remaining 4 electrons gather on one side. They push the water bonds away from the horizontal by ≈ 38°.

Bond angles are not due to "orbits" of electrons around an atom. They are caused by the mutual repulsion of paired electrons from each other. Electrons paired between two atoms stay paired to exchange angular momentum as they move, maintaining the bond.

7.7. Electron Spectra

Having established why electrons have discrete energy levels we need to see what happens when electrons are excited either by heat or by an electromagnetic field. If there is sufficient energy, the electron will be knocked out of its energy level. Immediately another electron will drop into that vacant energy position, emitting a photon to get rid of the additional energy as it drops to the lower energy level. In that way atoms can be excited to give off photons.

Photons emitted that way have a fixed energy, and hence, a fixed wavelength. They are called the emission spectra and are characteristic of the energy levels in individual elements. Because of the fixed values of energy levels mentioned above, it is not surprising that the electron "orbits" appear to be quantized.

The reverse also occurs. When photons of a particular energy impinge upon an atom, an electron can absorb it and be ejected out of, or to a higher level in, the atom. That is the origins of the absorption spectra. The energies of the emitted and absorbed photons will be the same for the transition between the same energy levels.

Figure 7.8 shows the different energy levels. The lower energy levels are filled first. Whenever an electron is removed from a lower level and another electron drops down to fill the vacancy, the level from which it drops down determines the frequency of the photon that is emitted.

For the above reasons, the binding energies of all the electrons are discrete and have fixed values. It is not, as quantum specialists point out, because electron orbits are quantized. Their discrete energy values are due to the limited number of protons in the nucleus, the number of inner electrons screening out some of the protons' positive charges, and the arrangement of those inner electrons.

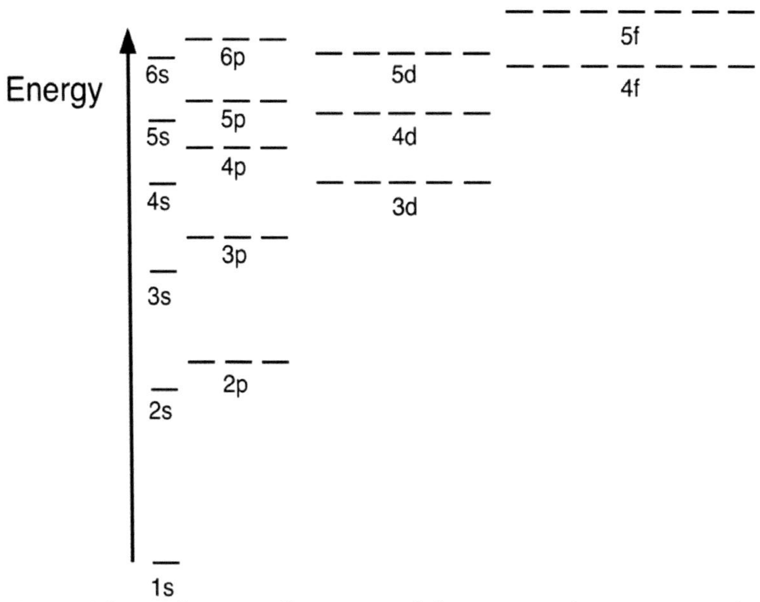

Figure 7.8: *Schematic illustration of the various electron energy levels associated with atoms. Each bar represents a pair of electrons.*

The unique nature of these values enables particular elements to be detected. Sometimes the frequency of a transition in one element corresponds to the frequency of a transition in another element.

Fortunately, there is more than one transition for each element, and the combination of two or more transitions can be used to distinctly identify any element. The complete set of photons emitted or absorbed by atoms are characteristic to each type of element, allowing them to be positively identified from two or more emission or absorption spectral lines. The lists of tables and calculations, together with the assortment of instruments available, enables each element to be identified through spectroscopy. Electron spectroscopy is a thriving analytical technique.

7.8. Quantum Mechanics

Classical mechanics, i.e., Newtonian mechanics, Maxwell's equations and their derivatives like the Carnot cycle, Bernoulli's principle and similar, work well when there is a continuum, relativistic effects excluded. Consider water as an example. Each molecule is made of two atoms of hydrogen around 1 atom of oxygen. When sufficient molecules provide a continuum, classical mechanics works well. When the volume is so small that the effect of individual molecules is noticeable, classical mechanics breaks down as each molecule is a single, or quantized entity in its own right. Individual molecules don't behave like they do in bulk water.

Within the molecule, individual atoms don't behave anything like bulk water. The electrons and nuclei have none of the properties of bulk water. The protons and neutrons within a nucleus behave entirely different from bulk water.

Because these individual components are quantized, they need a different mechanics to explain them. That is the origins of quantum mechanics. Within the mechanical world, the smallest elements are molecules, atoms, electrons, nuclei, protons, and neutrons respectively. In the electromagnetic field, the smallest electromagnetic waves are photons. Individual photons behave differently from Maxwell's equations. They apply to a continuum of photons. Quantum mechanics handles this world of individual particles.

When Individual particles were discovered, they were known to exist. Their structure, particularly photons, electrons, protons, neutrons, and neutrinos, were unknown. They were known to have wave properties. Quantum mechanics developed around treating particles as waves. Initially that focused on electrons with wave equations developed by Schrödinger (1926) and Dirac (1928).

Electrons seemed to have so many apparently conflicting properties that the best way to handle them seemed to be with the mathematically versatile wave equations. Physical properties were added mathematically to match observation. Quantum mechanics has largely developed into wave mechanics.

It is not only electrons that have been treated this way. The same wave techniques have been applied to protons, neutrons and "elementary particles. There are very few photon specialists who agree on the nature of photons. The standard model describes photons and electrons as point particles with properties attached mathematically.

Theoretical progress in nuclear physics has also been slow. Almost endless variations of quantum chromodynamics have stalled progress. Particle physicists are pursuing ever larger and larger machines to find an order that has eluded them for decades. Theoreticians are going nowhere in their efforts to unite Einstein's relativity and gravitational theories with their quantum field theories.

All the physics in this presentation has been based in the real world. Particles are treated as particles. Descriptions are given of what is occurring and why. It is suggested that an understanding of the physical mechanisms involved make it easier to understand processes and predict new possibilities.

Protons, electrons, neutrons, and neutrinos are particles. It is time to treat them as such. Planck's constant and quantization are an essential part of particles. The above has shown that wave equations are not needed to explain what have previously been considered quantum properties.

The previous chapters have shown that knowledge of the structure of those particles enables classical to be applied to their structure to describe their behaviour. The biggest difference is in chapter 5. It shows that classical concepts give an accurate description of nuclear physics, solving problems that have bothered nuclear physicists for many decades. It also suggests that visualizing quantum processes may give a better understanding of how to make more meaningful manipulations.

7.9. Summary

Table 7.3 summarizes some of the differences between electron properties according to particle versus wave theories. Although the results may appear to be the same in several cases, their explanations are very different. The quantized orbits model requires artificial constraints. The rotating photon model shows that all the quantized properties can be explained as a natural consequence of the structure and properties of electrons and nuclei.

Electrons Travel in Fixed Orbits	
Rotating Photon Model	**Bohr's Quantized Orbitals**
Electron energy levels are determined by the potential well generated by the number of protons in the nucleus and the way electrons in inner potential wells screen out the protons within the nucleus. It is the potential wells that determine atomic spectra. Electrons will not collapse into the nucleus because they are far too large to fit into it.	Postulated to explain why electrons travel around the nucleus in fixed energy levels and do not spiral into the nucleus. They are artificially constrained to multiples of orbital angular momenta or de Broglie wavelengths.
Electrons are free to move all around the nucleus, screening out the positive charge.	Additional postulates are necessary to make the electron's orbit rotate completely and uniformly around the nucleus.
This model predicts, and Berry has observed, that outer electrons travel in pairs or multiply paired pairs. That is a necessity because of the electron's very high angular momentum. They need to exchange photons so their angular momentum direction can change, and they can travel in a different direction.	Offers no explanation for this observation. Electrons traveling in pairs was postulated to match observation.

Electrons act as powerful gyroscopes and must exchange angular momentum to change direction. They do so by exchanging photons with another electron, or multiple electrons in paired groups. In that way they do not lose energy when they change direction. They move all-round the nucleus typically at $>10^{15}$ oscillations per second.	Electrons have a property called spin. Theoreticians seem unclear about what it is. They insist it does not have any classical analogy and reject the concept of spin being angular momentum.
Paired electrons cannot occupy the same point in space at the same time because their like charge will repel each other. That occurs irrespective of their spins.	Paired electrons having opposite spins was postulated to satisfy Pauli's exclusion principle and to match observation.
The bonding of atoms to form molecules, crystals, and other structures is due to their need to form pairs so they can exchange photons and change their angular momentum.	The bonding of atoms to form molecules, crystals, and other structures is due to their postulated need to form pairs.
Electrons are treated as particles with their properties are due to their structure.	Electrons are treated as probability waves with their properties mathematically attached.

Table 7.3: *A comparison list of some of the differences between the Bohr quantized orbits model and the rotating photon model.*

Electrons do not spiral into a nucleus because they are too large to fit into them. In most cases nuclei do not have a free energy level to accommodate them. Electron energy levels in atoms are fixed by the number of protons in the nucleus and the screening effect of inner electrons on those outside them.

The particle property called spin Is considered to be a peculiar quantum mechanical property that has no classical analogy. Apart from considering its conservation and using it in calculations, it does not receive much other attention.

According to this model, spin is angular momentum, as first proposed by Uhlenbeck and Goudsmit to explain the Zeeman effect. A magnetic field split spectral emission lines. The spin of an electron is the angular momentum of its whole mass travelling at c in a circle of rest radius 193 fm. That constitutes a very powerful gyroscope. It is why photon exchange is required for electrons to change direction.

Electron pairing and the number of electrons in an atom are the keys to the properties of atoms and how they bind to other atoms. The role of nuclei is restricted to providing over 99.9% of the atoms mass and acting as a platform to attract a particular number of electrons, which determines all other properties.

It Is the high angular momentum to mass ratio of the electrons that makes them bind in pairs or multiples of pairs. Next time you see a tall building, remember it is the angular momentum of electrons that keeps it so high. When you go into the wild and see trees tens of metres tall, remember that it is the angular momentum of the electrons that enables them to draw water up even the tallest trees, some over 100 metres tall.

The conservation of the intrinsic $\hbar$ angular momentum of the rotating photons that are electrons is responsible for their existence and stability. Spin is angular momentum, $I\omega$. It is so strong that, when combined with an electron's magnetic field, it keeps the rotating photon as a particle.

There is a lot more to learn about the behaviour of electrons in atoms. The important feature here is that they can be treated as particles, and to obtain adequate descriptions of their properties and behaviour. They do not have to be treated as waves. It is possible that understanding their physical behaviour may give some scientists a better understanding of how they can be manipulated to give improved electronic or material properties.

As mentioned at the beginning of this book, the natural way events occur is that everything follows its path of least resistance to reach a stable state. It is humanity's role to discover these paths of least resistance and stable states, and then to work with them for the well-being of all. Quantized orbits are an artificial constraint that can be misleading for people, and upon this foundation, quantum mechanics practitioners have struggled to unite it with Einstein's relativity theories and gravity.

Chapter 2 showed how the structure of matter automatically led to Einstein's special relativity corrections and $E = mc^2$. The following chapters will investigate the properties of gravity, how it operates, how it originates, and how it influences the universe.

7.10. List of References

Berry RS, (1989); "How good is Niels Bohr's atomic model?", Contemporary Physics, **30**, pp 1 – 19

Bohr N, (1913); "On the Constitution of Atoms and Molecules, Part 1", Phil. Mag., **26**, pp 1 – 24

de Broglie L, (1925); "PhD Thesis", published in Annalen der Physik. (Paris), **111**, p 22+

Dirac PAM, (1928); "The Quantum Theory of the Electron", Proceedings of the Royal Society of London, **117**, pp 610 – 624

Geiger H and Marsden E, (1913); "The Laws of Deflexion of α Particles Through Large Angles", Philosophical Magazine, **25**, pp 604 – 623

Gerlach W and Stern O, (1922): "Der experimentelle Nachweis des magnetischen Moments des Silberatoms". Zeitschrift für Physik. **8,** pp 110 – 111 doi:10.1007/BF01329580

Mendeleev DI, (1863); "The Natural System of Chemical Elements"

Pauli W, (1925); "Uber den Zusammenhang des Abschlusses der Elektronengruppen im Atom mit der Komplexstruktur der Spektren" Zeitschrift für 287hysic, **31**, pp 765 – 783

Rutherford E, (1911); "The Scattering of α and β Particles by Matter and the Structure of the Atom", Philosophical Magazine, **21**, pp 669 – 688

Rydberg JR, (1889); "Recherches sur la Constitution d'emission des elements chimique", Den Kungliga Svenska Vetenskapsakademiens Handlingar, **23**, pp 11

Schrödinger E, (1926); "An Undulatory Theory of the Mechanics of Atoms and Molecules". Physical Review, **28**, pp 1049 – 1070

Thomson JJ, (1897 a); "On the Cathode Rays", Proceedings of the Cambridge Physical Society, **V 9**, pp 243 – 267

Thomson JJ, (1897 b); "Cathode Rays", Philosophical Magazine, **44**, pp 293 – 317

Uhlenbeck GE and Goudsmit SA, (1925); "Spinn Elektronen und die Struktur der Spektren" Naturwissenschaften, **13**, pp 953-954

Uhlenbeck GE and Goudsmit SA, (1926); "Spinning Electrons and the Structure of the Spectrum", Nature, **117**, pp 264 – 26

8. Einstein's Gravity

Abstract:

Gravity attracts matter to massive objects. The moon and everything on planet Earth are attracted to it, planets are attracted to their stars, and stars are held within their galaxies, all by gravity. This presentation describes the relationship between Newton's gravity and Einstein's gravity theory derived from his general relativity theory. It expands on Einstein's work on gravity to show its origins and its close relationship to Newtonian gravity. Along the way, it shows the physics behind Einstein's gravitational field equations from his general relativity theory.

8.1. Introduction

We now understand all the matter in the universe, along with a reasonable knowledge of its properties. We can now look at its distribution within the universe and explain why it is the way it is. The assumption is that matter is not uniformly distributed. This is supported by the large variations in the density of matter observed throughout the universe. The first thing that would happen with this non uniform distribution would be that gravity would exert its influence over matter. We therefore need to look at the properties and influence of gravity. After all, it was gravity that bound the particles together into the structures we now see in the planets, stars, and galaxies.

In the modern era, Copernicus (1543) suggested that the Sun was the centre of motion. Brahe (1572) made detailed studies of the motion of the planets. His assistant, Kepler (1609, 1619), used Brahe's observations in his work that showed the planets were orbiting the Sun. He determined three laws of planetary motion around the Sun. Their contributions were summarized by Sagan (LOC).

With his construction and use of a telescope, Galileo Galilei (1619, 1632) was able to observe the motions of the planets with greater accuracy. He also observed that Jupiter had four moons orbiting it. His work confirmed the idea that the planets moved around the Sun and that gravity was responsible for those heavenly motions. Together, they firmly established the solar centric model for the motion of the planets.

The first person to give a useable explanation of gravity was Isaac Newton (1687, 1729). He developed his universal law of gravitational attraction. Any two particles with mass would exert an attractive force, F_G, on each other that was proportional to the product of their masses, M_1 and M_2, and inversely proportional to the square of the distance, r, between their centres of mass. It was expressed mathematically as

$$F_G = \frac{GM_1M_2}{r^2} \qquad (8.1$$

where G is Newton's universal gravitational constant. It has been so widely used and so successful that equation 8.1 was regarded as Newton's universal law of gravity. Along with his other laws of motion, equation 8.1 allowed the acceleration due gravity to be calculated, as shown in equation 8.2.

$$g = \frac{GM}{r^2} \qquad (8.2$$

Equation 8.1 and 8.2 were not the full extent of Newton's work on gravity. He also worked out that a planet gravitationally attracted to the sun would orbit the sun along an elliptical path. If the law of attraction was inverse-square, as given in equation 8.2, the ellipse would have its perihelion, its closest point to the sun, and its aphelion, its furthest distance from the sun, always returning to the same place in space. In his Proposition 45, Newton went on to show that, if gravity was stronger than inverse-square, its aphelion and perihelion would occur slightly earlier in each orbit. The ellipse would then regress as the planet orbited the sun. If gravity were slightly weaker than inverse-square, the planet's perihelion and aphelion would advance in the direction of its orbit. The ellipse would then precess as the planet orbited the sun.

Newton's work predicted every gravitational observation for nearly 200 years. Its weakness was that it gave no physical reason for the origins of gravity. Some two hundred years later, it was shown to have an error of about 1.22 x 10^{-8} in its prediction of Mercury's orbit.

Einstein (1915a, 1915b, 1916, 1952c) modified Newton's gravitational work when he derived his gravitational field equations from his general relativity theory. In his theory, Einstein suggested that mass distorted the fabric of space–time, producing gravity. He gave a mathematical description of his calculations through his gravitational field equations. He solved the difference in Mercury's orbit and indicated that Newton's gravity was a first approximation to his theory. He went on to make other predictions, some of which have been verified.

Other mathematicians have extended his calculations to make predictions such as the existence of black holes. Their proponents acknowledge that there is no physics behind their existence. They are a mathematical prediction from a successful theory. Opponents point out that their existence violates other physical laws and breaks down in other mathematical approaches.

Einstein's mathematics were too complex for most mathematicians to follow, let alone a layman. So, although he gave space–time distortion as the cause of gravity, there was never an adequate description of the fabric of space–time or the physics that causes it. In an attempt to shed further light on the origins of gravity, this study uses Newtonian gravity and Einstein's "packets of electromagnetic energy," now referred to as photons, that are given a wave function, Ψ. Those results will be used to add gravity to the matter we have already described. We are then better placed to see how the universe is built. Some of the following work has been published, Robinson (2021).

8.2 Space–Time Outside Matter

Consider the situation of a mass, m, at distance, Δr, above the surface of mass, $M \gg m$. It has a potential energy $mg\Delta r$ for $\Delta r \ll r$, the distance between their centres of mass. The first section of this chapter follows the approach used by Einstein (1911). As derived by (Einstein,

1905, 1952a) photons have mass. That has been experimentally verified by Pound and Rebka, (1960), Pound and Snider (1961), Pound (2000) and used by van der Mark and t'Hooft (2000) and others. Photons have mass, m_p, given by

$$m_p = \frac{hv}{c^2} \tag{8.3}$$

where h is Planck's constant, v is the photon's frequency of oscillation, and c is the speed of light. Photons are unique among all particles in that they are constant velocity particles. When their energy state changes, their mass, and hence frequency, changes.

Substituting equation 8.3 into equation 8.1 shows that mass, M, exerts a gravitational attraction on photons given by

$$F_G = \frac{GM}{r^2}\frac{hv}{c^2} \tag{8.4}$$

A photon moving a distance, Δr, vertically away from M will increase its potential energy since it has mass and moves against gravity, as shown in figure 8.1.

A photon's energy, which is its kinetic energy (KE), is equal to hv. As it gains potential energy it loses kinetic energy. A photon is a constant velocity particle that always travels at c. A change in its kinetic energy therefore results in a change in its frequency, not its velocity.

Consider a photon moving distance, Δr, from r_1 to r_2, against mass M, as in figure 8.1. Its potential energy increase, ΔPE, over distance Δr is given by

$$\Delta PE = \frac{hv}{c^2}.\frac{GM\Delta r}{r^2} \tag{8.5}$$

Conservation of energy means that $KE_2 = KE_1 - \Delta PE$. That gives

$$hv_2 = hv_1 - \frac{hv_1}{c^2}.\frac{GM\Delta r}{r^2} \tag{8.6}$$

Equation 8.6 can be re-arranged and written as

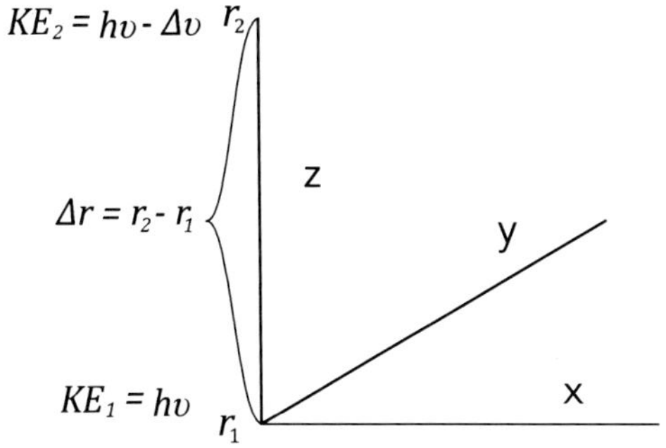

Figure 8.1: *Schematic illustration of a photon's increasing potential energy as it moves distance Δr travelling from r_1 to r_2.*

$$h(\upsilon_2 - \upsilon_1) = h\upsilon_1 . \frac{GM}{c^2}\frac{\Delta r}{r^2} \quad (8.7$$

Re-arranging and dividing both sides of equation 8.7 by *h* gives

$$\frac{\Delta\upsilon}{\upsilon} = -\frac{GM\Delta r}{c^2 r^2} \quad (8.8$$

G, M, and *c* are constants that can be combined as $\alpha = \frac{2GM}{c^2}$. Substituting α into equation 8.8 and setting the limit $\Delta r \to 0$ gives

$$\frac{d\upsilon}{\upsilon} = -\frac{\alpha dr}{2r^2} \quad (8.9$$

Integrating equation 8.9 from r_1 to r_2 gives

$$ln\upsilon_2 - ln\upsilon_1 = -\frac{\alpha}{2r_2} + \frac{\alpha}{2r_1} \quad (8.10$$

or

$$\upsilon_2 = \upsilon_1 \, e^{\left(\frac{\alpha}{2r_2} - \frac{\alpha}{2r_1}\right)} \quad (8.11$$

We can apply equation 8.11 to the frequency change of photons travelling from the sun's surface to Earth, knowing that α for the sun is 2.954 km, the sun's surface radius is 7 x 10^5 km, and Earth's orbital radius is 1.49 x 10^8 km from the sun. Substituting these into equation 8.11 gives a redshift of

$$e^{\frac{2.954}{1.4x10^6}} - e^{\frac{2.954}{2.98x10^8}} \quad (8.11a$$

or 2.1 x 10^{-6}. That is the same value reported by Einstein (1915 a, 1916, 1952 c), and agrees with experimental observation (Adams, 1910, Adam, 1948, Gonzalez Hernandez et al., 2020). That was as far as Einstein took the calculations, however. What follows is an expansion of the above.

Because c is constant for all local observers and $c = \upsilon\lambda$, the relationship between wavelengths and distance can be expressed as

$$\lambda_2 = \lambda_1 \, e^{\left(\frac{\alpha}{2r_2} - \frac{\alpha}{2r_1}\right)} \quad (8.12$$

A change in wavelength is called a redshift and denoted by z. It can be due to either a Doppler effect or gravity, and is given by

$$z = \frac{\lambda_r - \lambda_0}{\lambda_0} \quad (8.13$$

where λ_0 is its original wavelength and λ_r is its wavelength at distance, *r*, from its origin. From equations 8.12 and 8.13, setting $\lambda_1 = \lambda_0$ and $\lambda_2 = \lambda_r$ gives

$$z = e^{\frac{\alpha}{2r}} - 1 \quad (8.14$$

When $r \gg \alpha$, equation 8.14 becomes

$$z = \frac{\alpha}{2r} \quad (8.14a$$

The only way we have of measuring the distance and time associated with distant objects is by using photons. We use Euclidean space–time — well away from any gravitationally attracting mass — as a reference for distance, *r*, and time, *t*. They are assigned distance, r_0, and time, t_0, for distance and time intervals respectively. Newtonian distance, *d*, is determined by $d = n\lambda_0$, the number of wavelengths, *n*, of a particular frequency, λ_0. Equation 8.12 shows that the relationship between instantaneous radial distance, d_r, from a massive object with mass, $M = \frac{\alpha c^2}{2G}$, as

$$d_E = d_N . e^{\frac{-\alpha}{2r}} \qquad (8.15$$

The distance, d_E, is the distance under Einstein's gravity. D_N is the distance under Newtonian inverse square law gravity, otherwise known as the Euclidean distance. Applied to a planet orbiting the sun, equation 8.13 relates the distance, d_E, under Einstein's gravity to the distance, d_N, under Newton's inverse square law. In the same manner, time is defined by $t_0 = n v_0$. Time is measured by the number of oscillations of a known frequency. Equation 8.11 means the relationship between Einstein's relativistic time and Newton's Euclidean time at distance, r, axially away from the centre of mass of a massive object becomes

$$t_E = t_N . e^{\frac{\alpha}{2r}} \qquad (8.15a$$

Equations 8.15 show that distance under Einstein's gravity is equal to Newtonian or Euclidean distances multiplied by $e^{\frac{\alpha}{2r}}$. In the same manner, time under Einstein's gravity is equal to Newtonian or Euclidean time divided by $e^{\frac{\alpha}{2r}}$. The distance, r, applies to the distance from the centre of mass of the object of mass $M = \frac{\alpha c^2}{2G}$. It should be noted that r can only be positive. As such, distance measurements under Einstein's gravity can only be longer than Newtonian time. Einstein's time will therefore always be slower than Newtonian time.

Equations 8.15 can be used independently, as for example equation 8.11a. To use them together we must follow the work of Minkowski (1908, 1952) and Schwarzschild (1916). Minkowski determined that space and time were related through the space–time co-ordinate, s, given by

$$s^2 = c^2t^2 - x^2 - y^2 - z^2 \qquad (8.16$$

Following the work of Schwarzschild, the differential of equation 8.16 was modified to consider the distortion due to mass, giving

$$ds^2 = Fdt^2 - H(dx^2 + dy^2 + dz^2) - J\{xdx + ydy + zdz\}$$

where F, H, and J are functions of $r = \sqrt{x^2 + y^2 + z^2}$. This Cartesian co-ordinate expression was changed to Polar co-ordinates giving

$$ds^2 = Fdt^2 - (H + Jr^2)Jdr^2 - Jr^2(d\theta^2 + sin^2\theta d\varphi^2)$$

The above is the standard change of the differential form of the Cartesian co-ordinates of x, y, and z in equation 8.16 to polar co-ordinates of *r*, θ, and φ. It is a standard mathematical expression.

The distortion of space and time are measured by the variations in the wavelength and frequency of photons respectively. It follows from equations 8.15 that $F = e^{-\frac{\alpha}{r}}$ and $H + Jr^2 = e^{\frac{\alpha}{r}}$ and $J = 1$, yielding

$$ds^2 = dt^2 e^{\frac{-\alpha}{r}} - dr^2 e^{\frac{\alpha}{r}} - r^2(d\theta^2 + sin^2\theta d\varphi^2) \qquad (8.17$$

A Maclaurin/Taylor series expansion of $e^{\frac{-\alpha}{r}}$ yields

$$e^{\frac{-\alpha}{r}} = 1 - \frac{\alpha}{r} + \frac{\alpha^2}{r^2 2!} - \frac{\alpha^3}{r^3 3!} + \cdots \left[\frac{(-\alpha)^n}{r^n n!}\right]$$

For r >> α, a first approximation of equation 8.15 becomes

$$ds^2 = dt^2\left(1 - \frac{\alpha}{r}\right) - \frac{dr^2}{1-\frac{\alpha}{r}} - r^2(d\theta^2 + sin^2\theta d\varphi^2) \qquad (8.18$$

Equation 8.18 is one of the accepted formats of the Schwarzschild solution to Einstein's field equations, see for example Droste (1917) and Misner et al. (1973). It is called the strong field solution, predicting the existence of black holes with an event horizon within which emitted light cannot escape.

Other accepted alternative equations include c^2 associated with the dt^2 term and a reversal of polarity. Exact solutions from Minkowski's space–time continuum should include c^2. Normalization of constants to 1 is often used in mathematical physics. The reversal of signs is associated with calculations from different directions.

8.3. Calculating Some Effects

The principle of Newtonian gravity altering the wavelength and frequency of photons can be used to calculate the effects of Einstein's gravity, commonly referred to as general relativity effects. Equation 8.11 demonstrated how the change in photon frequency results in a change in time between the sun and earth. That change in frequency arises due to the distortion of time by gravity. Two other effects are gravitational lensing, where light rays are bent by gravity, and the precession of Mercury's perihelion as it orbits the sun.

8.3.i. Gravitational lensing

Figure 8.2 shows the passage of a photon travelling past a massive object of Schwarzschild radius α, making its closest approach at distance r_0 from the centre of mass.

At distance r from the centre of mass, O, it will experience an instantaneous space distortion $d\Delta_{dist} = \frac{\alpha}{2r}$. The total space distortion it experiences is given by

$$\Delta_{dist} = \int_{-\infty}^{+\infty} \frac{\alpha}{2r} dr.$$

Since $r = \frac{r_0}{cos\theta}$, that becomes

$$\Delta_{dist} = \int_{-\frac{\pi}{2}}^{+\frac{\pi}{2}} \frac{\alpha cos\theta}{2r_0} d\theta = \frac{\alpha}{r_0} \qquad (8.19$$

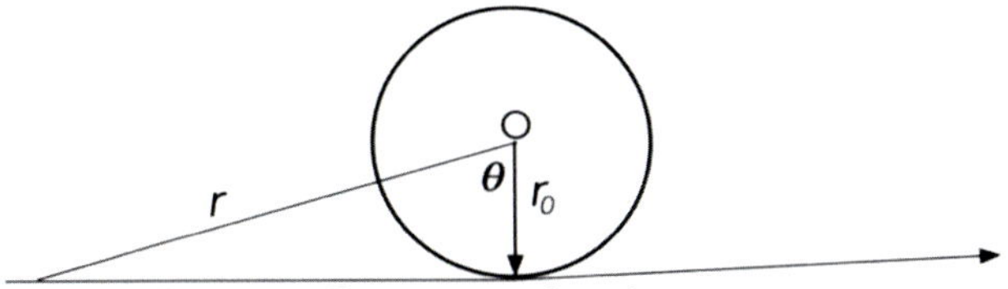

Figure 8.2: *Schematic illustration of a photon passing a massive object centred on O, with its distance of closest approach being r_0.*

Equation 8.17 shows that when there is a change in space distortion, there is a corresponding change in time distortion. For small distortions, i.e., $r >> \alpha$, that distortion is of equal magnitude and opposite sign. The total space–time distortion experienced by the photon is double that given in equation 8.19. That gives

$$\Delta s = \frac{2\alpha}{r_0} = 4z \qquad (8.19a$$

For the sun, α = 2.954 km and r_0 = 700,000 km. Inserting these into equation 8.19a gives Δs = 4.22 x 10^{-6} radian = 1.74 arc sec. That agrees with Einstein's (1911) prediction and matches the observations made by Eddington (1920) and Dyson, et al. (1920).

Equations 8.19 apply for distortions around masses that do not distort space–time around the observer. Figure 8.3 shows that the space–time distortion at the sun's surface is ≈ 2.1 x 10^{-6} and at Earth it is ≈ 0.987 x 10^{-8}. For observations of light ray trajectories altered by the sun, equation 8.19a should be changed to $\Delta s = 4z_s - 2z_E$, where z_s is the sun's redshift at its surface and z_E is its redshift at Earth. This small difference changes the observed deflection from 1.74 arc sec to 1.73 arc sec.

It should be noted that $\frac{\alpha}{2r_0}$ is the redshift of photons from their distance of closest approach to the massive object. The mathematics of gravitational lensing is that the total deflection, in radians, of a photon passing close to a massive object is four times the redshift of photons at their position of closest approach to the centre of mass.

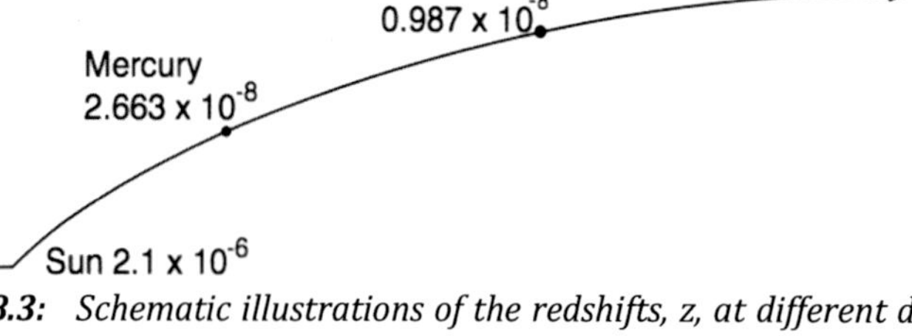

Figure 8.3: *Schematic illustrations of the redshifts, z, at different distances from the sun.*

8.3.ii. Mercury's Anomalous Orbital Precession

Observed from Earth, Mercury's total orbital precession is 5,600 arc seconds per century (as/c) (Matzner, 2001). Newtonian mechanics explained all but about 43 as/c. Einstein (1915b) explained this difference in his equation, $\varepsilon = 24\pi^3 \frac{a^2}{T^2c^2(1-e^2)}$. More accurate measurements now give his prediction of Mercury's anomalous precession as 42.98 as/c.

In Einstein's work, equations 8.14a and 8.15 indicates that Newtonian distances become relativistic distances when multiplied by (1+z), where z is the gravitational redshift produced by the Sun at the distance of interest. Figure 8.3 schematically illustrates the sun's redshift at different distances from its centre of mass. Those figures were calculated from equation 8.14, using the data given in table 8.1.

Mercury's corrected orbit radius becomes $r_M(1 + z_M)$. That increase in radius can only occur if the Sun's gravitational attraction is less than inverse square.

It means that Mercury has to travel a greater distance to complete an orbit from perihelion to perihelion. That increased circumference is given by $\Delta_{Circ} = 2\pi r_M(1 + z_m) - 2\pi r_M = 2\pi r_M z_M$. It is the extra distance Mercury travels to complete an orbit.

Physical Property	**Value**
Speed of light, ***c***	2.9979×10^5 km/sec
Newton's universal gravitational constant, ***G***	6.6743×10^{-11} Nm^2/kg^2
Sun's mass, *M*	1.989×10^{30} kg
These values of *c*, *G* and *M* give	2.954 km
Mercury's semi major axis, r_0	57.91×10^6 km
Mercury's orbital eccentricity, *e*	0.2056
These values give redshift at Mercury orbit, z_M	2.663×10^{-8}
Earth orbital radius	149.6×10^6 km
These values give redshift at Earth orbit, z_E	0.987×10^{-8}
Mercury orbital period, M_{Orb}	88 Earth days
Earth orbital period, E_{Orb}	365.25 Earth days

Table 8.1: *The physical values needed to calculate Mercury's anomalous orbital precession.*

Its angular precession per orbit, ε_M, is given by $r_M . \varepsilon_M = \Delta_{Circ,}$ or:

$$\varepsilon_M = 2\pi z_M \qquad (8.20$$

Mercury's orbital period is 88 Earth days. In 100 earth years, it will make 415 orbits around the Sun. During that time, Earth makes 100 orbits around the Sun. From Earth, Mercury will only be seen to make 315 orbits. As observed from Earth, equation 8.20 becomes

$$\varepsilon_M = 2\pi(z_M - z_E)\left(1 - \frac{M_{Orb}}{E_{Orb}}\right) \text{ radian/orbit} \qquad (8.21$$

Inserting the appropriate values from table 8.1, converting from radians to arc seconds and multiplying by the number of Mercury orbits per Earth century gives $\varepsilon_M = 42.99\ as/c$. The most up to date figures applied to Einstein's calculations gave 42.98 as/c. There are suggestions that the value observed by Park et al. (2017) differs from the predicted value by ≈1.2 as/c, a value considerably larger than the 0.01 difference between this approach and Einstein's approach.

This indicates that the approach in this model explains the physics that underpins Einstein calculations. Namely, mass distorts distances, increasing lengths by the redshift, z, of photons. It also distorts time by decreasing photon frequency by redshift. These are both indicated in equations 8.15. It is suggested this calculation is easier to follow than the derivations by Einstein and others, and it has the added advantage that it can be extended into other situations.

Einstein's theories are called relativity because all measurement results depend upon the position of the observer. There is no absolute reference point or frame. His theories enable observers in different places and reference frames to obtain the same results when the speed of light is constant for all observers.

Equation 8.21 can be modified to obtain the general equation for Mercury's anomalous precession per orbit for observers at different positions within the solar system. It becomes

$$\varepsilon_M = 2\pi(z_M - z_O)\left(1 - \frac{M_{Orb}}{O_{Orb}}\right) \text{radian/orbit} \qquad (8.22$$

where z_0 is the Sun's redshift at the observer's position and O_{Orb} is the orbital period of the observer around the Sun.

Table 8.2 gives the expected results for some different positions associated with the solar system. These results are additional to the 532.2 as/c due to the precession caused by the gravitational pull of the other planets on Mercury.

Equation 8.22 can be modified to obtain the general equation for Mercury's anomalous precession per orbit for observers at different positions within the solar system. It becomes

$$\varepsilon_M = 2\pi(z_M - z_O)\left(1 - \frac{M_{Orb}}{O_{Orb}}\right) \text{ radian/orbit} \qquad (8.23$$

where z_0 is the Sun's redshift at the observer's position and O_{Orb} is the orbital period of the observer around the Sun.

Notable features of table 8.2 are that an observer from beyond the solar system would see Mercury's anomalous orbit precess an additional 90 arc seconds when they had seen Earth make 100 complete revolutions around the Sun.

Observers accelerated in retrograde Earth orbit and held in place by a large solar sail, would see Mercury's anomalous orbit precess by 56.63 arc seconds per Earth century.

Observer Position	z_0 (x 10 $^{-8}$)	$\frac{M_{Orb}}{O_{Orb}}$	**ε/Mercury Orbit** (x 10^{-8})	**Per Earth Century** (arc sec)
Flat Minkowski space-time	0		2π x 2.663	89.99
Stationary at Earth Orbit	0.987		2π x 1.676	56.63
Earth Orbit	0.987	0.241	2π x 1.676x.759	42.99
Mars Orbit	0.648	0.128	2π x 2.015x0.872	59.38
Saturn Orbit	0.103	0.0082	2π x 2.562x0.9918	85.86

Table 8.2: *The precession of Mercury's orbit in arc sec per Earth century as measured from different positions. Earth Orbit was calculated from equation 8.23.*

The further away the observer is from the Sun, the larger the observed anomalous precession will be. Table 8.2 demonstrates that observations are relative to the position of the observer. In the same manner, the general equation for the precession of a satellite orbiting the sun and measured from Earth, in radians per satellite orbit, becomes:

$$\varepsilon = 2\pi(z_S - z_E)\left(1 - \frac{S_{Orb}}{E_{Orb}}\right) \qquad \text{(8.23a}$$

where z_S is the sun's redshift at the satellite's corrected orbit, its semi major axis, and S_{Orb} is its orbital period. The closer the satellite orbits the sun, the greater are other effects such as the sun's oblateness and its radiation and solar wind.

The general expression for the precession, in radians per orbit, of any object about a mass M external to the solar system is given by

$$\varepsilon = 2\pi\left(\frac{GM}{c^2(1-e^2)r_0}\right) = 2\pi\frac{\alpha_M}{2r_{OC}} = 2\pi z_{OC} \qquad \text{(8.24}$$

where r_0 is the orbiting body's semi major axis, α_M is the massive body's Schwarzschild radius, and z_{OC} is its redshift at the orbiting body's corrected semi major axis, r_{OC}.

In his Principia, Newton predicted that, if gravity is exactly inverse square, a planet would prescribe an exact ellipse around the Sun, with its aphelion and perihelion always returning to the same points, as shown in figure 8.4A.

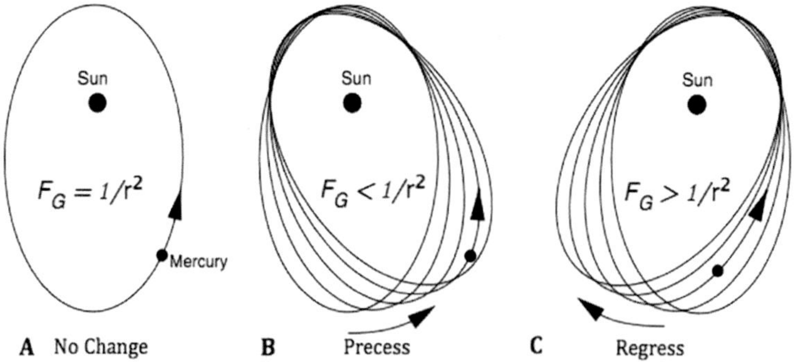

***Figure 8.4** Schematic illustrations of change in planetary elliptical orbits depending upon the nature of the attraction.*

If gravity were weaker than inverse square, the planet travels further away from the Sun and under the influence of a weaker force. That greater distance causes it to return to its aphelion and perihelion positions a little later. It will therefore precess in its direction of orbit, as shown in figure 8.4B.

If gravity were stronger than inverse square, the planet would orbit a little closer to the Sun. It would have to travel a shorter distance under the influence of a stronger force to return to its aphelion and perihelion positions. They would be reached a little earlier, and its perihelion would regress in the opposite direction to its direction of orbit, as shown in figure 8.4C.

It should be noted that the accepted Schwarzschild metrics, equation 8.18 and its variants, all contain the term $1 - \frac{\alpha}{r}$. They are recognized as the strong field solution to Einstein's gravitational field equations. They all predict that gravity will be stronger than inverse square. According to the mechanism worked out by Newton and presented above, it is not possible for an orbit's perihelion to precess in its direction of travel if gravity is stronger than inverse square. This poses a problem for those who accept the Schwarzschild metric as an exact solution to Einstein's gravitational field equations.

Einstein did not calculate Mercury's total orbit and derive the difference between his and Newton's calculations to be 43 as/c. He calculated only the difference his theory introduced to Newton's calculations. The approach used in this derivation, based on photon redshift, is similar. It starts with Newton's gravity and uses it to calculate its effect on photons. It shows the new orbital distance Mercury has to travel to complete a perihelion-to-perihelion orbit is $2\pi r_M(1+z)$. Mercury's Newtonian orbit of $2\pi r_M$ is subtracted to arrive at its anomalous angular precession of $2\pi z = 42.99\ as/c$. Both approaches have a close relationship to Newton's calculations.

Even if Mercury's orbit were circular, this difference would have been picked up because changes in its orbit were first determined by its transits across the Sun between 1677 and 1881 (Le Verrier, 1859; Newcomb, 1882). The elliptical orbit calculation makes it easier to recognize precession in other orbits. A change in orbital radius of a distant object would not be recognized.

8.4. Comments on Einstein's Calculations

The majority of Einstein's 1916 paper on "The Foundations of the General Theory of Relativity" is taken up with tensor studies, the purpose of which includes establishing the relationship between the four-dimensional space–time co-ordinates under the influence of gravity. Whether tensors clarify or confuse the reader depends upon the reader's understanding of them.

The above redshift calculations have matched Einstein's predictions of photon redshift, gravitational lensing, and orbital precessions. They indicate that the process by which mass distorts space–time is through changes in the wavelength and frequency of photons. It becomes a philosophical question as to whether mass distorts space–time to produce gravity that distorts photons, or whether gravity is produced by mass distorting photon wavelengths, which equates to space–time distortion. In either event, the calculations remain the same.

The redshift metric of equation 8.17 and its associated calculations do not match predictions under the Schwarzschild metrics of equation 8.19 and its equivalents. The latter predicts that gravity is stronger than expected under Newton's inverse square law. At the same time, the calculations for equation 8.21 show that, for Mercury's orbit to precess, gravity must be weaker than inverse square. That was predicted by Newton. Did Einstein therefore make a mistake in his calculations?

With many people not understanding the complexity of his work, some suggest he did. In his publication on The Foundations of the General Theory of Relativity, Einstein introduced approximations. After deriving his field equations, (47), he points out that "*There is only a minimum of arbitrariness in the choice of these equations*". He goes on to explain the reasons for the equations not being exact, and in that passage, he acknowledges his use of approximations.

In section 22 of that paper, Einstein (1916, 1952c) derives that $g_{11} = -\left(1 + \frac{\alpha}{r}\right)$. From this it follows that, when $r \gg \alpha$, $dx = \frac{1}{\left(1+\frac{\alpha}{2r}\right)}$. He indicates that "*... correct to a first order of small quantities* $dx = 1 - \frac{\alpha}{2r}$". When $\frac{\alpha}{2r} \approx 3 \times 10^{-8}$, see figure 8.3, that is a valid approximation. Equations 8.14 show that $\frac{\alpha}{2r} = z$.

After his equation (71), Einstein states *"The unit measuring rod thus appears a little shortened in relation to the system of co-ordinates by the presence of the gravitational field, if the rod is laid along a radius"*. That could appear to conflict with equation 8.15, which points out that relativistic lengths are longer than Newtonian lengths by a factor of (1+z). In its full context, it *"appears a little shortened"* by (1–z) because the background against which it is observed is lengthened by (1+z).

There is nothing in and around Einstein's equation (71) that could be construed as gravity being stronger than inverse square, thus shortening lengths.

Additionally, the present work agrees with his calculation that lengths are increased by (1+z). Both approaches find that, *"correct to a first order of small quantities"*, mass increases length by the redshift, z. This can only occur if gravity is weaker than inverse square, a pre-requisite condition for the precession of an orbiting body's perihelion.

Apart from his approximations, it could be suggested that there is one other problem associated with his work. It appears so complex that most of those who try to follow it are confused. That is not, however, an error on Einstein's part. His approximations appear to have caused problems for those who have attempted to provide exact solutions to his field equations.

Two examples of field equations are given below, Einstein's original, his equation (47), and a modern Ricci tensor format, equation 25.

$$\left.\begin{array}{l} \dfrac{\partial \Gamma^{\alpha}_{\mu\nu}}{\partial x_{\alpha}} + \Gamma^{\alpha}_{\mu\beta}\Gamma^{\beta}_{\nu\alpha} = 0 \\ \sqrt{-g} = 1 \end{array}\right\} \qquad \text{E (47}$$

$$R_{\mu\nu} - 0.5Rg_{\mu\nu} = \frac{8\pi G}{c^4} T_{\mu\nu} \qquad (8.25$$

Their generally accepted solutions take one of the variations in equation 8.18. The problem posed by these solutions is that they are exact solutions to the **approximations** mentioned above. **Exact** solutions to approximations are still **approximations**.

8.5. Comparing the Different Gravitational Predictions

The above has shown that Mercury's anomalous orbital precession is caused by gravity being weaker than inverse square. This extends the circumferential distance Mercury has to travel further to return to its perihelion position in each orbit. That additional distance is determined by the redshift, z. Under Newtonian gravity, an object of mass, m, at a distance, r, from the centre of mass of a body of mass, M, experiences a gravitational attraction $F_G = \frac{GMm}{r^2}$ and travels a distance $2\pi r$ to make one complete revolution. For the circumference to increase from $2\pi r$ to $2\pi r(1 + z)$, the gravitational attraction must decrease to

$$F_G = \frac{GMm}{[r(1+z)]^2} \qquad (8.26$$

Inserting equation 8.14 into equation 8.26 gives

$$F_G = \frac{GMm}{r^2 e^{\frac{\alpha}{r}}} \qquad (8.27$$

from which the acceleration due to gravity becomes

$$g_z = \frac{GM}{r^2 e^{\frac{\alpha}{r}}} \qquad (8.28$$

The gravitational field equations derived by Einstein, plus the variants by others following his work, are general equations from which the gravitational acceleration due to the presence of the mass, M, can be calculated. Equation 8.28 gives the direct expression for the acceleration due to gravity at different distances, r, under this redshift study. In the same manner, equation 8.2 gives the acceleration due to gravity for any distance, r, under Newtonian gravity.

Equations 8.2 and 8.28 enable the direct calculation of the gravitational field strength at any distance, r, from the centre of mass of any object of any mass, M. Using the approximations introduced by Einstein, the gravitational force and acceleration due to gravity can be determined for the Schwarzschild metric by replacing r with $r(1 - z)$. Einstein's approximations mean it is only accurate for $r \gg \alpha$, and even then, only if no other errors are introduced. The equations for the three different theories are shown in table 8.3.

The terms for the acceleration due to gravity can be used to calculate the strength of gravity at any radial distance, r, from the centre of mass. Table 8.3 has the advantage over gravitational field equations in that the values can be calculated for any value of M and r using simple pocket calculators. Examples of some calculations for M and r expressed in terms of α, the Schwarzschild radii, are given in table 8.4. The results are plotted in figure 8.5.

The first column shows various distances expressed as a function of $\frac{r}{\alpha}$. The other columns show the results of the relative gravitational strengths under the three theories described in table 8.3. In all cases, the stated results are for theories described in table 8.3. In all cases, the stated results are for bodies moving directly towards the centre of mass of M. The actual accelerations are determined by multiplying the calculated values by GM. At $r = 10\alpha$, the differences between the redshift and Schwarzschild results are within about 10% of the Newtonian acceleration.

At $r \approx 5\alpha$, the accelerations of the three metrics begin to differ significantly. Newtonian acceleration tends to infinity as $r \longrightarrow 0$. Gravitational acceleration under the Schwarzschild metric tends to infinity at $r = \alpha$. Both are as expected.

Gravity Theory	Acceleration due to Gravity
Redshift	$g_Z = \frac{GM}{[r(1+z)]^2} = \frac{GM}{r^2 e^{\frac{\alpha}{r}}}$
Newton	$g_N = \frac{GM}{r^2}$
Schwarzschild	$g_S = \frac{GM}{[r(1-z)]^2} \approx \frac{GM}{\left(1 - \frac{\alpha}{r}\right) r^2}$

Table 8.3: *The gravitational attraction and accelerations due to gravity for different distances from the centres of mass of objects of any mass M.*

Distance	**Redshift**	**Newton**	**Schwarzschild**
$\frac{r}{\alpha}$	$g_Z \propto \frac{1}{r^2 e^{\frac{\alpha}{r}}}$	$g_N \propto \frac{1}{r^2}$	$g_S \propto \frac{1}{\left(1-\frac{\alpha}{r}\right) r^2}$
10	9.048 x 10^{-3}	1.00 x 10^{-2}	1.11 x 10^{-2}
9	1.101 x 10^{-2}	1.23 x 10^{-2}	1.263 x 10^{-2}
8	1.377 x 10^{-2}	1.56 x 10^{-2}	1.783 x 10^{-2}
7	1.899 x 10^{-2}	2.04 x 10^{-2}	2.380 x 10^{-2}
6	2.353 x 10^{-2}	2.78 x 10^{-2}	3.366 x 10^{-2}
5	3.618 x 10^{-2}	4.00 x 10^{-2}	5.000 x 10^{-2}
4	4.867 x 10^{-2}	6.25 x 10^{-2}	8.333 x 10^{-2}
3	7.963 x 10^{-2}	0.111	0.167
2	0.1516	0.250	0.500
1.5	0.2282	0.444	1.337
1.0	0.3679	1.000	∞
0.9	0.4062	1.234	
0.8	0.4475	1.562	
0.7	0.4890	2.041	
0.6	0.5245	2.778	
0.5	0.5413	4.000	
0.4	0.5130	6.250	
0.3	0.3956	11.11	
0.2	0.1648	25.0	
0.1	4.54 x 10^{-3}	100	
0.05	8.35 x 10^{-7}	400	

Table 8.4: *The accelerations due to gravity for different distances under the different theories, with distances expressed as functions of the Schwarzschild radius* α*.*

Gravitational acceleration under this redshift metric gets stronger closer to the origin, although it is weaker than inverse square and Schwarzschild acceleration. At $r < 0.5\alpha$, space–time distortion, given by $e^{\frac{\alpha}{r}}$, dominates the inverse square law, and gravity becomes weaker as r tends further towards 0.

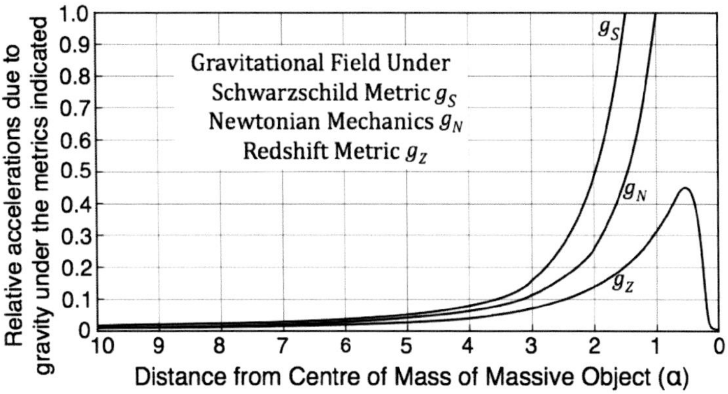

Figure 8.5: *Plots of the relative accelerations due to gravity under the different equations given in table 3.*

This may seem like a surprising result. It should be considered from the perspective of Newton's shell theorem. A body anywhere inside a spherical shell of uniform density and any thickness will not experience any gravitational attraction from the shell surrounding it. When applied to a large body like the Sun, it means that there will be no gravitational attraction at the Sun's centre, even though the pressure will be high. This simply means that mass has distorted space and time so much that gravity has become very weak.

Table 8.3 offers a simple way of determining the acceleration due to gravity under any of the three theories. It requires knowledge of the mass, M, of the attracting body to calculate its Schwarzschild radius, α, and the distance, r, from the centre of mass to the position where it is desired to determine the acceleration. Figure 8.5 displays the strengths of these gravitational fields in one dimension. To appreciate the field in the three spatial dimensions, it is necessary to rotate the curves.

Figures 8.6 show a two-dimensional slice through a three-dimensional representation of the accelerations due to gravity under the different theories. It was obtained by rotating the curves in figure 8.5 through 180° about the horizontal axis.

That was followed by rotating the compound curves about the vertical axis. They show the accelerations due to gravity on the vertical axis against distance from the centre of mass on the horizontal axis.

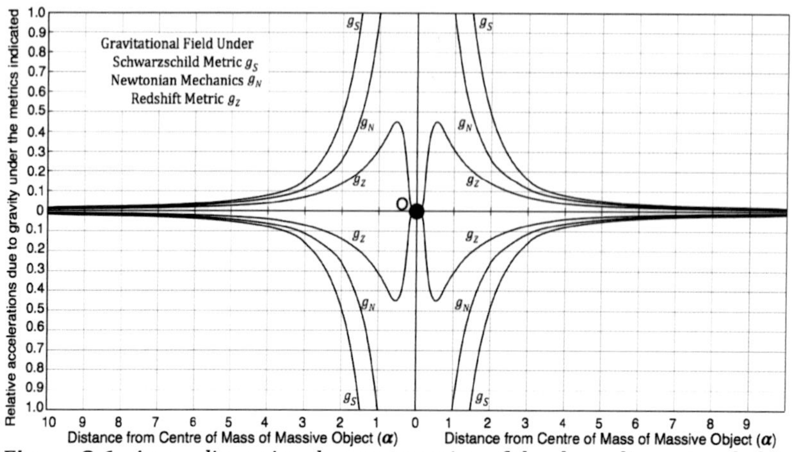

Figure 8.6 *A two-dimensional representation of the three-dimensional plots of the relative accelerations due to gravity under the different equations given in table 8.3.*

The vertical axis thus represents the distortions in space–time caused by the mass at centre, O. What are called the distortions of space–time are the accelerations due to gravity at that point in space. Any particle that experiences a distortion in space–time will be accelerated by that distortion in the direction of increasing distortion.

It is suggested that the g_S equation and its curve provide a better understanding of the nature of the Schwarzschild metric and its 'gravitational infinity' than do field equations such as those in equations 8.29 and 8.30. It is irrelevant because the Schwarzschild metric is not accurate.

$$G_{\mu v} + g_{\mu v}\Lambda = \frac{8\pi G}{c^4}T_{\mu v} \tag{8.29}$$

$$R_{\mu v} - 0.5Rg_{\mu v} = \frac{8\pi G}{c^4}T_{\mu v} \tag{8.30}$$

As mentioned earlier, the approximations Einstein used in developing his field equations mean that, in the absence of other errors, the Schwarzschild metric equations are only valid for $r \gg \alpha$. This is the conditions under which $1-\frac{\alpha}{r}$ is observationally indistinguishable from $\frac{1}{\left(1+\frac{\alpha}{r}\right)}$.

As is shown in section 8.7, other errors were introduced in the accepted solutions to Einstein's field equations. They result in the Schwarzschild metric calculations being mathematically incorrect and physically impossible. The failure of Newton's inverse square law to make accurate predictions leaves the redshift metric as the best representation of gravitational effects for space outside matter.

Table 8.4 and figures 8.5 and 8.6 indicate that the gravitational field strengths of the redshift metric are not much different from Newtonian mechanics for $r > 5\alpha$. As such, reasonably accurate predictions of reshift gravitational values can be made using Newtonian mechanics, with a slight end correction for the redshift metric equation.

Einstein's gravitational field equations and their variants did not allow mathematicians to calculate the strength of gravity at any point in space. They were calculations of the distortion of space–time away from Euclidean space and time. That distortion had to be added to Newton's inverse square law to calculate the strength of the gravitational field. The equations in table 3 allow direct calculations of the field.

To understand the effects of the redshift gravity, consider the situation of a ball rolling along the redshift gravity profile at the bottom of figure 8.6. It will be subject to stronger gravitational attraction as it gets close to $r = 0.5\alpha$. It will reach a maximum speed at that distance. From there it will continue on, rising against the gravitational attraction. If it is captured by the centre, O, it will stay put. Otherwise, it will roll back with increasing speed. The same applies to particles under the influence of this redshift metric.

The approximate speed an incident particle will reach as it approaches $r = \alpha$ can be determined by the reverse of the escape velocity. The Newtonian equation for the escape velocity, V_{e_N}, for distance, *r*, from a gravitationally attracting mass, *M*, is given by

$$V_{e_N} = \sqrt{\frac{2GM}{r}} = \sqrt{\frac{\alpha c^2}{r}} \qquad (8.31$$

Adjusting for the redshift metric gives

$$V_{e_Z} \approx \sqrt{\frac{\alpha c^2}{re^{\frac{\alpha}{2r}}}} \qquad (8.31a$$

Setting $r = \alpha$, equations 8.31 give the escape velocity for a particle at the Schwarzschild radius of $\approx 0.78c$. That also means that particles falling inward under the influence of gravity could achieve a velocity up to $\approx 0.75c$. That shows relativistic velocities can be achieved if incoming particles do not collide with other particles before reaching $r = \alpha$. Electrons reaching speeds $\approx 0.75c$ have to be accelerated through $\approx$ 1 MeV. That corresponds to temperatures in the vicinity of $\approx 10^{10}$°K. These figures depend upon the object being sufficiently massive that its Schwarzschild radius is somewhat larger than double its physical radius.

Particles do not travel along a radius as they approach the centre. Their different trajectories will cause them to collide with other particles close to the centre. With those collisions and other competing factors, it is difficult to determine a definite value for the speed of particles that reach $r = 0.5\alpha$. The best that can be indicated is that they will be relativistic. Collisions will result in significant energy release. These could correspond to temperature up to $\approx 10^{10}$°K. Without further corrections for relativistic effects and the distribution of mass at a radius of 0.5α, that energy release will be concentrated around the radius $r = 0.5\alpha$.

The maximum gravitational strength at $r = 0.5\alpha$ is not a solid surface. Particles moving through it from outside would be attracted back to it if they were not captured by the central mass. Particles moving away from the centre would also be attracted back towards $r = 0.5\alpha$. That would result in the accumulation of high energy particles in the vicinity of $r = 0.5\alpha$. They would move at relativistic velocities in random directions. Rapidly moving particles could move outside that radius. From $g_Z = \frac{GM}{r^2 e^{\frac{\alpha}{r}}}$, we can replace GM by $0.5\alpha c^2$ and r by 0.5α to get $g_Z = \frac{2c^2}{\alpha e^2}$. That attraction is sufficiently strong that particles with relativistic speeds would be reduced to classical speeds at $r > \alpha$. In the absence of other effects, it appears unlikely that high energy collisions would generate a significant signal at $r > \alpha$.

Figure 8.7, a reproduction of g_z taken from figure 8.6, shows the expected result would be the emission of high energy radiation in a torus shape if the incoming particles were preferentially aligned.

The insert shows the expected extent of the emissions from within the torus, when viewed perpendicularly to its preferential plane. Its predicted central radius is at $r \approx 0.5\alpha$. The intensities are entirely due to the emissions. They do not require any light bending around the massive object. A feature of the escape velocity equations and figure 8.7 is that particles can attain high energies, equivalent to temperatures up to $\approx 10^{10}$°K, without the need for the strong gravity predicted by the Schwarzschild solution.

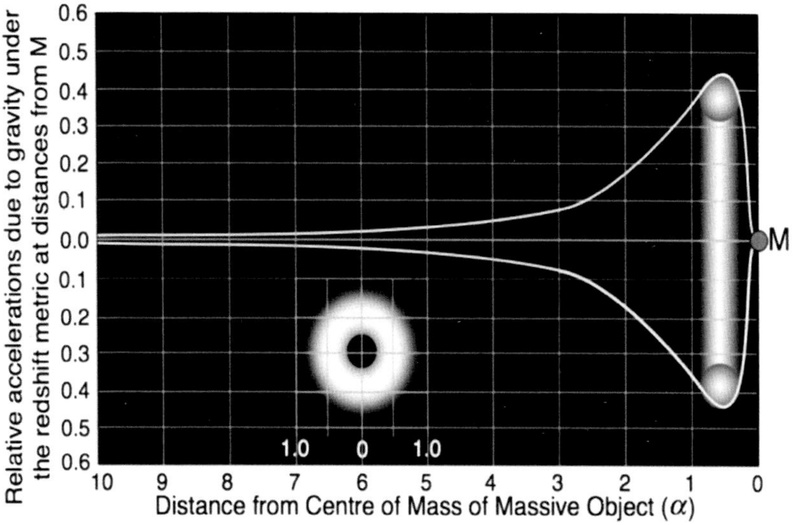

Figure 8.7: *A two dimensional representation of the relative accelerations due to gravity under the redshift metric. The insert shows the expected torus distribution of radiation emitted from an accretion disk that would accumulate around* $r \approx 0.5\alpha$.

8.6. Exact Solution to Einstein's Gravity Field Equations

There are several formats to Einstein's gravitational field equations. They all provide solutions that contain the term $\left(1-\frac{\alpha}{r}\right)$.

Equation 8.18 was derived as a first approximation to an exact solution to the redshift of photons caused by the presence of a gravitational field. Other derivations of equation 8.18 and its variants are complex. The first solution to Einstein's field equations was provided by Schwarzschild (1916). He introduced an approximation as he derived:

$$ds^2 = -dt^2\left(1-\frac{\alpha}{R}\right)+\frac{dr^2}{\left(1-\frac{\alpha}{R}\right)}+R^2(d\theta^2+sin^2\theta d\varphi^2)$$

$$R = \sqrt[3]{r^3+\alpha^3} \qquad (8.32$$

When $r \gg \alpha$, equation 8.32 merges into equation 8.18. Equation 8.32 did not predict a singularity at $r > 0$. It did not predict black holes, in which Einstein never believed. It was still an approximation and was rejected in favour of "exact" solutions.

The equation below is Einstein's equation 53.

$$\frac{\partial}{\partial x_a}\Gamma^{\alpha}_{\mu\nu}+\Gamma^{\alpha}_{\mu\beta}\Gamma^{\beta}_{\mu\alpha} = -K\left(T_{\mu\nu}-0.5g_{\mu\nu}\right) \qquad E(53$$

Like equations 8.29 and 8.30, it contains the terms $g_{\mu\nu}$. All "exact" solutions by different authors have the term $\left(1-\frac{\alpha}{r}\right)$. The mode of derivation of the solution from the various equations is not as important as the origins of the term $\left(1-\frac{\alpha}{r}\right)$. That can be determined by evaluating the term $g_{\mu\nu}$. It is common in gravitational field equations and is shown in equation 8.33.

$$g_{\mu\nu} = \begin{matrix} g_{11} & g_{12} & g_{13} & g_{14} & = & g_{xx} & g_{xy} & g_{xz} & g_{xt} \\ g_{21} & g_{22} & g_{23} & g_{24} & = & g_{yx} & g_{yy} & g_{yz} & g_{yt} \\ g_{31} & g_{32} & g_{33} & g_{34} & = & g_{zx} & g_{zy} & g_{zz} & g_{zt} \\ g_{41} & g_{42} & g_{43} & g_{44} & = & g_{tx} & g_{ty} & g_{tz} & g_{tt} \end{matrix} \quad \text{(8.33}$$

The numbers 1, 2, and 3 in $g_{\mu\nu}$ represents the orthogonal directions x, y, and z respectively. The number 4 represents time, t. The symbol g represents aspects of the gravitational field. For example, g_{11} represents its rate of change in x direction with respect to x. g_{12} represents its rate of change in the x direction with respect to y. g_{14} represents its rate of change in the x direction with respect to time, t, and so on.

Just before his equation (71), Einstein derived $g_{11} = -\left(1 + \frac{\alpha}{r}\right)$. In radial travel towards or away from the centre of mass, g_{11} can be used as the measure of the change of the g factor towards or away from the centre of mass. Because that change is due to the change in the photon's wavelength, λ, it follows that g_{44}, which is due to the change in frequency, must be the negative inverse of g_{11}. That makes:

$$g_{44} = \frac{1}{\left(1+\frac{\alpha}{r}\right)} \quad \text{(8.34}$$

Einstein noted that "$-1 = g_{11}dx_1^2$". Setting $dx_1 = dx$, because x is the only direction being considered, it follows that $dx = \sqrt{\frac{1}{\left(1+\frac{\alpha}{r}\right)}}$. As an approximation, when $r \gg \alpha$, that gives $dx = 1 - \frac{\alpha}{2r}$, which is what Einstein stated in his equation 71. Near his equation 70, Einstein stated, "*... to the first approximation ...* $g_{44} = 1 - \frac{\alpha}{r}$". Again, because of equation 8.34, that is a good first approximation when $r \gg \alpha$.

Table 8.5 lists the terms as derived by Einstein, as he approximated them, and as others incorporated them into the Schwarzschild metrics. Einstein's derivations involving the term $\left(1 + \frac{\alpha}{r}\right)$ were exact within the approximations he used. It appears his approximations involving $\left(1 - \frac{\alpha}{r}\right)$ were incorporated into what are now

Term	Derived by Einstein	Einstein's Approximation	Used in Schwarzschild Solutions
g_{11}	$-\left(1+\frac{\alpha}{r}\right)$	$-\frac{1}{\left(1-\frac{\alpha}{2r}\right)^2} \approx -\frac{1}{\left(1-\frac{\alpha}{r}\right)}$	$-\frac{1}{\left(1-\frac{\alpha}{r}\right)}$
g_{44}	$\frac{1}{\left(1+\frac{\alpha}{r}\right)}$	$\left(1-\frac{\alpha}{r}\right)$	$\left(1-\frac{\alpha}{r}\right)$

Table 8.5: *The terms g_{11} as derived by Einstein, as he approximated and as used in the Schwarzschild metrics.*

called the Schwarzschild metrics. It is believed that this led to the "strong field" solutions of Einstein's field equations.

All forms of the Schwarzschild metric derived solutions use the term $\left(1-\frac{\alpha}{r}\right)$. As shown in tables 8.3 and 8.4 and figures 8.5 and 8.6, this choice can only produce a gravitational attraction that is stronger than inverse square law. As shown in figure 8.4, this cannot give the observed orbital precession.

In these calculations, g_{11} was referenced to the *x* direction. In a radially symmetric situation, there is nothing special about the direction of x. The other orthogonal directions, y and z would have gravitational fields of the same strength as g_{11} at the same radius, r. Figure 8.8 is a schematic illustration of that effect.

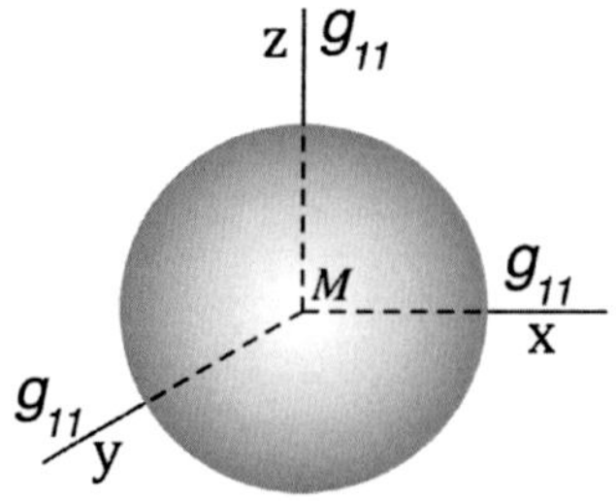

Figure 8.8: *Schematic illustration of how gravity in the x direction has an equal value in the orthogonal y and z directions.*

To determine the full metric, it is necessary to label the x direction as r, convert the y and z contributions to r, θ and φ and add them. When that is done, the general expression for the space–time metric becomes

$$ds^2 = g_{44}dt^2 + |g_{11}dr^2 + r^2(d\theta^2 + sin^2\theta d\varphi^2)| \quad (8.35$$

The g_{11} and r^2 terms are added because they deal with the spread of the gravitational field through all three orthogonal directions. Because g_{11} is negative, equation 8.35 becomes

$$ds^2 = \frac{dt^2}{\left(1+\frac{\alpha}{r}\right)} - dr^2\left(1+\frac{\alpha}{r}\right) - r^2(d\theta^2 + sin^2\theta d\varphi^2) \quad (8.36$$

It is herein designated as the Einstein metric.

Einstein's gravitational work has led to three gravitational metrics. His 1916 paper on "The Foundations of the General Theory of Relativity" leads directly to equation 8.36. When others expanded his work, they included the approximations he used, as shown in table 8.5. When they are inserted into either of equations 8.28, they give

$$ds^2 = dt^2\left(1-\frac{\alpha}{r}\right) - \frac{dr^2}{\left(1-\frac{\alpha}{r}\right)} - r^2(d\theta^2 + sin^2\theta d\varphi^2) \quad (8.18$$

The expansion of Einstein's paper presented above, derived equation 11, repeated here for ease of comparison.

$$ds^2 = dt^2e^{-\frac{\alpha}{r}} - dr^2e^{\frac{\alpha}{r}} - r^2(d\theta^2 + sin^2\theta d\varphi^2) \quad (8.17$$

The differences between the three metrics make no difference for situations such as accurate global positioning systems and international timing. It also makes no difference to the magnitude of the precession of Mercury's perihelion. It does, however, result in a change in the direction of that precession. The "strong field" solution predicts orbital regression. Only orbital precession has been observed.

8.7. Differentiating the Theories

The accuracy of Einstein's gravity field equations from his general theory of relativity in predicting gravitational effects should not be overlooked. Einstein (1918) made a further prediction that non-symmetrical rotating massive bodies would radiate gravity waves and hence lose power. That loss of power would be detected by the slowing down of the rotating body. Hulse and Taylor (1975) reported a binary pulsar, PSR B1913+16, two neutron stars rotating about each other, which seemed to fit the situation of a non-symmetrical rotating body.

Observation by Weisberg and Taylor (1980, 2002, 2005) showed it was indeed slowing down according to Einstein's predictions. Consequently, it is generally believed that Einstein's general theory of relativity has been verified to an accuracy of approximately 1 part in 10^{14}. The decay rate is independent of whether the metric used is equation 8.17 or equation 8.18. The same situation applies for other verifications of Einstein's field equations. The detection of Mercury's anomalous orbital precession and that of star S2 in Sagittarius A can only occur with equations 8.17 and 8.36. They predict that gravity is weaker than inverse square.

The above has demonstrated that treating photons as particles of variable mass and constant velocity while calculating the effect of Newtonian gravity upon them has produced the space–time metric for space outside matter. This suggests there are now three theories of gravity: Newton's inverse-square law, Einstein's gravitational field equations from his general relativity theory, and this redshift presentation.

Newtonian mechanics is based upon the speed of light being infinite and/or the mass of a photon being zero. Newton had no alternative knowledge available to him. It accurately describes local situations where experimental error is not sufficient to distinguish between c and ∞, or to measure the change in frequency and hence redshift of photons. As this presentation shows, giving photons a mass and a finite velocity enables Newtonian mechanics to predict relativistic effects when conservation of energy is applied to photons.

A good comparison of the differences between the three approaches is shown in figure 8.9. It shows that the denominator of

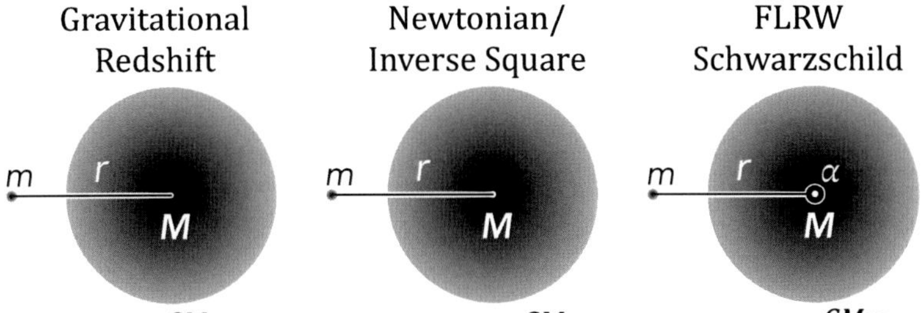

$$F_{GZ} = \frac{GMm}{[(1+z)r]^2} \qquad F_{GN} = \frac{GMm}{r^2} \qquad F_{GS} = \frac{GMm}{(1-2z)r^2}$$

$$g_Z = \frac{GM}{r^2 e^{\alpha/r}} \qquad g_N = \frac{GM}{r^2} \qquad g_S = \frac{GM}{(1-\frac{\alpha}{r})r^2}$$

Figure 8.9: *Schematic illustrations of some gravity properties under the three metrics.*

gravitational attraction varies from Newton's inverse square law by the redshift, z. Apart from that, gravity acts through the centre of mass in the redshift and Newtonian metrics, and at the Schwarzschild radius in the Schwarzschild metrics.

It should be restated. There is no physical reason for gravity to act through the Schwarzschild radius. It is a mathematical prediction that was incorrectly "derived" from Einstein's gravitational field equations. As indicated above, those derivations used flawed mathematics.

Figure 8.10 shows the difference between gravity under Newton's inverse square law, g_N, the metric derived from Einstein's field equations without the Schwarzschild approximations, g_E, and the redshift metric, g_Z. The Einstein metric was derived in the same manner as shown in table 8.4. It shows Einstein's approximations were good for $r > \approx 3\alpha$.

Einstein's gravitational field equations do not determine the strength of gravitational fields. They determine the difference between his predictions and Newton's predictions. The equations in figure 8.9 predict the gravitational field strengths, and they also show the relationship to Newton's gravity.

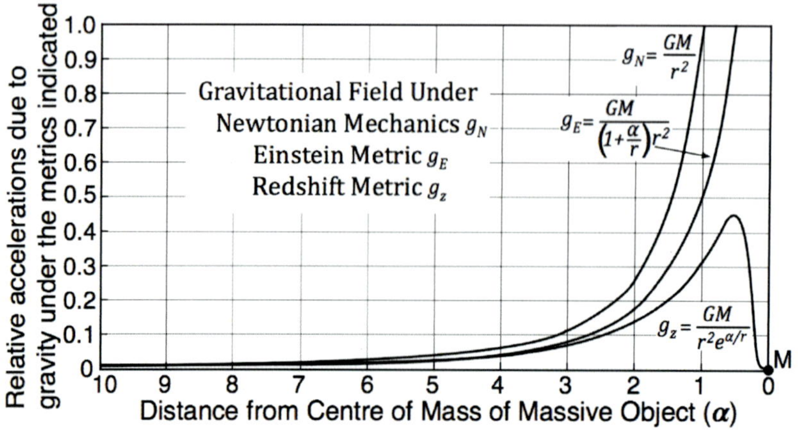

Figure 8.10: *Plots of the relative accelerations due to gravity under Newton's theory, Einstein's metric, and this redshift study metric.*

8.8. Discussion

Table 8.6 lists some properties of the different metrics. The field status is relative to Newton's inverse square law. As far as gravitational attraction is concerned, when $r \gg \alpha$, equations 8.17, 8.18 and 8.36 all trend to Newtonian gravity. The g_{11} terms separate the gravitational fields of equations 8.17, 8.18 and 8.36 from Newton's inverse square law of gravity. Einstein acknowledged that his equations were approximations.

At $r \gg \alpha$, it is not possible to distinguish between the Einstein and redshift metrics because $e^{\frac{\alpha}{2r}} \approx 1 + \frac{\alpha}{2r}$. They predict the same values within experimental error. They both predict orbital precession. Figure 8.10 shows that the only difference occurs in the region where $r \approx \alpha$.

The only results available in that region are from The Event Horizon Telescope Collaboration (2019) images, one of which is shown in figure 8.11A. Figure 8.11B is extracted from figure 8.7 above and presented side by side at approximately the same scale. It is suggested that their similarity is more than simply coincidence.

Metric	Gravity	Field Status	Orbit Status	Black Hole
Newton	$\frac{GM}{r^2}$		No Change	No
Einstein	$\frac{GM}{\left(1+\frac{\alpha}{r}\right)r^2}$	Weak	Precess	No
Schwarz-schild	$\frac{GM}{\left(1-\frac{\alpha}{r}\right)r^2}$	Strong	Regress	Yes
Redshift	$\frac{GM}{r^2 e^{\frac{\alpha}{r}}}$	Weak	Precess	No

Table 8.6 *Some gravitational properties under the different gravitational theories and metrics.*

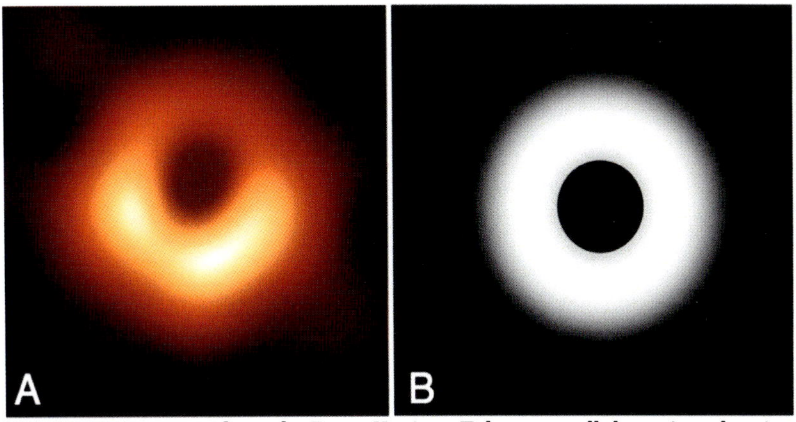

Figure 8.11: *Image from the Event Horizon Telescope collaboration, showing the bright torus shaped structures detected in the centre of galaxy M87, A. B shows the expected torus shape of emissions from a massive object under this redshift metric. B was adjusted to the same image size as A.*

The Event Horizon Telescope Collaboration images were presented as depicting a black hole at the centre of galaxy M87, approximately 55 million light years away. It was suggested they have a mass of about 5 to 7 billion solar masses, based upon the Schwarzschild metric.

Predictions under the redshift metric would give a higher mass. There is nothing in the above that excludes mass accumulating to any amount. A feature of this work is that the redshift metric is the only one that predicts both the torus shape for a massive object and the observed orbital precessions.

These predictions are a good indication that the redshift metric applies for all values of $\frac{r}{\alpha}$ in space–time outside matter. Apart from its inability to predict orbital precession, the Schwarzschild metric is only accurate for $r \gg \alpha$. It has only been tested up to $r \approx 10^{-6}\alpha$. As shown in figure 8.10, the Einstein metric is good for $r > 3\alpha$.

Figures 8.5, 8.6 and 8.10 and their associated equations, show that when $r \gg \alpha$, the gravitational field strengths all trend towards Newton's inverse square law values. As all gravitational waves are detected at distances where $r >>> \alpha$, their detection is independent of the metric used for their generation.

Einstein's great contribution was to show that gravity worked by mass distorting space–time. That removed all concern about gravity acting at a distance for no apparent reason. This work shows that space–time distortion is measured by photon redshift. Einstein's work involved determinations of the differences caused by detectable redshift effects.

In the above calculations, the effect was to introduce a redshift term into the denominators in equations 8.1 and 8.2. Einstein's field equations approximations changed Newton's r^2 to $r^2\left(1+\frac{\alpha}{r}\right)$. The doubly approximated Schwarzschild metric changed it to $r^2\left(1-\frac{\alpha}{r}\right)$. This redshift study changed it to $[r(1+z)]^2 = r^2e^{\frac{\alpha}{r}}$. Beyond these and the extensions that come from them, Newtonian gravity is otherwise unchanged.

It is possible that some authors may have derived the accepted Schwarzschild metric without reference to Einstein's field equations. In that case, they are not solutions to his field equations. They all face the problem that gravity stronger than inverse square physically cannot generate the observed precession of bodies orbiting massive objects that create detectable redshifts.

Knowledge of Einstein's equation, $ds^2 = g_{\mu\nu}dx_{\mu}dx_{\nu}$, his calculation of g_{11} , from which g_{44} is automatically derived, plus equations 8.33 and 8.35, which include the conversion from Cartesian to Polar coordinates, are sufficient to derive the metric.

Equations 8.14 shows that, under these conditions, $z = \frac{\alpha}{2r}$. That changes the denominators for the gravitational acceleration equations derived from equations 8.36, 8.18 and 8.17 to $(1+z)r^2$, $(1-z)r^2$ and $[(1+z)r]^2$ respectively. When there is no detectable effect of redshift, Newton's inverse square law of gravity will match all observations.

An important feature of table 8.6 is that the exact solutions to Einstein's gravitational field calculations, equation 8.36, and the exact extension of his 1911 paper, equation 8.17, both predict that gravity will be weaker than inverse square law. It follows that those equations predict orbital precession and do not predict the existence of black holes.

The prediction of black holes requires gravity to be stronger than inverse square – "*the strong field solution*". When gravity is stronger than inverse square, the orbiting body will travel closer to the attracting body and complete a perihelion-to-perihelion orbit slightly earlier. That would cause its perihelion position to regress.

The accurate prediction of Mercury's anomalous perihelion precession, and the observation of the precession of star S2 orbiting the massive object in Sagittarius A at the centre of our galaxy, confirm that orbital precession has been detected. They require the gravitational attraction to be less than inverse square. Those observations physically exclude the possibility that mass has distorted space–time to make gravity stronger than inverse square, and this excludes the possibility of black holes.

8.9. Summary

This chapter has shown that relativistic gravitational effects can be calculated from the effects of Newton's inverse-square law of gravity upon photons that have mass $m = \frac{hv}{c^2}$, as assigned to them by Einstein. It shows that mass distorting space-time are the changes to a photon's wavelength and frequency respectively. Space distortion is the change in the wavelength of photons, measured through redshift, z, changing distance r_0 to $r_0(1+z)$. Time distortion is the change in the frequency of the photon, changing υ_0 to $\upsilon_0/(1+z)$.

They are the space and time distortions to which Einstein referred. They were used to match some of his calculations such as the anomalous precession of Mercury's orbit, gravitational lensing, and photon redshift. It is believed the calculations were simpler to follow than derivations used by Einstein and others. Their match shows that it is the same theory. It also shows that Einstein's gravity is founded upon the distance and time changes associated with photon redshift. These space and time distortion equations suggest redshift, z, is the key to understanding gravity.

Einstein's mathematical approach and his use of approximations made it difficult for many people to follow his work. By his own admission, Einstein's field equations were an approximation. Exact solutions to approximations always remain approximations. This study suggests his approximations were valid for $r > 3\alpha$.

An exact expansion of Einstein's 1911 study was derived. It predicted that the strongest gravitational field is at $r \approx 0.5\alpha$, if the object's mass is so great that its physical radius is smaller than half its Schwarzschild radius. Particles of relativistic velocity would swirl around near that radius. Their collisions would cause them to emit high energy radiation from a torus shape. Its prediction has the shape and approximate dimensions of the phenomenon observed by the Event Horizon Telescope collaboration. That strongly suggests the exact redshift metric, equation 8.17, is applicable for all values of $\frac{r}{\alpha}$.

It is believed this redshift metric will match all observations for space–time outside matter. It shows that Einstein's mass distorting space–time is measured by photon redshift. When the effects of redshift are not physically detectable, Newton's inverse square law of gravity will match all observations. It is suggested that this vindicates the approximations he used. The redshift metric will hold for all values of $\frac{\alpha}{r}$, including when $r < \alpha$.

The Schwarzschild metrics fail because they are only valid for $r \gg \alpha$ and they predict gravity to be stronger than inverse square. They only allow for the perihelion of orbiting objects to regress against their direction of travel. Appropriate orbiting bodies show a perihelion precession that can only be achieved when gravity is weaker than inverse square. Predictions of black holes have no physical foundation. They also have no foundation in Einstein's gravitational field equations from his general theory of relativity. Einstein never believed in them, suggesting he was right yet again.

Standard model cosmology accepts that equation 8.18 is an exact solution to Einstein's field equations and will apply. Based upon it, they have announced that black holes, objects so massive that not even photons can escape their immense gravity, are part of the universe. Mathematicians believe that equation 8.18 is an exact solution to Einstein's gravitational field equations derived from his general theory of relativity. The above has shown his field equations were not exact. He acknowledged they were approximations.

Einstein's expression for the precession of Mercury's perihelion predicts 42.98 arc sec/century. The expression derived herein gives 42.99 arc sec/century, being essentially equal to Einstein's term. The equality of the answers indicates these approaches are addressing the same topic from a different perspective. Both answers differ from the measured value of 41.77 ($\pm$ 0.65) arc sec/century. That suggests there is another reason for the difference.

An important feature of table 8.2 is that all observations are relative to the observer. Further, all aspects of the observer's position and motion must be taken into consideration. That is why Einstein's theories were called "Relativity".

The derivation of equation 8.17 was an exact solution to Einstein's (1911) work. Equation 8.36 is the direct solution to the approximations Einstein used in deriving his field equations. They both predict Mercury's perihelion precession to within 0.01 arc sec per century. The closeness of the predictions and the exact nature of the derivation of the equations suggest that equations 8.17 and 8.36 will give a closer match with observation than will equations 8.18. That brings into question the very concept of black holes and worm-hole paths to other parts of the universe or to parallel universes.

Astronomers have detected torus shaped objects at the centres of galaxies. Mathematicians believe they are black holes, suggesting this presentation could be wrong. Astronomers have detected massive objects. Under current standard model theory, they can only be explained as black holes. Every observation that supports the existence of black holes is nothing more than the detection of a very massive object. The nature of that massive object is subject to conjecture because they have not measured anything close enough to its centre to determine exactly what it is. Figures 8.7 and 8.11, as well as this redshift metric, suggest they could be images of massive objects under this redshift metric.

The difference between the prediction shown in figure 8.11 and the black hole explanation is a factor of two (2) over a distance of 55 million light years, if no second order effects are taken into consideration.

Gravity acts through the distortion of space–time, just as Einstein predicted. That is discussed further in the next chapter. It shows the origins of gravity and why it varies as inverse square with Einstein's correction. It shows why it is not physically possible for mass to distort space–time in a manner that can increase gravitational attraction beyond inverse square.

Additional to the above, it is respectfully suggested that the derivations of space–time distortion in this presentation are easier to follow than Einstein's derivation of his gravitational field equations from his general relativity theory. Further, these equations have been reduced to the redshift of photons produced by massive objects. It is suggested that this makes it easier to extend space–time distortion effects to other situations.

8.10. List of References

Adam MG, (1948); "Interferometric Measurements of Solar wave-lengths and an Investigation of the Einstein Gravitational Displacement," Monthly Notices of Royal Astronomical Society, **108**, pp 446 – 464

Adams WS, (1910); "An Investigation of the Displacement of Spectrum Lines at the sun's Limb," Astrophysical Journal, **31** pp. 30 – 61

Brahe T, (1572); "Astronomiae Instauratae Progymnasmata"; https://en.wikipedia.org/wiki/Tycho_Brahe

Copernicus N, (1543); De revolutionibus orbium coelestium libri VI ("Six **Books** Concerning the Revolutions of the Heavenly Orbs"), J Petreius, Venice

Droste (1917) "The field of a single centre in Einstein's theory of gravitation and the motion of a particle in that field" Royal Netherlands Academy of Arts and Science **19** pp 447 – 455

https://www.dwc.knaw.nl/DL/publications/PU00012325.pdf

Dyson FW, Eddington AS and Davidson C, (1920); "A Determination of the Deflection of Light by the Sun's Gravitational Field, from Observations made at the Total Eclipse of May 29, 1919," Philosophical. Transactions of the Royal Society of London, **220A**, pp 291 – 333

Eddington AS, (1920); "Space Time and Gravitation," Cambridge University Press, Ch 7, 8

Einstein A, (1905); "Ist die Trägheit eines Körpers von Seinem Energiehalt Abhängig?" Annalen der Physik, **18**, pp 639 – 641

Einstein A, (1911) "Uber den Einfluss der Schwerkraft auf die Ausbreitung des Lichtes", Annalen der Physik, **35**, pp 898 – 908

Einstein A, (1915 a); "Zur Allgemeinen Relativätstheorie," Sitzungsberichte der Königlich Preussischen Academie der Wissenschaften, Part 2, pp 778 – 786

Einstein A, (1915 b); "Erklärung der Perihelbewegung des Merkur aus der Allgemeinen Relativätstheorie," Sitzungsberichte der Königlich Preussischen Academie der Wissenschaften, Part 2, pp 831 – 839

Einstein A, (1916) "Die Gründlage der Allgemeinen Relativätstheorie," Annalen der Physik, **42**, pp 769 – 822

Einstein A, (1918) "Über Gravitationswellen," Sitzungsberichte der Königlich Preussischen Academie der Wissenschaften, pp 154 – 167

Einstein A, (1952 a); "Does the Inertia of a Body Depend upon its Energy Content?" in Lorentz HA, Einstein A, Weyl H and Minkowski H, The Principle of Relativity, translated by Perret W and Jeffery GB, Dover Publications, New York, NY, pp 69 – 71

Einstein A, (1952) b; "On the Influence of Gravitation on the Propagation of Light," in Lorentz HA, Einstein A, Weyl H and

Minkowski H, The Principle of Relativity, translated by Perret W and Jeffery GB, Dover Publications, New York, NY, pp 99 – 108

Einstein A, (1952 c); "The Foundation of the General Theory of Relativity," in Lorentz H.A., Einstein A., Weyl H. and Minkowski H., The Principle of Relativity, translated by W. Perret and G. B. Jeffery, Dover Publications, New York, NY, pp 109 – 164

Galileo G, (1610); "Siderius Nuncius (Starry Messenger)" (Venice: Galileo G)

Galileo G, (1632); "Dial Sopra I due Massimi Sistemi del Mondo (Dialogue Concerning the Two Chief World Systems)" (Florence: Galileo G):- https://en.wikipedia.org/wiki/Dialogue_Concerning_the_Two_Chief_World_Systems

Gonzalez Hernandez, J.I. et al., (2020); "The solar gravitational redshift from HARPS-LFC Moon spectr*a.* A test of the general theory of relativity*", Astronomy and Astrophysics, 643 A146,* https://doi.org/10.1051/0004-6361/202038937

Hulse RA and Taylor J H, (1975); "Discovery of a Pulsar in a Binary System," Astrophysical Journal, **195**, pp 51 – 53

Kepler J, (1609); "Astronomia Nova (New Astronomy)", Rudolph II, Prague: https://archive.org/details/Astronomianovaa00Kepl/page/n5/mode/2up?view=theater

Kepler J, (1619); "Harmonices Mundi (Harmony of Worlds)", Linz:- https://en.wikipedia.org/wiki/Harmonices_Mundi

Le Verrier U, (1859); "Lettre de M. Le Verrier à M. Faye sur la théorie de Mercure et sur le mouvement du périhélie de cette planète", Comptes rendus hebdomadaires des séances de l'Académie des sciences (Paris), **49,** pp. 379–383.

Matzner RA, (2001); "Dictionary of Geophysics, Astrophysics and Astronomy," CRC Press, London, p 360.

Minkowski H, (1908); "Raum und Zeit," Jahrescherricht der Deutschen Mathematiker-Vereinigung, **18**, pp. 75-88. (Address delivered at the 80th Assembly of German Natural Scientists and Physicians, Cologne, September.)

Minkowski H, (1952); "Space and Time," in Lorentz HA, Einstein A, Weyl H and Minkowski H, The Principle of Relativity, translated by Perret W and Jeffery GB, Dover Publications, New York, NY, pp 75 – 91

Misner CW, Thorne KS and Wheeler JA, (1973); "Gravitation," W. H. Freeman and Co., New York, NY, Ch 31

Newcomb S, (1882) US Nautical Almanac Office Astronomical paper 1 (Washington) p 473 https://www.relativitycalculator.com/pdfs/mercury_perihelion_advance/S.Newcomb.pdf

Newton I, (1687); "Philosophiae Naturalis Prinicipia Mathematica", London

Newton I, (1729) Mathematical Principles of Natural Philosophy **1** ed. A Motte (London: Motte B) Propositions1 – 32, 45

Park RS, Folkner WM, Konopliv AS, et al., 2017; "Precession of Mercury's Perihelion from Ranging to the Messenger Spacecraft", The Astronomical Journal, **153**, pp 121 – 127

Pound RV and Rebka, GA, (1959); "Gravitational Red-shift in Nuclear Resonance," Physical Review Letters, **3**, pp 439 – 441

Pound RV and Snider JL, (1961); "Effect of Gravity on Nuclear Resonance," Physical Review Letters, **13**, pp 539 – 540

Pound RV, (2000); "Weighing Photons," Classical and Quantum Gravity, **17**, pp 2303 – 2311

Robinson VNE, (2021; "Physical Explanations of Einstein's Gravity", Journal of Physics Communications, **5** 035013:

https://doi.org/10.1088/2399-6528/abee2f

Sagan C, (LOC) "Whose Revolution? Copernicus, Brahe & Kepler", https://loc.gov/collections/finding-our-place-in-the-cosmos-with-carl-sagan/articles-and-essays/modeling-the-cosmos/whose-revolution-copernicus-brahe-and-kepler

Schwarzschild K, (1916); "Aus von Gravitationsfelf eines Massenpunktes Entsprechend Einsteinschen Theorie", Sitzungsbereichte der Königlich Preussischen Akademie der Wissenschaften, Phys. Math., **K1**. Pp 189 – 196. (English language: https://arxiv.org/pdf/physics/9905030.pdf

The Event Horizon Telescope Collaboration *et al.* (2019), Astrophysical Journal Letters, **875** L1 – L6) Further details can be found at: https://iopscience.iop.org/article/10.3847/2041-8213/ab0e85

van der Mark, MB, and 't Hooft, GW, (2000) "Light is Heavy", van Atom tot Quantum, Nederlandese Natuurkundige Vereninging, pp 1 – 6

Weisberg JM and Taylor JH, (1980; "Gravitational Radiation from an Orbiting Pulsar", General Relativity and Radiation, **13**, pp 1 – 6

Weisberg JM and Taylor JH, (2002); "General Relativistic Geodetic Spin Precession for Binary Pulsar B1913+16: Mapping the Emission Beam Two Dimensions", Astrophysical Journal, **576**, pp 942 – 949

Weisberg JM and Taylor JH, (2005); "The Relativistic Binary Pulsar B1913+`6: Thirty Years of Observation and Analysis", in Rasio FA and Stairs JH, Binary Radio Pulsars, Aspen Conference, ASP Conference Series, **328**, pp 25 –

9. The Physical Origins of Gravity

Abstract:

This chapter shows that, if mass induces changes in the electric permittivity of space, photon trajectory, wavelength and frequency would be altered in the same manner as they are altered by gravity. The radial differential of changing electric permittivity gives the single equation that derives Newton's inverse square law of attraction as a first approximation and the solution to Einstein's description of mass distorting space–time to produce gravity as a second approximation. The exact solution predicts the torus shaped objects observed at the centre of galaxies It indicates that mass distorts space–time by generating variations in the electric permittivity of space.

9.1. Introduction

By the 16th century CE, the study of gravity was changing from Earth at the centre of the universe, to planets orbiting the Sun (Sagan, LOC). Using his telescope, Galileo (1610, 1633) observed that Jupiter had four moons orbiting it. His work confirmed the idea that the planets moved around the Sun and that gravity was responsible for those heavenly motions. The solar centric model for the motion of the planets around the Sun was firmly established.

Newton (1687, 1729) placed gravity and gravitational effects on a firm foundation. He introduced his universal law of gravitational attraction between two bodies of masses M and m, through the equation

$$F_G = \frac{GMm}{r^2} \quad (9.1$$

where F_G is the gravitational attraction between the two bodies, G is Newton's universal gravitational constant, and r is the distance between their centres of mass. Newton also indicated that gravity was also responsible for holding everything onto the surface of Earth, and other planets and heavenly bodies. Gravity was universal and affected everything. He gave no mechanism by which gravity caused two masses to attract each other.

His inverse square law successfully predicted all gravitational observations for almost 200 years. Towards the end of the 19th century CE, it was established that Mercury's orbit had an anomalous precession of 43 arc sec per century (as/c) (Le Verrier, 1859, Newcomb, 1882). Based upon the gravitational field equations associated with his general theory of relativity, Einstein (1915a, 1916, 1952) predicted that Mercury should have an anomalous orbital precession of 43 as/c (Einstein 1915b). He also made other gravitational predictions including gravitational lensing, and gravity waves.

Many other attempts were made to explain gravity. A good summary was given in Krasnoholovets (2021). Most, like Einstein's theory, were based upon mathematical calculations. Quantum gravity introduces gravitons as the force carrying particles. Apart from suggesting they are spin 2 particles, very little else is known about them. They have never been detected. Quantum field theory suggests particles are a subset of the larger field. They, along with others, are very much mathematical work in progress.

Einstein's theory of mass distorting space-time was the most successful. Space–time distortion gave a mechanism by which gravity could attract He gave no indication of the physical properties of space and time that mass distorted.

Through his equation

$$s^2 = c^2t^2 - x^2 - y^2 - z^2 \quad (9.2$$

Minkowski (1908, 1952) suggested that space and time were interlinked in a single continuum by the speed of light. In equation (9.2, s is the space–time co-ordinate, c is the speed of light in vacuum, t is time and x, y and z are the orthogonal space dimensions.

The complexities of Einstein's mathematics made it difficult to understand what the physical principle by which mass distorted space–time was. For over 100 years after Einstein published the gravitational field equations associated with his general theory of relativity, people knew the mathematics worked, but not the physical principle of mass distorting space–time.

Chapter 8 (Robinson, 2021) solved that by pointing out that space–time distortion was determined by the redshift of photons. That considerably simplified the mathematics of some of the effects Einstein calculated. It also corrected some misconceptions that had developed in those who followed and extended Einstein's gravity calculations. Among those was the prediction of black holes. Einstein never believed they could be predicted from his field equations.

As chapter 8 pointed out, Einstein introduced approximations when he derived his gravitational field equations. Exact solutions to approximations are still approximations. Using tensor notations, chapter 8 also derived an exact solution to Einstein's field equations. That solution did not allow for black holes. In Proposition 45 volume 1 of his Principia, Newton pointed out that, if gravity was stronger than inverse square, a planet orbiting the sun would have its perihelion and aphelion positions regress against its orbital direction. If gravity was weaker than inverse square, a planet's perihelion and aphelion positions would precess in its orbital direction. The physical reasons were given by Newton and in chapter 8.

Mercury's anomalous orbital motion, i.e., that not predicted by Newtonian gravity, was a precession of $\approx$ 43 arc seconds per century (as/c). The orbit of star S2 about the massive object in Sagittarius A precesses in its orbital direction (Hees et al., 2017; Abuter et al., 2020). For those to occur, gravity must be weaker than inverse square. Chapter 8 also showed that the accurate solution to Einstein's gravitational field equations predicted, along with exact solution to the redshift produced by mass distorting space–time.

The above points out the advances made by knowing that Einstein's mass distorting space–time is determined by photon redshift.

It does not point out how mass distorts space–time to produce that redshift.

Apart from pointing out that mass distorted space–time through photon redshift, no indication was given of the physical properties of space and time that mass distorted to generate gravity. One objective of this chapter is to point out that property. This is done in several stages.

1) Establish a relationship between changing electric permittivity and properties of gravity.
2) Describe the mechanism by which mass can change the electric permittivity of space.
3) Show how that mechanism matches nucleons to known gravity properties through the electric permittivity of space.
4) Show how particles respond to the changing electric permittivity.

9.2. Variations in Electric Permittivity with Distance

In his first attempt at calculating the bending of light rays by the sun, Einstein (1911) suggested they could be bent by a change in the refractive index of matter free space. Roy and Sen (2019) summarised the work of others who also suggested that changes in the refractive index of matter free space could be responsible for some gravitational effects. None continued the calculation to determine how such a change would give rise to gravity.

Maxwell (1865) determined that the speed of light in vacuum, c, was given by

$$c^2 = \frac{1}{\varepsilon_0 \mu_0} \qquad (9.3$$

where ε_0 and μ_0 are the electric permittivity and magnetic permeability of matter free space. Huygens (1690) suggested that the diffraction of light was caused by changes to the speed of light in different media. It has since been established that the speed of light through a medium m, c_m, was always slower than that in free space. The refractive index, n, of medium m, n_m, was given by $n_m = \frac{c}{c_m}$, where c_m is the speed of light through medium m. From that we get $c_m^2 = \frac{1}{\varepsilon_m \mu_m}$, which gives

$$n_m = \sqrt{\frac{\varepsilon_m \mu_m}{\varepsilon_0 \mu_0}} \qquad (9.4$$

The refractive index, n_m, of a medium is always greater than 1. Light passing through a medium with an increased refractive index is slowed and deflected towards the material with the higher n.

Magnetic permeability, μ, is defined as $4\pi \times 10^{-7}$ Henry/m. The speed of light in vacuum, c, is defined as 299,792,458 m/sec. The electric permittivity, ε_0 , is derived from them. With magnetic permeability fixed, equation 9.4 becomes

$$n_m = \sqrt{\frac{\varepsilon_m}{\varepsilon_0}}. \qquad (9.5$$

The speed of light through medium m is given by

$$c_m = \frac{c}{n_m} = c.\sqrt{\frac{\varepsilon_0}{\varepsilon_m}} \qquad (9.6$$

The wavelength of light changes inversely with its refractive index. An increase in refractive index slows down the speed of photons, decreasing their wavelength. An increase in wavelength when a photon goes from ε_m to ε_0 is denoted by redshift, z, defined as

$$z = \frac{\lambda_0 - \lambda_m}{\lambda_m} = \frac{\lambda_0}{\lambda_m} - 1 \qquad (9.7$$

where λ_m and λ_0 are its wavelengths at its origin in medium m and at matter free space respectively.

$$\frac{\lambda_0}{\lambda_m} = n_m = 1 + z \qquad (9.8$$

Equations 9.5 and 9.8 combine to give $z = n - 1 = \sqrt{\frac{\varepsilon_m}{\varepsilon_0}} - 1$, which simplifies to

$$\frac{\varepsilon_m}{\varepsilon_0} = (1 + z)^2. \qquad (9.9$$

In equation 9.9, z is the redshift between ϵ_M and ϵ_0. That is the redshift from the massive object to effective infinity, a long way away. From equation 8.14, we get

$$(1 + z) = e^{\alpha/2r} \qquad (9.10$$

where $\alpha = \frac{2GM}{c^2}$, the Schwarzschild radius.

The redshift, z, is the same in both calculations. Equation 9.9 came from changes to the refractive index caused by changes in the electric permittivity of a medium. It is the same mechanism that applies in optics and refractive index studies that started with Huygens. Equation 9.10 came from the redshift caused by mass. The two can be equated when the medium is matter-free space. Then equations 9 and 10 combine to give

$$\varepsilon_m = \varepsilon_0 e^{\alpha/r} \qquad (9.11$$

Equations 9.3 and 9.4 show that the absolute values of electric permittivity determine the speed of light through a medium. It has been well known since Huygens' work, that a change of refractive index changes the direction light travels. A change in refractive index changes the speed of photons travelling perpendicular to that change. It deflects photons travelling non-perpendicularly to the change.

Since Maxwell's work, it has been known that the refractive index of a medium is governed by its electric permittivity when the value of magnetic permeability is fixed. Electric permittivity determines the refractive index of a medium. Changes in electric permittivity determine the speed and deflection of photons.

Most of the universe is matter free space, where the electric permittivity is ε_0. Equation 9.11 shows that ε_0 changes to ε_m as photons approach a mass. The rate of change of electric permittivity from ε_0 to ε_m, is given by the differential of $\frac{\varepsilon_0}{\varepsilon_m}$ with the distance r, namely

$$\frac{d}{dr}\left(\frac{\varepsilon_0}{\varepsilon_m}\right) = \frac{d}{dr}\left(e^{-\alpha/r}\right) = \frac{-e^{-\alpha/r}}{r^2} = \frac{-1}{r^2 e^{\alpha/r}} \qquad (9.12$$

Equation 9.12 gives the variation of the rate of change of the electric permittivity with distance from the centre of mass for distances outside the mass. In turn, the rate of change of electric permittivity with radius is proportional to the rate at which photons speed changes, or they are deflected. For r >> α, the $e^{\alpha/r}$ term becomes 1 for all practical purposes. That means the speed of photons is constant and their deflection proportional to the inverse square of the distance from the centre of mass.

Figure 9.1 shows an object of mass M with a field of varying electric permittivity surrounding it. The white background represents the value ε_0. The intensity of the grey shades represents the increasing value of ε_m as it approaches M. The dotted lines surrounding it are lines of equal electric permittivity. Photons are represented by γ.

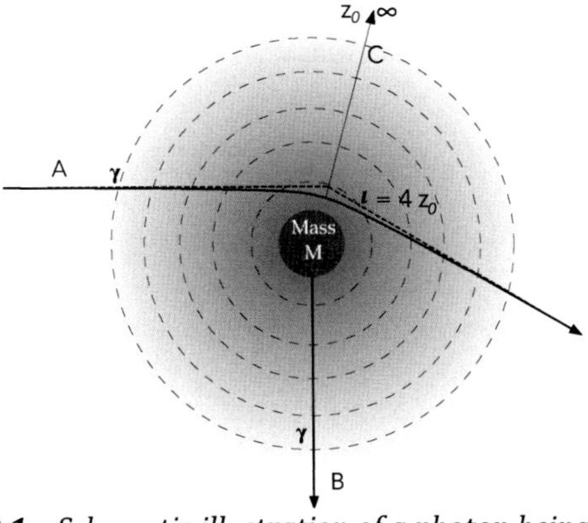

Figure 9.1: *Schematic illustration of a photon being deflected by an angle ɩ (iota) as it approaches and departs a massive object composed of N nucleons.*

Photon A travels across the lines of equal electric permittivity. Its trajectory becomes increasingly deflected the closer it gets to the mass M. After it reaches its point of closest approach, its deflections diminish as it returns to ε_0.

This is illustrated by the differences in the photon's trajectory, full curve, and the dashed lines that extend the photon's initial and final directions back to their crossover at its position of closest approach.

Using the Schwarzschild radius calculations, the total deflection, ι (iota) in radians, is 4 times the redshift of photons leaving the position of closest approach and reaching ε_0, the effective infinity, see equations 8.19. That is shown by the line C.

Photons travelling perpendicular to the lines of equal permittivity are not deflected. That is shown by photon B. Because they start out in a high electric permittivity zone and travel to a lower electric permittivity zone, they travel a little faster. The faster speed stretches their wavelength. They are redshifted. That has also been observed. Using the Schwarzschild radius calculations, the predicted redshift between the sun and Earth is 2.1 x 10^{-6}. That was the same as Einstein (1911) calculated. It has been observed (Adams 1910, Adam, 1948, Gonzalez Hernandez et al., 2020.

Those examples indicate that the alterations to photons by a mass induced change in electric permittivity in space matches the trends of the effect of gravity upon photons. Referring to equation 9.13, the minus sign $(-)$ plays a similar role to the – sign in Einstein's gravitational field equations, which were conditional upon $\sqrt{-g} = 1$. It indicates gravity acts in the direction towards the centre of mass, *M.* Measurements of r are made away from the centre of mass.

Equation 9.12 also shows that the mass induced changes in electric permittivity are directly proportional to the gravitational attraction under the redshift metric, as derived in chapter 8. Using Newton's G as the constant of proportionality gives the alteration, alt, to photons due to the mass induced electric permittivity of space as

$$\text{alt} = \frac{GM}{r^2 e^{\alpha/r}} \qquad (9.13$$

Equation 9.13 should be compared to the redshift calculation for gravity, equation 8.28, reproduced in equation (9.14.

$$g_Z = \frac{GM}{r^2 e^{\alpha/r}} \qquad (9.14$$

The equality of the two gives a strong indication that the changes to the electric permittivity induced by mass M are the cause of alterations to photon trajectories. Both equations derive the Newton's inverse square law of gravitational attraction when r >> α. When α is a noticeable fraction of r, equation 9.14 shows that photon redshift will have a noticeable effect beyond Newton's inverse square law. Both closely match Einstein's gravity of

$$g_E = \frac{GM}{r^2\left(1+\frac{\alpha}{r}\right)} \qquad (9.14a$$

for values of r > ≈ 3α, see figure 8.10.

Developments from Huygens' diffraction studies and Maxwell's equations show that variations in the electric permittivity of a medium affect the wavelength, frequency, and trajectories of photons by changing the refractive index of the medium through which photons travel. Since Einstein's work, it was known that gravity induced by mass changed the properties of photons. Chapter 8 showed that mass distorting space–time to produce gravity, was its redshift of photons. The objective of this presentation was to determine if there was a possibility of mass inducing changes to the electric permittivity of space which had the same effect on photons as did gravity.

Equation 9.9 expressed the redshift of photons in terms of changes in the electric permittivity between ε_0 and ε_m. Equation 9.10 expressed the redshift in terms of gravitational effects based upon the Schwarzschild radius α. The two were combined into equation 9.11. One of the effects of gravity is to change the trajectory of photons. Electric permittivity varies photon trajectories when it changes. The expression for $\frac{\varepsilon_0}{\varepsilon_m}$ was differentiated with respect to the distance r from the centre of mass of M.

The differential produced equation 9.12. It shows the same variation of effects like changes to photon trajectories and their wavelength. Using Newton's universal gravitational constant, G, as the constant of proportionality gave equations 9.14 as the term that expressed the alteration to photons at different distances r. Equation 9.14 is equation 8.28, the same term derived from gravity considerations for the influence of gravity upon photons.

Both terms derive Newton's inverse square law and the redshift corrections in the one equation. When r >> α, Newton's inverse square law works well. When r > ≈ 3 α, Einstein's field equations work quite well when correctly interpreted, see figure 8.10.

Einstein indicated that gravity was caused by mass distorting the space–time continuum. He gave no indication of the physical nature of space–time or how mass distorted it. Chapter 8 indicated that space–time distortion was mass redshifting photons. Changes in photon wavelength change space dimensions. Changes in photon frequency was responsible for time distortions. No indication was given of the physical mechanism by which mass changed photon frequencies and wavelengths.

The above gave a very strong Indication that those changes were caused by mass altering the electric permittivity of free space. The constant of proportionality between the predictions under the varying electric permittivity induced by mass and the gravitationally observed effects was Newton's universal gravitational constant G.

The above deals with the effect of photon redshift due to changes in the electric permittivity of free space. It shows that the photon redshift and trajectories due to changes in the electric permittivity of space are the same as that due to gravity. The next step is to show how mass can distort the electric permittivity of matter free space.

9.3. Mass Changes the Electric Permittivity of Space

The standard model (Schumm, 2004; Oerter, 2006) uses nucleons composed of quarks held together by gluons (Halzen and Martin, 1984). It has had some successful predictions through the theory of Quantum Chromodynamics (Greiner et al., 2007). It has many failures. Being based upon mathematics associated with undetected particles, they are hidden behind a wall of great complexity.

Failures show up by their inability to predict much about nuclear physics and the structure of nuclei. Electrons, photons, and neutrinos are treated as point particles to which properties are mathematically attached. All predictions are made mathematically, mostly being to match known properties. Very few predictions were made in advance and subsequently verified. There is almost no physical Indication of the processes Involved.

Another model was proposed in which all matter particles were considered as being composed of confined photons. The original presentation was by Williamson and van der Mark (1997). They proposed a model in which an electron was considered as a photon of the appropriate frequency confined by making two complete revolutions per wavelength. That gave electrons the appearance of toroidal electromagnetic field. Their study was followed by similar proposals by Qui Hong (2007) and Robinson (2011). Details are given in chapters 2 and 3.

Chapters 2 and 3 continued that study to include the other particles, namely protons, neutrons, and neutrinos. They included a physical description of how the rotating photon model of the structure of particles gave rise to many known particle properties. These included the origins of the special relativity corrections when particles moved, why $E = mc^2$, the de Broglie wavelength, chirality, parity, the distribution of charge and magnetic field within particles and more, see chapter 2.

Chapters 2 and 3 used the model that all matter particles were composed of photons of the appropriate frequency making two revolutions per wavelength. The frequency, υ_N, of the photons that are the nucleons, was given by

$$\upsilon_N = \frac{m_N c^2}{h} \qquad (9.15$$

where m_N is the mass of the nucleons and h is Planck's constant.

Figure 9.2 shows two representations of a proton, extracted from chapter 3. A show a plan view of the primary rotating photon, P/1,

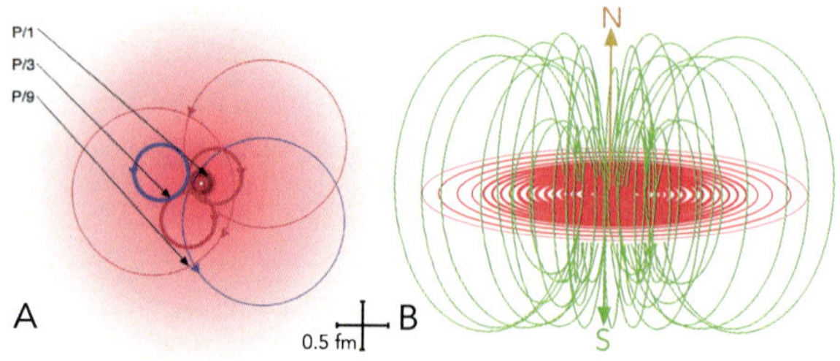

Figure 9.2: *Schematic representations of the structure of protons under the rotating photon model of matter.*

with one third, P/3, and one ninth, P/9 frequency harmonics. Those harmonics are needed to match the proton's charge distribution and magnetic moment. The P/9 frequency harmonic has the same properties as a muon, with slightly higher mass. Its central P/1 rotation has the same dimensions as the measured central core of protons.

The three P/3 harmonics are close to It. Two are positive and one is negative, the same ratio as quarks. They all rotate in the plane, generating the electric charge distribution. It is shown as the variations in the red, for positive charge, colour. The blue circles represent oscillations that generate negative charge. The magnetic field is in the third dimension, shown schematically in 2 B.

Through equation 9.15, a proton's mass gives its primary photon a frequency of 2.269 x 10^{23} Hz. Its P/3 and P/9 harmonic oscillations have frequencies of 7.56 and 2.52 x 10^{22} Hz respectively.

Figure 9.3 shows similar representations of a neutron. Its N/1, N/3, and N/9 harmonic oscillation frequencies are almost the same as those of a proton. Its central core diameter is the same as the proton's central core. It has three N/3 frequency harmonics surrounding it, two negative and one positive. Neutrons have an additional N/27 harmonic oscillation of frequency 8.4 x 10^{21} Hz.

The mass and other properties of pions are very similar to a combination of N/9 and N/27 frequency harmonic oscillations. Blue represents the distribution of negative charge. The charge distribution of positive core surrounded by a negative and a second weak positive charge matches observation. The additional very weak outer negative charge is required to match the neutron's negative magnetic moment.

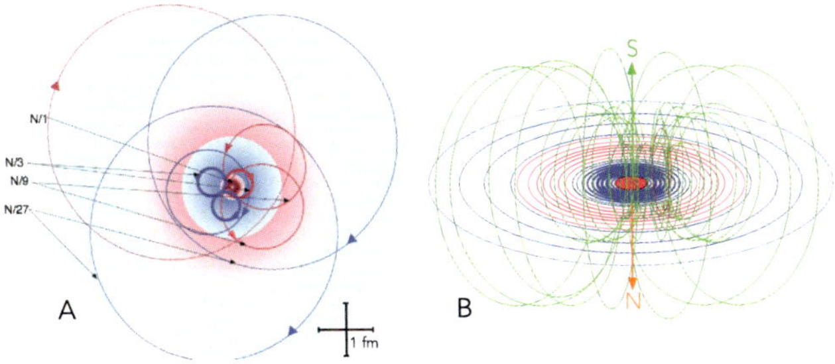

***Figure 9.3:** Schematic representations of the structure of neutrons under the rotating photon model of matter.*

Of particular note is that the nucleons are entirely electromagnetic. Nowhere is there anything that can be considered as solid matter. Chapters 2 and 3 show how this model explains many of the unknowns associated with nucleon properties. Chapter 5 shows how they combine to form nuclei. It also shows how to calculate the structure and many of the properties on any nucleus of any A and Z), answering many of the difficulties associated with nuclear physics.

Neutrally charged mass has equal numbers of electrons as protons. Through equation 9.15, an electron's mass gives it a fundamental frequency of 1.24 x 10^{20} Hz, the frequency associated with its Bohr wavelength. They too can generate electric field oscillations. They would be counted as part of the nucleon oscillations because of their close association with them.

The Important feature Is that there are high frequency variations in the electric and magnetic fields. The above has illustrated that the nucleons' oscillating electric fields gives them an equivalent to the electrons' zitterbewegung. The electric field varies as the inverse square law. The magnetic field varies as the inverse cube law. Once away from a nucleus, the magnetic field diminishes much faster than the electric field. For all practical purposes we may consider only the effect of the varying electric fields. There is a range of frequencies in the vicinity of $\approx 10^{21}$ to 10^{24} Hz. Their effects are cumulative.

9.4. Inducing Changes in Electric Permittivity

In his determination of the general expression for the refractive index of a medium with n close to 1, Feynman (1963) indicated that the fields affecting refractive index in a medium such as glass, were produced by the motion of all the charges in that medium. As the positive charges in nuclei are effectively at rest, that left the motion of the electrons in the material as the source of the field.

Feynman indicated the motion of the charges, i.e., electrons, (quote) *"makes them behave like little oscillators, with a resonant frequency ω_0"*. Their varying electric fields altered the electric permittivity of the space within the medium. Feynman went on to calculate an approximate general expression for the refractive index of a medium. His equation 31.19,

$$n = 1 + \frac{N_\rho q_e^2}{2\varepsilon_0 m(\omega_0^2 - \omega^2)} \qquad (9.16$$

for refractive index was determined by using the electric field generated by moving charges. In it, N_ρ is the number of moving charges per unit medium volume, q_e and m are an electron's charge and mass respectively, ω_0 is the resonant frequency of the electrons bound in the atom and ω is the frequency of the radiation. He pointed out that it was an approximation and some of the properties he used were not easy to determine. It was useful to point out the principles of refractive index were based upon the resonant frequency associated with oscillating charges. His derivation also pointed out some of its properties, such as a variation with different wavelengths.

Equation 9.16 applies to the variations in the electric field within a medium, induced by the motion of electrons within that medium. They have no effect beyond the medium. In this situation, the nucleon oscillations within a mass have an impact on the medium of space surrounding them. It is suggested the same principle applies, without a resonant frequency.

Applying equations 9.8 and 9.9 to equation 9.16 gives

$$z = \sqrt{\frac{\varepsilon_m}{\varepsilon_0}} - 1 = \frac{N_\rho q_e^2}{2\varepsilon_0 m(\omega_0^2 - \omega^2)} \qquad (9.16a$$

Space outside matter requires electric field oscillations to induce changes in the electric permittivity of matter free space. Because there is nothing in empty space to resonate, it requires only the oscillating frequency of one unit of electric charge per nucleon.

It must be noted that it is the oscillation of electric charge that changes the electric permittivity of space. It is independent of the charge polarity. Oscillating adjacent opposing charges still alter the electric permittivity, even if their charges cancel to zero.

The oscillations are of variable frequency and random. They will accumulate until they reach a maximum frequency. That maximum frequency is the inverse of Planck's time. Planck's time of ≈ 5 x 10^{-44} sec is the smallest possible time in which an event can occur. It follows that its inverse, the Planck frequency, ω_P is the highest possible frequency, namely ≈ 2 x 10^{45} Hz. Once that frequency is reached, further oscillations will increase the amplitude, and hence value of the electric permittivity.

The Planck time, $t_P = \sqrt{\frac{\hbar G}{c^5}} \approx 5.4 \times 10^{-44}$ sec, is derived from considering the energy required for a photon to be so massive that it generates a black hole. The black hole derivations as solutions to Einstein's field equations result from a misuse of the approximations Einstein used when he published his gravitational field equations from his general relativity theory. In that regard, Planck's time and hence frequency are not valid limits.

It is suggested the oscillations will average to be the same per nucleon. There are no other charges upon which the oscillating nucleon charges can interact. The effect is that of an oscillating high frequency charge on the electric permittivity of space, a linear relationship will apply. To a first approximation, equation 9.16a, can be rewritten as

$$z \approx fn.N.\left(\frac{e}{\omega_M}\right) \qquad (9.17$$

where ω_M is the maximum frequency at which electromagnetic radiation can oscillate. It will be lower than the fictitious Planck's frequency. The fn term is a function that takes all other factors into consideration.

The important feature of equation 9.17 is that the redshift is a function of the number of oscillating charges, and the magnitude of the electric charge, e, divided by the maximum frequency of their combined oscillations. Avogadro's number of ≈ 6 x 10^{23}/mol gives 6 x 10^{26} nucleons per kg. Equation 9.17 is converted to mass in kilograms, by replacing *N* by 6 x 10^{26} *M*, and inserting the value of e to give:

$$z = fn\left[6\,x\,10^{26}M.\frac{1.6\,x\,10^{-19}}{\omega_M}\right] \qquad (9.18$$

It indicates the redshift outside matter is proportional to the number of oscillators, i.e., nucleons, inside the matter, as well as a function of the unit electric charge divided by the maximum frequency. Equation 9.17 applies for space outside matter, commencing at its surface.

The nucleon's frequency is ≈ 5 x 10^{23} Hz. In random oscillations, it would require less than 1 gm, i.e., 10^{23} nucleons, to reach a maximum frequency of less than 10^{45} Hz. Gravitating masses are much larger, typically > 10^{20} kg. The maximum frequency will always less than the Planck frequency. In the first instance we shall set $\omega_M = 10^{44}$ Hz.

Equation 9.18 can be rewritten as

$$z = fn(10^{-36}.M) \qquad (9.19$$

In equations 9.18, z is the redshift at infinity from the mass of *N* nucleons. From equation 8.14, we have

$$z^* = \frac{\alpha}{2r} = \frac{GM}{rc^2} \qquad (9.20$$

Equations 9.19 and 9.20 can be used independently on known masses within the solar system. The results for equation 9.20 are shown in table 9.1. It lists the major bodies within the solar system. The diameters have been measured directly by observation. The redshifts have been measured, either directly, or by inference from satellites orbiting them.

Determinations of G have been made on Earth involving known masses of up to 10^2 kg. A recently recommended value for G is 6.673 ± 0.010 x 10^{-11} $m^3.kg.s^{-2}$. Higher accuracies have been reported within a range of ≈ 1500 ppm (Grundlach and Merkowitz, 2018).

The masses in table 9.1 were determined from bodies orbiting them. The orbital distances and speeds of satellites, natural or artificial, are accurately measured. They are proportional to GM. The assumption is that G, Newton's universal gravitational factor, is universal and constant.

Body	Mass (kg)	Diameter (m)	α (m)	$z^* = \alpha/2r$
Sun	2×10^{30}	1.4×10^{9}	2.95×10^{3}	2.1×10^{-6}
Mercury	3.3×10^{23}	4.88×10^{6}	5.5×10^{-4}	1.1×10^{-10}
Venus	4.9×10^{24}	12.1×10^{6}	7.3×10^{-3}	6.0×10^{-10}
Earth	6×10^{24}	12.7×10^{6}	9×10^{-3}	7.1×10^{-10}
Mars	6.4×10^{23}	6.78×10^{6}	$9 .1 \times 10^{-4}$	1.5×10^{-10}
Jupiter	1.9×10^{27}	1.4×10^{9}	2.84	2.0×10^{-9}
Saturn	5.68×10^{26}	1.1×10^{9}	8.4×10^{-1}	7.3×10^{-10}
Uranus	8.68×10^{25}	5.1×10^{7}	1.3×10^{-1}	2.55×10^{-9}
Neptune	1.02×10^{26}	4.9×10^{7}	1.5×10^{-1}	3.1×10^{-9}

Table 9.1: *Some properties of the major bodies within our solar system.*

Individual measurements claim to have accuracies lower than 20 ppm. The properties of G are not necessarily fixed. Table 1 has been derived using a fixed value of G. The large range of masses and distances means all measurements are rounded to two or three significant figures only.

The results for equations 9.19 are shown in table 9.2. Equation 9.18a was simplified to $fn = z/10^{-36}.M$, where z is the redshift calculated in table 9.1.

The masses of the different bodies vary by about 6.3 orders of magnitude. The required fn terms vary by about 3.5 orders of magnitude. It represents an error of ≈ 1000 ppm over that range. That is similar to the errors in the measurement of G on Earth.

The table 9. 2 results divide into four groups, the gaseous planets and sun, the ice planets, the small rocky planets and the large rocky planets. The gaseous planets and the sun have large differences in mass, but similar densities and fns ≈ 1. In the absence of other factors, if the maximum frequency they could generate was 10^{40} Hz, the equality of equations 9.19 and 9.20 suggests G ≈ 6.7×10^{-11} $m^3.kg.s^{-2}$.

Body	**Mass (kg)**	**z (Eqn 19)**	fn **(z*/z)**	**Density (kg/m³)**
Sun	2×10^{30}	2.1×10^{-6}	1.00	1.4×10^{3}
Mercury	3.3×10^{23}	3.3×10^{-13}	3×10^{3}	5.43×10^{3}
Venus	4.9×10^{24}	4.9×10^{-12}	1.24×10^{2}	5.24×10^{3}
Earth	6×10^{24}	6×10^{-12}	1.2×10^{2}	5.5×10^{3}
Mars	6.4×10^{23}	6.4×10^{-13}	4.3×10^{3}	3.93×10^{3}
Jupiter	1.9×10^{27}	1.9×10^{-9}	1.1	1.32×10^{3}
Saturn	5.7×10^{26}	5.7×10^{-10}	1.3	0.69×10^{3}
Uranus	8.7×10^{25}	8.7×10^{-11}	29	1.27×10^{3}
Neptune	1.0×10^{26}	1.0×10^{-10}	31	1.64×10^{3}

Table 9.2: *Some results determined from equation 9.19.*

The ice planets, Uranus and Neptune, have similar sizes, masses, densities and fns, ≈ 30. If their maximum frequency, ω_{MIP} was reduced to ≈ 3.3×10^{38} Hz, the equality of equations 9.19 and 9.20 suggests G ≈ 6.7×10^{-11} m³.kg.s⁻².

The large rocky planets, Venus and Earth, have similar size, mass density and fn, ≈ 120. If their maximum frequency, ω_{MLR}, was 8×10^{41} Hz, they would also have G ≈ 6.7×10^{-11} m³.kg.s⁻².

Mercury and Mars have masses differing by ≈ 2, and fns differing by ≈ 40%. They require a maximum frequency, ω_{MSR}, of ≈ 3×10^{40} Hz to have G ≈ 6.7×10^{-11} m³.kg.s⁻².

The unexplained variations in measured values of G, ≈ 1500 ppm or 0.1% (Grundlach and Merkowitz, 2018) are similar to the ≈ 0.1% difference between the fn terms for the sun and Mars, the masses of which differ by ≈ 6.3 orders of magnitude, see table 9.2.

One difference could be in their densities, listed in table 9.2. The redshifts calculated by $\alpha/2r$ start at the body's surface. The densities can be normalized by calculating the redshift from the radius all bodies would have if their density was the same as Earth. The difference is related to the cube root of their density ratios. It does not change much. That does not mean their densities do not alter frequencies.

Another possible difference is the effect of the sun's redshift at planetary orbits on that generated by the planets. Table 9.3 compares each planet's redshift with that of the sun at their orbital radius. Could the sun's strong electric field oscillations set up a resonance with those of the inner planets? The sun's redshift, and through that its electric permittivity and oscillations, is ≈ 250 times stronger than is Mercury's.

It is not until Jupiter's orbit and its large mass that it matches the sun's local strength.

Feynman's derivation of his equation 31.19 (9.16), indicates resonances could be induced. They could change the effect of the planetary oscillations, giving the appearance of a difference mass.

The sun's large mass means their mutual effect on the sun would be so low as to not register. Other possibilities include G is not constant. A variation in G requires a change in ω_M in equation 9.18. If G is not constant, the product GM for the body remains the same. That would mean mass varies according to equation 9.20. The large frequency variations required to match the accepted value of G require significant changes.

It is suggested the above results indicate there is a trend between the cumulative *fn*s of the various bodies and their mechanical properties.

Body	**Planet** $z = \alpha/2r$	**Sun at Orbit** $z = \alpha/2r$
Mercury	1.1×10^{-10}	2.6×10^{-8}
Venus	6.1×10^{-10}	1.4×10^{-8}
Earth	7.1×10^{-10}	1×10^{-8}
Mars	1.5×10^{-10}	6.5×10^{-9}
Jupiter	2.0×10^{-9}	1.9×10^{-9}
Saturn	7.3×10^{-10}	1×0^{-9}
Uranus	2.55×10^{-9}	5×10^{-10}
Neptune	3.1×10^{-9}	3.3×10^{-10}

Table 9.3: *Comparisons of planet and solar redshifts*

In the approximations used in this model, that variation could be due to changes in the maximum frequency ω_M due to internal effects. From herein it becomes conjecture as to the nature of the relationship. There are some features to note.

The redshift $z = \alpha/2r$ has been measured and is accurate. That is because it is determined outside the body. It is the effects inside the body that generate the magnitude of the oscillations and hence the change in electric permittivity that gives rise to gravity.

All measurements of the strength of the gravitational field associated with a large mass are of GM. G is assumed to be constant, and hence the measurement directly indicates mass.

Equations 9.16 and 9.17 indicate that a reduction in frequency or an increase in its magnitude will increase the redshift, and therefore electric permittivity of space surrounding it.

Gravity effects determined by orbiting bodies measures the product GM. That does not mean that G is necessarily fixed, allowing M to be determined.

Measurements of G made on Earth use low mass weights, typically less than 100 kg. Extending those results by over 20 orders of magnitude for Mars and over 28 orders of magnitude for the sun is not necessarily accurate when its origins are unknown.

In deriving his equation 31.19, equation 9.16, Feynman introduced approximations. Equations 9.17 to 9.20 are further approximations. The calculated values above show there are differences between calculated and observed results.

Despite those differences, it is suggested the above indicates a form of Feynman's idea of refractive index being caused by a vibrating electric field is applicable to gravity and space outside matter. The properties it calculates match known properties of gravity.

The very high frequency of the denominator in equation 9.18 means a permittivity change from ε_0 to ε_m requires a large number of nucleons. Equations 9.18 are consistent with gravity being very weak and requiring a large number of nucleons and mass respectively to have a detectable effect.

It is suggested that a refinement of the calculations would give a better match with observation. That is left to others.

9.5. The Effect on Matter

The high value of the maximum frequency, $\omega_M \approx 10^{44}$, Hz is much higher than the frequencies of all particles and photons. When Feynman's principle, given in equation 9.16, indicates that photons of frequencies significantly different from the oscillating frequency will respond similarly. The combined oscillation frequencies of the nucleons is $\approx 10^{40}$ Hz. That of the nucleons is $\approx 10^{23}$ Hz, lower by $\approx 10^{17}$. Linear photons have even lower frequencies. Those differences are so great that the deflection is always the same for all objects.

That enables photons with masses less than 4 eV to have the same response to gravity as massive objects such as Jupiter's 2 x 10^{27} kg, equivalent to $\approx 10^{63}$ times its mass.

With all particles composed of rotating photons, those photons will respond to changes in the electric permittivity in the same manner. Figure 9.1 shows photon A travelling at all angles to the lines of equal electric permittivity, from perpendicular at great distance, to parallel at position of closest approach. Equation 8.19 applies to all angles. The response of the rotating photons that are all particles will always be the same, irrespective of their orientation to the changing electric permittivity.

The effect is the same for linear and rotating photons. Linear photons pass rapidly from one value of electric permittivity to another, undergoing small diffractions over nucleon distances of $\approx$ 1fm. They pass rapidly through the changed electric permittivity associated with a large mass. The sun, which has 10^{57} nucleons, deflects linear photons that pass close to its surface 1.74 arc sec, a small amount for such a large mass.

Rotating photons stay in the one place, undergoing the same diffraction over the same distance. They are repeatedly subject to the same deflection. An indication of the strength of the effect can be obtained by considering gravity near Earth's surface. It has a value of $g = 9.8$ m/sec^2. We can apply Newton's equation

$$s = ut + ½at^2 \qquad (9.21$$

to a nucleon. In equation 9.21, s is the distance travelled, u is the initial velocity at time $t = 0$ and $a = g$, the particle's acceleration under gravity, Nucleons have a frequency of ≈ 4.5 x 10^{23} Hz. In 2.3 x 10^{-24} sec, the time it takes a nucleon's photon to travel the two rotations it makes in one wavelength, that acceleration causes them to be deflected by ≈ 1.1 x 10^{-48} m.

Table 9.4 shows the results of applying equation 9.22 to that deflection over the different times indicated. It shows the displacement is exceedingly small per wavelength cycle. A nucleon has to make ≈ 10^{16} wavelength cycles before it has moved ≈ 1fm, a typical nucleon dimension. It makes ≈ 10^{20} revolutions before it has travelled 1 micron. The displacement per unit time is the same for nucleons and electrons.

It does not matter how many nucleons are in an object within the varying electric permittivity, they all undergo the same deflection towards the region with higher electric permittivity. Figure 9.1 and its associated equations show that the displacement is independent of the direction of travel of the photons. If the object containing the nucleons is unsupported, each time the rotating photons repeat their trajectory, they do so from a slightly different position and velocity towards the increased electric permittivity. That is what gives the free particles their acceleration under gravity. If the object is resting on a support, the deflection will exert a force on the support, converting its mass to its weight. Its mass is fixed. Its weight depends upon the gravitational field strength.

Time (sec)	Number of cycles	Displacement (m)
2.3 x 10^{-24}	1	1.1 x 10^{-48}
10^{-20}	4.5 x 10^{3}	4.9 x 10^{-40}
10^{-15}	4.5 x 10^{8}	4.9 x 10^{-30}
10^{-12}	4.5 x 10^{11}	4.9 x 10^{-24}
10^{-8}	4.5 x 10^{15}	4.9 x 10^{-16}
10^{-6}	4.5 x 10^{17}	4.9 x 10^{-12}
10^{-3}	4.5 x 10^{20}	4.9 x 10^{-6}
1	4.5 x 10^{23}	4.9

Table 9.4: *A list of the displacement individual nucleons will experience in free fall under the influence of Earth's gravity near sea level.*

Every object containing nucleons will generate changes in the electric permittivity of space. To do that, there must be sufficient nucleons to reach a maximum frequency of $\approx 10^{40}$ Hz. That requires less than 1 gm. Having generated a change in electric permittivity, that mass will deflect the nucleons in a neighbouring body.

9.6. Discussion

In operation, a body of mass M generates a change in the electric permittivity of the surrounding space. Each rotating photon in a body of mass m experiences a deflection towards M. Without a support, mass m moves towards mass M. In the same manner, body m generates a change in the electric permittivity in the space surrounding it. That causes nucleons in body M to move towards m. That gives the mutual attraction between the two bodies, as mentioned by Newton.

Equation 9.14 shows that Newton's inverse square law is a first approximation to Einstein's theory that mass distorts space–time to give gravity. Einstein's use of approximations to derive his field equations resulted in answers that are indistinguishable from equation 9.14 for all values of $r > 3\alpha$, see figure 8.10. Both equation 9.14 and exact solutions to Einstein's field equations, 9.14a, predict that gravity is weaker than inverse square law as mass gets larger. That is essential for the anomalous precession of Mercury's perihelion and the precession of star S2 about the massive object in Sagittarius A (Hees et al., 2017; Amorin et al., 2020).

As mentioned in chapter 8, the weaker gravity causes orbiting bodies to rotate at a longer distance under a lower strength gravity. That results in the orbiting body returning a little later each orbit to its aphelion and perihelion positions. That is why they precess.

Another feature shown in figure 9.1 is the role electric permittivity plays in the photon's trajectory. As it both approaches and departs the region of changing electric permittivity, it travels at the speed of light c_0 given by ε_0. The above has established that the electric permittivity, ε_m, of the medium that is space outside matter induces the same properties as gravity. Hereafter, ε_m will be referred to as ε_G, to indicate its relationship to gravity. In ε_G, the speed of photons, to an external observer, will be c_G. An internal observer in the ε_G region, will measure the local speed of light as c_0. The change in electric permittivity is the way that mass distorts space–time.

In ε_G the photon is slowed down. From standard refraction theory, its trajectory is changed. The energy difference between the photon travelling in ε_0 and ε_G is stored in ε_G. As it departs ε_G and returns to ε_0, that energy is returned to the photon. It continues its travel at c_0, with its changed trajectory. Einstein's space–time curvature is caused by changes to the electric permittivity of space, induced into it by the high frequency oscillations of the nucleons that make up the mass of the object that distorts space–time.

As photon A in figure 9.1 approaches the mass, it enters a region of higher refractive index where it is slowed down and deflected according to the normal rules of diffraction. As it passes the centre of mass, the refractive index diminishes, the photon increases its speed as it exits the masses gravitational attraction zone of influence with the same speed and frequency as it approached it. Energy is conserved even though the photon has been deflected.

The same applies to a comet approaching Earth. It is deflected by Earth's gravity but leaves its gravitational influence at the same speed as it approached. The deflection is caused by the changed refractive index induced into the electric permittivity of space by the high frequency oscillations of the nucleons within it.

It is that distortion of trajectories with the energy stored in the changed electric permittivity of space that differentiates gravity from other forces. In the classical sense, a force, F, applied to an object of mass, M, causes it to accelerate at a rate given by $F = Ma$, where a is the acceleration. As the moon orbits Earth, it is accelerated towards Earth, always moving at a sufficient speed that it matches Earth's curvature at its orbital radius.

If gravity was a conventional force, it would use energy to make that orbit. That would show up as a reduction in the strength of gravitational attraction. There is no evidence of that occurring. That has been mentioned many times in the past. It is a reason why many physicists have mentioned that gravity is not a force. The moon's slight increase in orbital distance from Earth is explained by its action in generating tides.

It is suggested that also shows up in Newton's universal gravitational constant, $G = 6.67 \times 10^{-11}\ m^3kg^{-1}sec^{-2}$. When mass M of the gravitationally attracting body and the distance r, are applied in equation 9.14, the result is an answer in $msec^{-2}$, the unit of acceleration. It indicates that whatever is causing gravity is a dimensionless quantity. A deflection angle is always expressed as a fraction of a complete revolution.

There appears to be nothing in the above that prevents the deflection caused by mass induced changes in electric permittivity from explaining other known properties of gravity. The effect of gravity travelling at the speed of light is governed by the electric permittivity, which, together with magnetic permeability, set the speed of light. It does mean that the speed of light, as observed by an external observer, varies with the strength of the gravitational field. That shows up in the gravitational redshift of photon. As photons go from ε_G to ε_0, their speed increases, stretching their wavelength.

All local measurements of the speed of light will get the same answer because of Maxwell's equation, 9.3, depends upon the local value of electric permittivity. Changes in electric permittivity automatically change lengths and time. Einstein's theories are called relativity because every observation is relative to the circumstances of the observer. That is a consequence of the speed of light being constant for all local observers. The mass induced changes in electric permittivity distort time by changing photon frequencies and length by changing photon wavelengths.

Gravitational waves (Einstein 1918), explained as distortions in space–time, become variations in electric permittivity. As two massive objects rotate about their mutual axis, their motions drag their changes in electric permittivity with them. Those changes spread through the surrounding space.

In this presentation, the properties of gravity were derived from sub-atomic and quantum properties of matter. It goes beyond the standard model, uniting quantum and sub-atomic particle properties with gravity.

The above has given a physical explanation of the role mass has in changing the direction of photons travelling in space surrounding it. The reasons relate to the rotating photon structure of nucleons. That causes them to generate high frequency electric field oscillations.

Feynman's principle is that oscillating electric fields generate a change in the electric permittivity of the space in which the electric field permeates. Through the work of Huygens and Maxwell, we know that introduces a change in the refractive index of space. A change in electric permittivity of space deflects the direction of travel of photons towards the region of higher electric permittivity.

Above a small number of nucleons, the random nature of the high frequency oscillations of individual nucleons' electric fields reaches a frequency limit, 10^{44} Hz. Further nucleons increase the amplitude of those oscillations. The two effects combine. The increased amplitude of the oscillations increases the changes in the electric permittivity, and hence deflection, in a linear manner.

An important feature from equations 9.12 to 9.14 is that they show the variation in electric permittivity with distance gives the properties of gravity. Newton's inverse square law is a first approximation that works well for all situations when $r >> \alpha$. Einstein's work is a second approximation which works when $1 + \frac{\alpha}{r}$ is indistinguishable from $e^{\frac{\alpha}{r}}$. For most practical purposes, that applies when $r > 3\alpha$. The exact equation predicts the toroid shape shown in figure 8.11.

This mechanism of gravity being the differential of the electric permittivity further shows there is no mechanism by which black holes can be generated. Gravity being stronger than inverse square requires mass to decrease the electric permittivity of free space. There is no physical mechanism by which a property can be removed when there is nothing to remove.

Section 9.4 only dealt with redshift produced by the high frequency oscillations of the particles within massive bodies. Section 9.2, particularly equation 9.12, show that changes in the electric permittivity of space control all other aspects of gravity.

Gravity and electromagnetic effects both depend upon the electric permittivity of space. This gives a smooth transition between electric, magnetic and gravity effects. Mass responds to whichever is the strongest in its vicinity.

Equation 9.17 is an approximation. However, it does suggest there is a trend for redshift to increase with lower frequency inducing a greater change in electric permittivity. The greater the change in electric permittivity, the stronger the attraction towards the charged source. This shows up with electrostatic dust attraction. A charged object attracts dust and other lightweight objects against gravity. It suggests a continuity between gravity and electric charge through changes in electric permittivity.

9.6. Summary

Photon speeds are controlled by the absolute value of electric permittivity of space. Deflections are due to changes in the electric permittivity. Expressions for the electric permittivity values were derived from the redshift due to gravity and the redshift due to changes in electric permittivity. The two were equated. The differential of the electric permittivity with distance was determined. It represents the distortion of photon trajectories at different distances. It showed the same proportionality as that obtained from redshift only.

Newton's universal gravitational constant, G, together with M, gives the equation for gravitational field strength that includes both Newton's inverse square law and Einstein's modification that gravity is caused by mass distorting space–time, see equations 9.12 to 9.15. It indicates that Einstein's mass distorting space–time, which shows up as photon redshift, is caused by changes to the electric permittivity of space caused by the high frequency oscillations of the nucleons' rotating photon structure.

This presentation also physically shows why variations in gravity move at the speed of light. That is the rate of passage of electric fields. Although that was an essential feature of Einstein's gravity, some perceptions of Newtonian gravity, often expressed in textbooks, were that its effect was instantaneous (Kenyon, 2002; Ohanian and Ruffini, 2013. Haug (2021) provided a mathematical reason why that interpretation was not correct. He did not give any physical reason for Newtonian gravity to be limited to the speed of light. This presentation of the physical origins of gravity shows why gravity propagates at the speed of light. That Newton inverse square law is a first approximation does not mean that it eliminated gravity's finite speed.

Gravity is a very live and dynamic phenomenon. A large number of nucleons, each generating a high frequency electric field, $\approx 10^{23}$ Hz, alter the electric permittivity in the space surrounding them. In turn, that variation deflects photons that experience it according to the differential of that variation with the distance from its centre.

With all matter particles composed of rotating photons, they are deflected by that differential. That deflection is experienced as gravity. It shows that the sum of the out of phase frequencies of the rotating photons that are the nucleons soon reaches the Planck frequency, the highest frequency possible. That results in changes to the electric permittivity of space increasing linearly with the number of nucleons in the massive object. The deflections increase linearly with mass.

The above has established the proportionality or physical trend of the effect described. It shows a proportionally related to Newton's universal gravitational factor, denoted G. It has neither derived G independently, nor shown that it is constant under all conditions.

The mechanism relies upon Feynman's principle that oscillating electric fields alter the electric permittivity of space. Feynman limited his calculations to space inside matter. Electric field resonances brought about by the mutual interaction between electron frequencies and photon frequencies gave rise to the refractive index of the medium. In this study on gravity, the oscillating electric fields within individual particles directly changed the electric permittivity of free space.

The rotating photon model of matter was used to derive the physical origins of gravity. The high frequency alternating electric fields of the nucleons generate a change in the electric permittivity of space. It was shown that proposed mechanism matched the known properties of gravity through the differential of the ratio of the electric permittivity of matter free space over that of space near matter. That differential in electric permittivity generates a differential in refractive index, which deflects in the trajectory of photons. With all matter particles composed of rotating photons, deflection gives rise to gravity. That differential explains Newton's inverse square law and the gravity from Einstein's field equations.

In this presentation, the properties of gravity were derived from sub-atomic and quantum properties of matter. It goes beyond the standard model, uniting quantum and sub-atomic particle properties with gravity. It shows that mass distorts space–time by varying the electric permittivity of space. Gravity is a deflection, not a force. Energy is stored in the electric permittivity as photons are slowed down approaching a massive object and returned to it as it moves away. With no energy loss, gravity does not lose its strength over time.

This presentation has shown all the properties gravity exerts. It is suggested this presentation matches more gravity properties than any, if not all, other gravity theories. It has no space for black holes. Even Einstein's calculations could not be used to derive them. It has not been able to derive independently derive G. The suggestion is that G may vary. That leaves the study of gravity under this mechanism with a lot more to be done.

9.8. List of References

Abuter A, Amorim A, Bauböck M, et al. (2020); "Detection of the Schwarzschild precession of the star S2 near the Galactic centre massive black hole", Astronomy and Astrophysics, 636, L5 (14pp)

Adam MG, (1948); "Interferometric Measurements of Solar wave-lengths and an Investigation of the Einstein Gravitational Displacement," Monthly Notices of Royal Astronomical Society, **108**, pp 446 – 464

Adams WS, (1910); "An Investigation of the Displacement of Spectrum Lines at the sun's Limb," Astrophysical Journal, **31** pp. 30 – 61

Einstein A, (1911); "Uber den Einfluss der Schwerkraft auf die Ausbreitung des Lichtes", Annalen der Physik, 35, pp 898 – 908

Einstein A, (1915 a); "Zur Allgemeinen Relativätstheorie," Sitzungsberichte der Königlich Preussischen Academie der Wissenschaften, Part 2, pp 778 – 786

Einstein A, (1915 b); "Erklärung der Perihelbewegung des Merkur aus der Allgemeinen Relativätstheorie," Sitzungsberichte der Königlich Preussischen Academie der Wissenschaften, Part 2, pp 831 – 839

Einstein A, (1916) "Die Gründlage der Allgemeinen Relativätstheorie," Annalen der Physik, 42, pp 769 – 822

Einstein A, 1952, "The Foundations of the General Theory of Relativity" in Einstein, A. et al. (ed) (New York: Dover Publications) pp 109–164

Einstein A, (1918), "Uber Gravitationswellen." Königlich Preußische Akademie der Wissenschaften (Berlin). Sitzungsberichte,154–167.

Feynman RP, (1963); "The Origin of the Refractive Index": The Feynman Lectures on Physics, Vol 1. Ch 31: https://www.feynmanlectures.caltech.edu/I_31.html

Galileo G 1610 Siderius Nuncius (Venice: Galileo G)

Galileo G, (1632); "Dial Sopra i due Massimi Sistemi del Mondo (Dialogue Concerning the Two Chief World Systems)" (Florence: Galileo G):- https://en.wikipedia.org/wiki/Dialogue_Concerning_the_Two_Chief_World_Systems

Gonzalez Hernandez, J.I. et al., (2020); "The solar gravitational redshift from HARPS-LFC Moon spectra. A test of the general theory of relativity", *Astronomy and Astrophysics, 643 A146,* https://doi.org/10.1051/0004-6361/202038937

Greiner W, Schramm S, Stein, E (2007). "Quantum Chromodynamics" Berlin Heidelberg: Springer. ISBN 978-3-540-48535-3.

Grundlach JH and Merkowitz SM (2018); "Measurement of Newton's Constant using a Torsion Balance with Angular Momentum Feedback" https://arxiv.org/pdf/gr-qc/0006043.pdf

Halzen F, Martin A, (1984), "Quarks and Leptons: An Introductory Course in Modern Particle Physics", John Wiley & Sons, NY ISBN 978-0-471-88741-6

Haug, EG, (2021); Demonstration that Newtonian gravity moves at the speed of light and not instantaneously (infinite speed) as thought! Journal of Physics Communications, 5 025005: https://doi.org/10.1088/2399-6528/abe4c8

Hees A, Do T, Ghez et al (2017); "Testing General Relativity with Stellar Orbits around the Supermassive Black Hole in Our Galactic Center" Phys Rev Lett 118 211101 2017 DOI: https://doi.org/10.1103/PhysRevLett.118.211101

Huygens C (1690) Treatise in Light: https://archive.org/details/treatiseonlight031310mbp/page/n3/mode/2up DOI: 10.1086/340578

Kenyon IR, (2002; General Relativity (Oxford: Oxford University Press)

Krasnoholovets V (2021); Ed. "The Origins of Gravity from First Principles", Nova, NY, ISBN: 978-1-53619-566-8

LeVerrier U 1859 Comp trend l'Acad de scie Paris 49 379 (https://archive.org/details/comptesrendusheb49acad/page/37

Minkowski H, (1908); "Raum und Zeit," Jahreschericht der Deutschen Mathematiker-Vereinigung, 18, pp. 75-88. (Address delivered at the 80th Assembly of German Natural Scientists and Physicians, Cologne, September.)

Minkowski H, (1952); "Space and Time," in Lorentz HA, Einstein A, Weyl H and Minkowski H, The Principle of Relativity, translated by Perret W and Jeffery GB, Dover Publications, New York, NY, pp 75 – 91

Maxwell JC, (1865); "A Dynamical Theory of the Electromagnetic Field", Philosophical Transactions of the Royal Society London, 155, pp 459 – 512

Newcomb S 1882 US Nautical Almanac Office Astronomical paper 1 (Washington) p 473 (http://www.relativitycalculator.com/pdfs/mercury_perihelion_advance/S.Newcomb.pdf)

Newton I, (1687); "Philosophiae Naturalis Prinicipia Mathematica", London

Newton I, (1729) Mathematical Principles of Natural Philosophy 1 ed. A Motte (London: Motte B) Propositions 1 – 32, 45

Oerter R, (2006); "The Theory of Almost Everything, the Unsung Triumph of Modern Physics", Penguin Group, ISBN 978-0-13-236678-6

Ohanian HC and Ruffini R, (2013); Gravitation and Spacetime (Cambridge: Cambridge University Press)

Qui-Hong H, (2007); "The Nature of the Electron" http://arxiv.org/pdf/physics/0512265v1.pdf

Robinson VNE (2011); "A Proposal for the Structure and Properties of the Electron, Particle Physics Insights, 4, pp 1 – 18

Robinson V N E (2021); "Physical Explanations of Einstein's Gravity", J. Phys. Commun. 5 035013 https://doi.org/10.1088/2399-6528/abee2f

Roy S and Sen AK, (2019); "Study of Gravitational Deflection of Light Ray", J. Phys. Conf. Ser. 1330 012002; doi:10.1088/1742-6596/1330/1/012002

Sagan C, (LOC) "Whose Revolution? Copernicus, Brahe & Kepler", https://loc.gov/collections/finding-our-place-in-the-cosmos-with-carl-sagan/articles-and-essays/modeling-the-cosmos/whose-revolution-copernicus-brahe-and-kepler

Schumm BA, (2004). Deep Down Things: The Breathtaking Beauty of Particle Physics. *Johns Hopkins University Press.* ISBN 978-0-8018-7971-5

Williamson JG and van der Mark MB, (1997); "Is the electron a photon with toroidal topology?" Annales de la Foundation Louis de Broglie, **22**, pp 133 – 178

10. Some Galaxy Properties

Abstract:

Astronomers have observed galaxies for well over a century. Along the way they have observed that some properties that theoreticians found difficult to explain. Included in these are galaxies rotating faster than they should under Kepler's law of gravity. General relativity does not provide the answer. Cosmologists currently explain that by the existence of belts of unknown dark matter around each galaxy's centre. Other observed galaxy properties include the emission of GeV gamma rays from the central galactic regions and Fermi lobes extending perpendicular from the central core for about half the galaxy's radius. This presents alternative explanations for those and other observations.

10.1. Introduction

There are billions of galaxies in the observable universe. Figure 10.1 gives two examples of spiral arm galaxies. There are many other types, including but not limited to globular, elliptical, and lenticular. There are so many different variations that no two galaxies are alike. Getting a single theory to cover them all is not possible. The best that can be done is to provide some information relating to some galaxy properties.

Galaxies typically contain between about 10^9 to 10^{12} solar masses as luminous matter, namely stars. They also average about 10 times that mass in a cold molecular gas and dust cloud surrounding their stars.

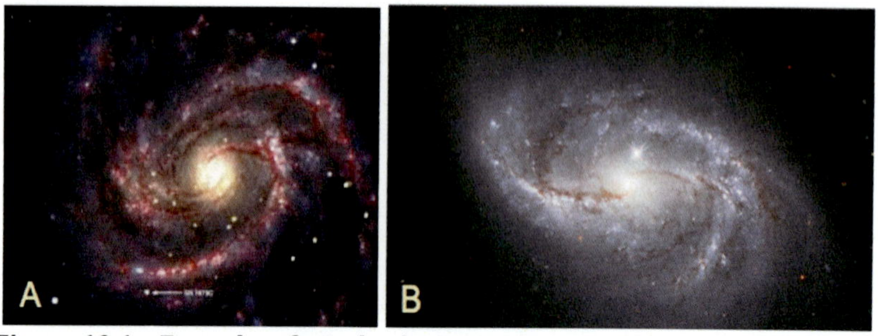

Figure 10.1: *Examples of spiral galaxies. Images courtesy ESA and Hubble.*

Within their stellar matter most galaxies have a few percent of the total number of stars as inter stellar particulate matter. With such huge mass, galaxies are heavily influenced by gravity.

Excluding the massive objects at the centre of galaxies, the differences between Newton's gravity and Einstein's gravity within galaxies is so small that Newtonian calculations work well.

Even for an object of mass 3 x 10^{10} solar masses, $m_{\odot}$, the difference between Newton's, Einstein's, and the redshift gravity, is only few percent at distances of about 10^{11} metres, $\approx$ 0.01 light years from the centre of mass. It makes no significant difference on the Newtonian calculations of gravity within galaxies.

10.2. Dark Matter?

Gravity is the force that binds galaxies together. Galaxy rotation, first detected by Slipher (1914), prevents their collapse. One feature that galactic studies revealed was that galaxies were observed to rotate faster than they should under the influence of gravity described by Kepler's (1609, 1619) laws. It treated the rotation of stars as if the mass of all the stars at a lesser orbit were concentrated at the centre of rotation of all the stars.

Kepler derived his laws of motion for planets revolving about the sun. His law of greatest interest is that all planets draw arcs of equal area when travelling for the same time around the sun. That area is determined by the difference in angle of a line between a planet and the sun's centre, as travelled for a given time. The further out are the planets, the slower they travel.

When it comes to galaxies, the same principle is applied using Newton's shell theorem. If mass is uniformly distributed within a radius, *r*, it can be treated as if all the mass is concentrated at its centre. Mass of uniform density and distribution outside that radius has no effect on the mass at *r*.

Figure 10.2 illustrates the expected rotational velocity of stars at different distances from the galactic centre, calculated using Kepler's law. It shows that the speed of star rotation should rise rapidly to a maximum over a short distance and then decline, getting slower as the stars are more distant from the centre of motion. The shape varies with the size of the galaxy.

Figure 10.3 shows some measured curves of the rotational speeds of stars at different distances from the galactic centre (Sofue et al., 1999; Rubin, 2000; Sofue and Rubin, 2001; Sofue, 2014).

All curves show the initial slow rotation near the centre that rises rapidly to a maximum. After that they all show the same trend of flattening of the rotation speed with increasing distance.

The expected decrease shown in figure 10.2 does not occur. That trend was documented by Rubin (1965) and has been observed many times. The experimental evidence is overwhelming. Galactic rotation does not follow Kepler's laws. Explanations have been sought.

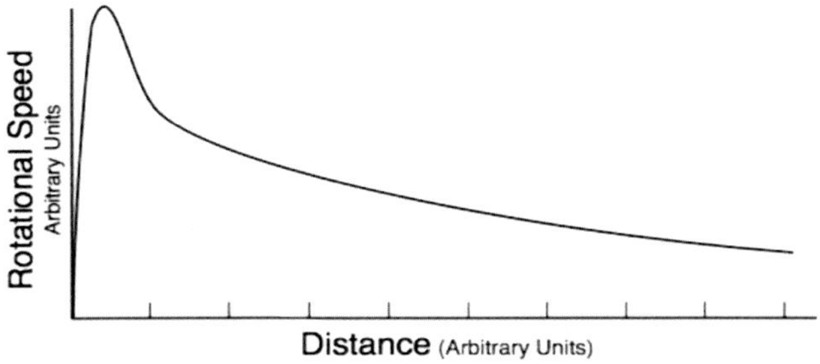

Figure 10.2: *Schematic illustration of the rotational speed of objects at their distance from the centre, showing what is expected when the mass is distributed throughout the galaxy rotating under its own gravitational field.*

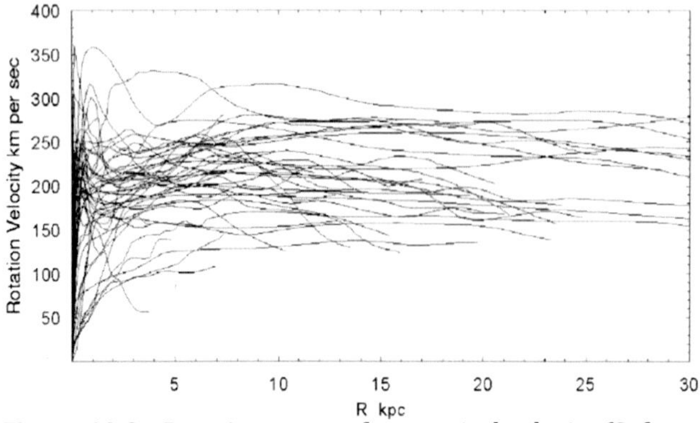

Figure 10.3: *Rotation curves of some spiral galaxies (Sofue et al., 1999; Sofue and Rubin, 2001).*

Galaxy rotation is a complex topic that has received much attention over the past one hundred years. A comprehensive review was given by Madore (2017).

One explanation forwarded was that there was additional matter in the galaxy that added extra gravitational attraction but had no other observable effect. It was called dark matter. Figure 10.4 A gives another example of a spiral galaxy. Like other spiral galaxies, it rotates in the direction indicated by the arrow. The spiral arms lag behind the rotation of the central regions. The circles indicate the approximate distribution required of the dark matter for the stars to rotate at the speeds indicated in figure 10.3. Figure 10.4 B shows that approximately six times as much dark matter over detected baryonic matter is required to match the figure 10.3 rotational speeds.

The dark matter only reacts gravitationally with the detected baryonic matter. Candidate particles have never been detected. No valid reason has ever been given as to why dark matter should exist in such quantities within a galaxy and only congregate near the indicated zone. The circumstances seem highly improbable. Other explanations have been sought.

A 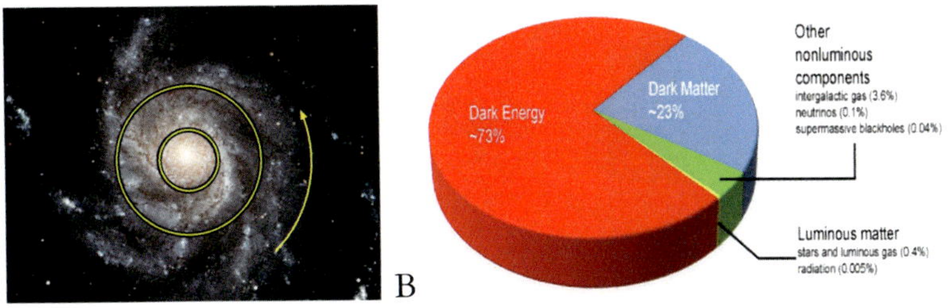 B

Figure 10.4: *A: Pinwheel galaxy: Yellow circles represent approximate distribution limits required of dark matter. (https://www.spacetelescope.org/images/archive/category/galaxies/) B: Ratio of luminous and non-luminous matter to dark matter required to match the observed rotation rates shown in figure 10.3. https://www.osti.gov/servlets/purl/1410794)*

The main problem with that theory is that, as figure 10.1 and other galaxy images show, galaxies are not of uniform density. Kepler's law and Newton's shell theorem do not apply to galaxies and their rotation speeds. Other explanations are needed.

10.3. Alternative Newtonian Explanations

Some attempts have been made to show the observed galactic rotation can be explained by Newtonian mechanics. A detailed study was undertaken by Pavlovich et al. (2014) and Sipols and Pavlovich (2020). They divided each galaxy into segments and assigned a mass to each segment based upon the luminosity of that segment. An example of the segmented sections is given in figure 10.5.

That was followed by a summation of the Newtonian attraction of each segment on all other segments. Some of their results, each for the named galaxy, are shown in figure 10.6. The blue circles represent experimentally measured points of rotational velocity in kilometres per second. The red line through them is their calculated plot for the surface mass density indicated in green.

The vertical axis on the left-hand side of each, is the rotational speed in kilometres per second for the measured and theoretically calculated for the distance, in kilo parsecs, from the centre.

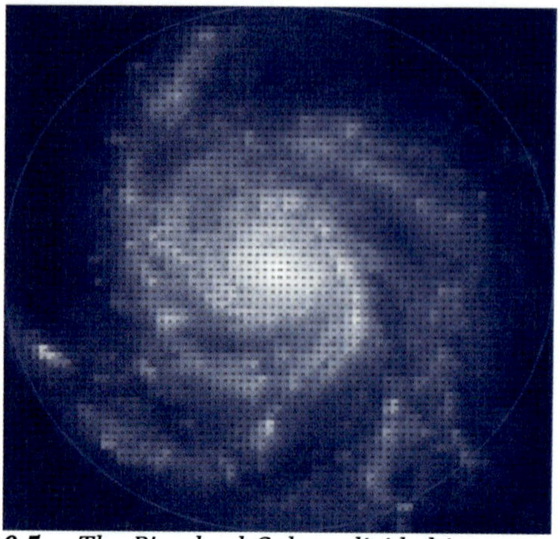

Figure 10.5: *The Pinwheel Galaxy divided into segments (Pavlovich et al., 2014).*

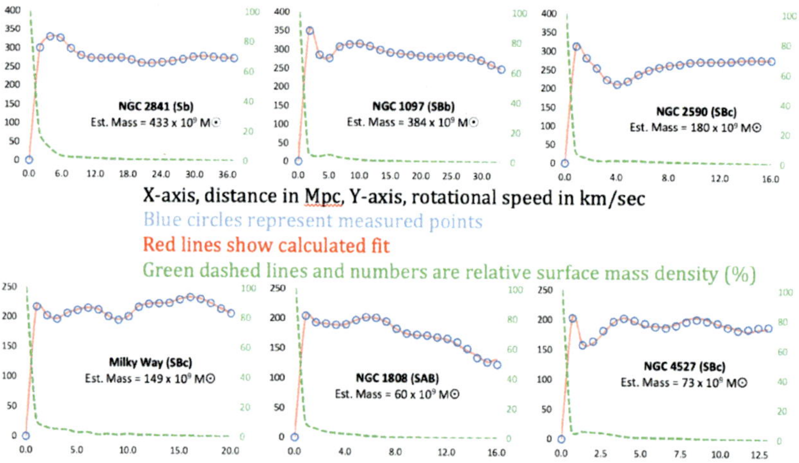

Figure 10.6: *Some results from the Pavlovich et al. (2014) study of the effect of segmented Newtonian gravity on other segments within the galaxy.*

The green numbers on the right-hand side of each graph are the relative mass densities for the estimated mass of each galaxy in the plot. It shows that the galaxy's centres have the greatest density, being 100% in each plot. The results display quite good agreement between prediction and observation.

Sofue (2014) has detected similar mass density distributions to those shown in the green curves in figure 10.6. The surface mass density for galaxies has a high core density followed by a rapid drop off in density to about 20% in core region of radius ≈ 1.5 kpc, ≈ 5,000 LYs. That is followed by a slow decline to edge of the galaxy.

Sipols and Pavlovich (2020) extended those calculations. They studied 214 galaxies and got good agreement in 132 of them. Their main difference was in mass to luminosity ratio. They have offered explanations for the remaining 82. In the absence of contradicting evidence and bearing in mind the huge variety of galaxy shapes and sizes, their explanation seems plausible.

Their work suggests the simplification of using Kepler's law, the inverse square law of gravity about a gravitational centre, was not a good approximation. Their use of the inverse square gravity of each segment of stars within a galaxy on all other galaxy segments, matches the observed galaxy rotation to their surface mass density. There is no need for additional dark matter to explain the observed galaxy rotation speeds.

Rourke (2020), like the Pavlovich/Sipols team, used Newtonian gravity's inverse square law and frame dragging to achieve the results shown in figure 10.7. The numbers along the axis lines, 20,000 and 40,000 represent the distances in light years from the centre of the galaxy. The density of the lines is proportional to the density of matter accumulating around those positions. Figure 10.8 shows an example of a galaxy that has a stellar mass distribution like that shown in figure 10.7.

Between them, the works of the Pavlovich and Sipols team and Rourke, indicate that Newtonian gravity is sufficient to explain many of the observed galaxy rotation curves. An important feature of both those studies is that the observed rotation and spiral arms of galaxies would automatically happen as the lowest stable state of the assembly of stars.

Their results in figures 10.7 and 10.8 need to be evaluated in terms of Newton's (1687, 1729) shell theorem. It relied upon the mass density being uniform. For Kepler's law to be valid, galaxies must have a uniform density. A quick glance at any galaxy image shows they do not have uniform density.

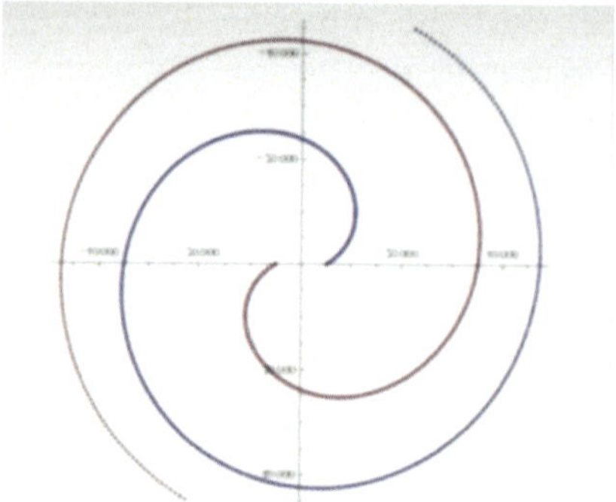

Figure 10.7: *Expected distribution of galactic matter on spiral arms, based upon frame dragging principles (Rourke, 2020)*

Figure 10.8: *Galactic spiral arms that show the expected mass distribution based on the frame dragging principles shown in figure 10.7 (Rourke, 2020).*

That reason suggests it is not a valid approximation to evaluate galaxy rotation in terms of Kepler's law. Neither the Pavlovich/Sipols nor Rourke approaches disagree with Newton's gravity or his shell theorem. It does raise the question of "Why was Kepler's law always used?" That is for others to answer.

Galactic rotation speeds are not the only unexplained phenomena associated with galaxies. The following introduces another effect, which also operates on the inverse square law. In addition to the other effects it introduces, its action would not be distinguishable from additional gravity effects.

10.4. Cosmic Ray Effects

Cosmic rays are high energy charged particles that originate from extra-terrestrial non-solar sources. Their earliest detection was made using William Gilbert's electroscope. All discharged naturally. That established they were due to charged particles passing through the electroscopes. More recent investigations established they originate from catastrophic events such as supernovae. Most originate from inside our galaxy.

Figure 10.9 shows the measured distribution of the number of cosmic rays, vertical axis, versus their energy, horizontal axis, as measured above Earth atmosphere (Delahaye, 2010a).

It shows that the greatest number of cosmic rays have a positive charge with an energy of above 1 GeV. Adding positive cosmic rays of energy between 0.1 and 10 GeV increases the positive cosmic ray flux to over 10^4 per m^2. That is 1 positive cosmic ray per second for every sq cm^2 of surface. Electrons have similar energies, with a flux of less than 1% of the proton flux.

The most obvious reason for the different flux densities of protons and electrons in cosmic rays is the effect of the galaxy's magnetic field on their propagation through it. In an explosion, most particles are ejected at about the same velocity. Their lighter rest mass, approximately 1837 times less than a proton, means they are affected much more by a galaxy's magnetic field.

All galaxies have magnetic fields (Beck and Wielebinski, 2013). A plot of the planar view illustrating our Milky Way galaxy's magnetic field, is shown in figure 10.10 (Delahaye 2010b). Although the magnetic fields are weak, the distances cosmic rays travel are large enough for them to affect charged particles.

The magnetic fields deflect and slow the lower mass electrons much more than the higher mass protons. Protons and charged nuclei will penetrate much further through the galaxy than will electrons. This has two effects. Electrons travel further to traverse the same linear distance. Their curved trajectory causes them to emit photons,

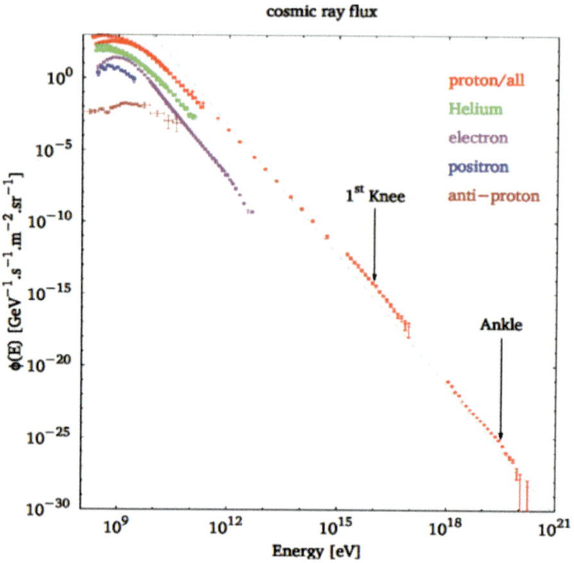

Figure 10.9: *Cosmic ray flux verses energy observed above Earth atmosphere (Delahaye, 2010a).*

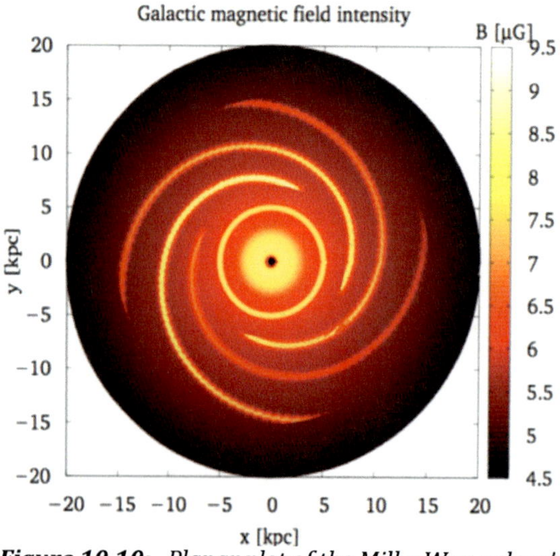

Figure 10.10: *Planar plot of the Milky Way galaxy's magnetic field.*

and thus lose energy. They combine to make electrons take longer to arrive at a destination than protons. That offers the possibility of a separation of the charges.

The various images of galaxies and figures 10.6 show there is a concentration of stars in the core. Although galactic cores contain a high density of stars, they represent ≈ 20% of the mass of galaxies. That leaves about 80% of the galactic stars outside of the core. With supernovae approximately uniformly distributed within the stars, about 80% of the cosmic rays generated would originate from outside the core. Within the galactic plane, more cosmic rays would head towards the central core than away from it.

Figure 10.9 shows the cosmic rays include both protons and electrons. Electrons are emitted at the same time and place as the much more massive protons and nuclei. They take longer to reach the central core region where there was a high density of stars. That creates a situation where there is a concentration of excess positive charge over electrons. That leads to a charge imbalance between the inner and outer galaxy regions.

10.5. Effects of Charge Imbalance

The electric repulsion between two protons is given by

$$F_E = \frac{q^2}{4\pi\varepsilon_0 r^2} \qquad (10.1$$

where q is the unit electric charge, ε_0 is the electric permittivity of free space and r is the distance between their centres. The gravitational attraction between two protons is given by

$$F_G = \frac{Gm_p^2}{r^2} \qquad (10.2$$

where G is Newton's universal gravitational constant, $m_p$is the mass of a proton and r is the distance between their two centers of mass. Within a galaxy, detectable gravitational redshift has almost no effect on the rotation of stars outside the central inner core surrounding a massive object of millions to billions of solar masses $M_\odot$. The object in the centre of galaxy M87 has an estimated mass of several billion solar masses. Non-Newtonian gravity effects are not detectable in galaxy rotation beyond a radius of $\approx 10^{-3}$ Lys. Galaxy radii are typically tens of thousands of light years. They can be ignored in these galaxy rotation calculations.

Combining equations 10.1 and 10.2 gives:

$$\frac{F_E}{F_G} = \frac{4\pi\varepsilon_0}{Gm_p^2q^2} \qquad (10.3$$

The electric force is about 8×10^{37} times stronger than the gravitational force. With both having an inverse square law of attraction, it would be difficult to detect an effect different from gravitational attraction if there was a means of separating the charges. Cosmic rays provide a source of charged particles. Magnetic fields provide a means of separating them.

A separated charge excess of one proton and 1 electron per 10^{18} nucleons is all it takes for electric attraction to rival and exceed gravitational attraction. All we need is a way that can occur. Similarly, an excess of 1 proton in 10^{18} nucleons associated with each of two bodies, is sufficient to overcome their gravitational attraction.

Reucroft (2015) investigated if such a mechanism was feasible. He concluded it was but gave no mechanism by which the charge could be either stored or acted. It was little more than a statement of those equations.

The overall effect of the cosmic rays, the mass difference between protons and electrons and a galaxy's magnetic field, accumulates positive charge in the galaxy's high mass density regions at the core and in spiral arms. That leaves a negative charge excess in the outer galactic regions. The opposing charge will attract each other with a force that is over 10^{36} times stronger than gravity for individual nucleons. It now needs to be determined if there is a possibility of a charge accumulation and distribution of sufficient magnitude that it could cause the outer galactic stars to rotate faster while not causing the inner galactic stars to rotate faster.

The above suggests that there will be an excess of positive charge associated with the inner galactic stars, leaving an excess of negative charge associated with the outer galactic regions. The following calculations demonstrate the effects such charge imbalances generate. They are divided into five parts:

i. The effect of protons on individual stars
ii. The cumulative effect of protons on all the stars in a galaxy's core
iii. The effect of electrons on individual outer stars
iv. The cumulative effect of electrons and all outer stars
v. The cumulative effect of the positive charge in a galaxy's core on the negative charge at the galactic edges.

Because of the vast distances and masses involved with galaxies, calculations will be order of magnitude only. If the effect depends upon exact calculations, it is considered that, with all the variables, it is unlikely to occur.

10.5.i. Cosmic Protons on an Individual Star

To simplify these calculations, each star will be assumed to have a mass $M_S = 10^{57}$ nucleons and a radius of 700,000 km, conditions that approximate our sun. The trend will be the same for stars of other dimensions. That gives the star a surface area of ≈ 6.16 x 10^{12} km^2 or 6 x 10^{22} cm^2 when rounded to whole numbers.

Cosmic rays making their way into the galactic centre have a probability of impinging upon a star. Cosmic rays striking a star with a radius of 700,000 km at the rate of ≈ $1/cm^2$ sec will add ≈ 6 x 10^{22} protons to the star every second. That causes the star to become positively charged. When it is sufficiently charged, individual free protons will be repelled from it when the electric repulsion force F_E is greater than the gravitational attraction force F_G. That occurs when

$$F_E = \frac{Qq}{4\pi\varepsilon_0 r^2} > F_G = \frac{GM_S m_p}{r^2} \qquad \text{(10.4}$$

where Q is the charge on the star, M_S is the mass of the star and r_s is the star's radius. When F_E for a proton is greater than F_G for the same proton, electric repulsion will cause lose protons to leave the star. From equation 10.3, that is given by:

$$\frac{F_E}{F_G} \approx \frac{10^{36}Qq}{M_S m_p} \qquad \text{(10.5}$$

For protons to be expelled from the star there needs to be a proton excess equal to the number of nucleons in the sun multiplied by $\approx 8 \times 10^{-37}$. A star, such as our sun, with $\approx 10^{57}$ nucleons, requires an excess of $\approx 1.25 \times 10^{21}$ protons $\approx$ 133 Coulombs (*C*) to expel single protons against its gravitational attraction. This would charge up the sun to a voltage, *V*, given by

$$V = Q/C \qquad (10.6$$

where Q and C are the star's charge and capacitance respectively. The capacitance of a sphere of radius *r* is given by

$$C = 4\pi\varepsilon_0 r \qquad (10.7$$

The capacitance of a star with a radius of 700,000 km is 7.77×10^{-2} F. Those values give the star a positive electric charge potential of $\approx$ 1,700 volts.

The sun's thermal energy and turbulent activity through solar mass ejections and other causes, result in proton ejection well before that value is reached. Even in periods of quiet solar activity, that turbulence is so great it ejects other atoms. There is what is called a solar wind of particles moving away from the sun. It is made up of about 92% hydrogen, 8% helium, by mass, and trace amounts of some higher Z elements. Their average speed varies between about 300 and 800 kilometres per second.

Data from www.spaceweather.com suggests there are approximately 10,000 hydrogen atoms thermally emitted from the sun for each proton emitted by this mechanism. The sun's ionizing ultra-violet light would ionize most of them, producing free electrons and protons. Those free electrons will be attracted towards the highest positive charge.

It is suggested that those electrons are captured by protons within the solar corona and emit X-rays of several hundred electron volts. One electron volt corresponds to a temperature of over 11,000 degrees Celsius. That would make it look as if the solar corona had a temperature of several million degrees C.

At a proton arrival rate of $\approx 6 \times 10^{22}$/sec, a charge accumulation of $\approx 1.25 \times 10^{21}$ protons take $\approx$ 50 mS. The ejected protons create a cloud of protons around the star, all being repelled from it. As the protons spread out from the star, there would be a charge within a volume. That would generate a positive voltage around that star. That cloud of protons would grow at the rate of $\approx 10^{22}$ per second. They are all repelled from the star. From Newton's third law of motion, they are also repelling the star.

Solar wind observations (www.spaceweather.com; Uprety 2016) show that there is a stream of protons flowing away from the sun at hundreds of km/sec. The approximations that $\approx 6 \times 10^{22}$ ejected protons flow away at an average of 500 km/sec is used in the following calculations. That represents $\approx$10,000 Amps (A), travelling away from the sun at $\approx$500 km/sec. The effect of neutralization by cosmic electrons can be added later. "Typical" solar wind conditions are $\approx$ 10 protons per cm^3 traveling at 500 km/sec at Earth radius.

The density of ejected protons at Earth radius represents less than $1{:}10^{10}$ of the total number ejected. The remainder appears to be ejected hydrogen atoms that have been ionized by the intense UV around the sun. They are accompanied by ejected electrons and play little part in the charge accumulation or neutralization.

That causes the star to accumulate a proton cloud of charge $Q = D.t$, where D is the dose rate of protons leaving the sun and t is time, around it. The capacitance of a sphere of radius $r + vt$ where v is the speed at which the photons are ejected and t is the time after starting the process, is given by $C = 4\pi\varepsilon_0(r + vt)$. Equation 10.6 can be rewritten as

$$V = \frac{Dt}{4\pi\varepsilon_0(r+vt)} \qquad (10.8$$

When v.t >> r, equation 10.8 reduces to

$$V = \frac{Dt}{4\pi\varepsilon_0.vt} = \frac{D}{4\pi\varepsilon_0 v} \qquad (10.9$$

Using D = 10,000 A and v = 500 km/sec gives V $\approx 1.8 \times 10^8$ volts if the star had a defined surface at that radius. There is no defined surface around a star. The above serves only to indicate that the accumulation of positive charge behind the emitted protons would accelerate them. Figure 10.11 illustrates the approximate charge accumulation rate around the sun if there was no charge neutralization. It shows charge accumulation at 700,000 km from the sun's surface would be approximately equal to half the maximum voltage accumulation, in this case 100 MeV from a maximum of 200 MeV.

The positive charge builds up away from the sun's surface. The solar corona is just outside the sun's surface. It has a higher potential than the sun's surface. Electrons released from ionized hydrogen atoms would be attracted to the higher potential of the solar corona than the sun's surface or where they were ionized. The electrons' small mass means they will be captured quickly, within a potential difference of a few hundred volts. As mentioned above, those X-rays would make it look as if the solar corona's temperature was several million °K.

With electrons neutralizing the charge, increased distance from the sun and the magnetic fields associated with the sun would result in the protons slowing down. Protons with energies in the range 2 – 20 MeV, have been observed in the sun's heliosphere (Kecskeméty et al., 2005). Their fluxes and energies were variable.

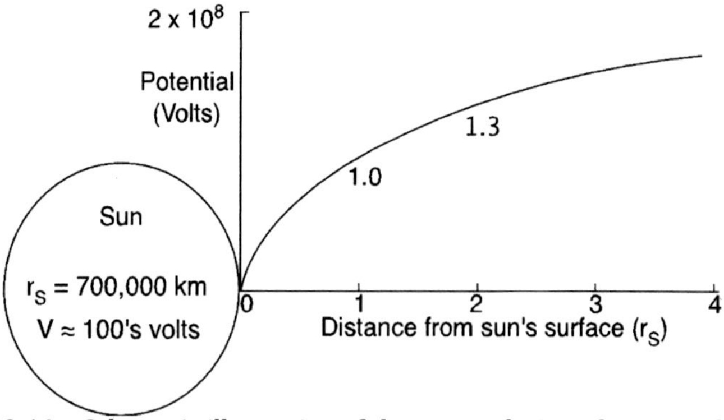

Figure 10.11: *Schematic illustration of the accumulation of positive charge around the sun.*

10.5.i.a. Possible indications of that effect

If this effect occurs, would it be detectable? Hydrogen and other atoms are emitted as the main part of the solar wind. As they move away from the sun, many are ionized by its UV radiation. Those ionized electrons, along with free cosmic electrons will be attracted to the sun's positively charged surface. Figure 10.11 shows the emitted protons provide a potential gradient away from the sun, including through the solar corona. That gradient combines with the ionizing effect of the sun's ultraviolet radiation, the existence of ions, the long mean free paths between particles in the corona and other effects.

Together they give the opportunity for events with potential differences of hundreds of volts to give the appearance of the solar corona having a temperature of millions of volts. The source of the high energy X-rays that indicate the solar corona's temperature is still the subject of debate. This offers another contender for its origin. It suggests there is no need for any additional heating to generate the ionization of iron and other trace elements detected in the solar corona. That would also be the source of X-rays of energies of hundreds of volts. With 1 eV representing ≈ 11,000°C, 100 eV would equate to a temperature of ≈ 1.1 x 10^6°C. Depending upon the incident electron energies, the corona would appear to have a temperature of several million °C.

There may well be other causes of the high temperatures detected in the solar corona. The influence of cosmic ray induced charge mentioned above should not be overlooked.

10.5.ii. Cosmic Protons and Stars in the Galactic Core

Estimates place the mass of galaxies such as the Milky Way or Andromeda at ≈ 1.2 x 10^{12} solar masses ($M_\odot$) (Watkins et al., 2010). The stars in the bulged central core make up typically 10% to 20% of the stars in the galaxy. That would give the galactic core a total of ≈ 10^{11} stars. They exist in a radius of ≈ 1 – 2 kpc, ≈ 5,000 LYs. The galactic bulge is somewhat larger. The Milky and Andromeda galaxies are a little larger than the "average" galaxy. The following calculations use the situation of 10^{10} stars in a diameter of 3 kpc. That approximately equates to the centre of a galaxy that is somewhat smaller than the Milky Way. A radius of 1.5 kpc has a surface area of ≈ 3 x 10^{43} cm^2 and a volume of ≈ 3 x 10^{40} km^3.

With the sun as an example, a single star has ≈ 10^{57} nucleons. For electric effects to rival gravitational effects, it requires a cloud of ≈ 10^{39} protons to accumulate around that star. At a cosmic proton arrival rate of ≈ 6 x 10^{22} per sec, it would take about 10^{16} sec for a single star to accumulate sufficient protons if they are not neutralized. That represents about 3 billion years. That is far too long to have any detectable effect. Other possibilities need to be considered.

Consider the galactic core as a whole. A radius of 1.5 kpc gives it a surface area of ≈ 3 x 10^{43} cm^2. 10^{10} stars, each with 10^{57} nucleons equates to 10^{67} nucleons. That requires 10^{49} protons before there will be sufficient charge accumulation to alter gravitational effects. A proton arrival rate of one tenth the previous rate could be expected if they came in along the galactic plane. Protons arriving at the rate of 10^{42} per second would only take 10^{7} seconds, or less than a year, to accumulate sufficient protons for gravitational force to be rivalled by the electric force. That depends upon no neutralization by incoming electrons. That only requires a time delay between proton and electron arrival times of 1 year. It is suggested the delay is much larger.

Figure 10.11 shows there would be a charge accumulation around each star. It would lead to a voltage build-up. It is cumulative for all stars. 10^{10} stars, each with a capacitance of 0.078 F gives a total capacitance of 7.8 x 10^{8} F. Although the low energy protons originated from the stars, they are not attached to them. To a first approximation, a positive charge potential of 10^{8} volts around each star would build up a total voltage of ≈ 10^{18} volts in the core. That greatly exceeds the average 1 GeV of the cosmic protons. Such a charge accumulation could not occur.

Storing an excess charge of 10^{49} protons at a voltage of ≈ 1 GeV requires a capacitance of ≈ 10^{21} Farad. The total capacitance of all 10^{10} stars is less than 10^{9} Farad. The capacitance of a sphere of radius 2.5 kpc is about 5 x 10^{4} F. These values are far too small to store such a huge charge. Another mechanism of charge storage is required.

To a first approximation, the self-capacitance of a dust particle of 10 nm diameter is ≈ 10^{-18} F. Capacitances in parallel are cumulative. To reach a total capacitance of 10^{21} Farad requires approximately 10^{41} dust particles, each of diameter ≈ 10 nm. To store $10^{\mathbf{49}}$ protons within a 1.5 kpc radius and have a charge build-up of 1 GeV would require ≈ 10^{41}

dust particles, with 10^{44} nucleons, $\approx 10^{-13}\ M_{\odot}$. A dust particle build-up of 1:10^{23} of the total mass of all the stars in the galaxy core is not considered an excessive figure. If the dust particles are smaller, such as molecules, the number required increases, but the number of nucleons required decreases.

That collective charge build-up will slow down incoming protons. They will be easier to capture or will simply come to equilibrium with the accumulated voltage. The lower energy cosmic protons would find themselves unable to pass through the voltage accumulated around all the stars in the central core. Their capture would increase the charge within the radius of the core, increasing its positive electric potential. In turn that would capture higher energy protons, further increasing the potential. The volume over which that would occur is within the core's radius of 1.5 kpc. The surface area available to capture incident cosmic protons increases from the total of the individual radii, $\approx 6 \times 10^{32}$ cm^2, to a radius of 1.5 kpc, or $\approx 3 \times 10^{43}$ cm^2.

A radius of 1.5 kpc equates to a volume of $\approx 4 \times 10^{50}$ km^3. That would result in 1 excess proton in every 40 km^3. Maintaining that charge requires comic electrons to take at least a year longer to arrive at the galactic core. A year in five thousand years – the approximate time it takes 1 GeV photons to travel 1,000 kpc – represents a speed difference of $\approx$ 0.003%.

A proton with energy of 1 GeV will, from special relativity correction, be travelling at $\approx$ 0.7 c. Cosmic rays can be emitted from supernovae exploding throughout a galaxy. A galaxy of the size used would have a radius of $\approx$ 20 kpc. Protons from a supernova exploding 3 kpc from the galaxy's core would take over ten thousand years between origin and arrival at the galactic core. A delay in electron arrival of 1% would have them arrive decades later. A delay of 1 year would see over six times the required charge for gravitational attraction to be matched by charge accumulation. That only requires that the electrons travel at about 99.9% of the proton's speed. Those figures indicate there is a real possibility of the accumulation of positive charge in and around a galaxy core.

Each star has a proton accumulation in it of $\approx 10^{21}$ protons. That is sufficient to repel individual protons within the star. Each individual star will repel protons within that cloud. If sufficient individual protons were repelled from another star, the repulsion they exert on the star would cause it to overcome the gravity attracting the two stars. 10^{10} stars in a volume containing 10^{49} protons equates to 10^{39} protons per star, $1:10^{18}$ of the nucleons in that star.

Under equation 10.5, that is sufficient for each star to have a charge accumulation around it large enough that it can have an electric repulsion from others that equals or exceeds the gravitational attraction from all the other stars. Those stars would not be attracted to each other by gravity. When the ratio of protons to nucleons exceeded $1:10^{18}$, the stars would repel each other.

The same assumption of each star repelling 6×10^{22} protons that are not neutralised by electrons at $\approx$ 500 km/sec are used. That equates to 6.2×10^{32} protons repelled from all those stars every second. At $\approx$ 500 km/sec it takes $\approx 10^{14}$ sec for protons from every star to travel 3 kpc, about 3×10^6 years.

During that time there would be $\approx 6 \times 10^{46}$ repelled protons. Of those, over half will travel outside the 1.5 kpc radius. The others will remain inside it. That would leave $\approx 3 \times 10^{46}$ ($\approx 5 \times 10^{28}$ C) of free repelled protons in a 1.5 kpc radius sphere ($r \approx 5 \times 10^{16}$ m). From equation 10.7, such a sphere has a capacitance of $\approx 5 \times 10^6$ F. If there was no charge neutralization by electrons, equation 10.6 gives $V \approx 10^{22}$ volts if the galactic core had a defined surface. That voltage is so high it will repel incoming protons long before that charge can be reached.

Each star has a capacitance of $\approx 7 \times 10^{-3}$ F. As the charge builds up around the star, it would require a larger charge accumulation to repel individual protons. From equation 10.6, setting V at 10^{10} volts gives $Q \approx 7 \times 10^7$ C. 10^{10} stars accumulating $\approx 7 \times 10^7$ C, for a total of 7×10^{17} C falls far short of the $\approx 1.5 \times 10^{30}$ C required to be accumulated. A voltage accumulation of 10^{10} V would repel further incoming protons, preventing the possibility of sufficient charge accumulation for it to have any effect. Other means must be found for the charge to be stored that did not repel the incident cosmic rays.

To a first approximation, equation 10.7 indicates the self-capacitance of a dust particle of diameter 10 nm is $\approx 10^{-18}$ F. A single proton has a charge of $\approx 1.5 \times 10^{-19}$ C. Using equation 6, $V = Q/C$, an excess of 1 proton appended to an $\approx$ 10 nm Ø dust particle would give it a potential of $\approx$ 0.1 volt. To store 10^{49} protons within a 1.5 kpc radius and have a charge build-up of 0.1 V would require $\approx 10^{49}$ positively charged dust particles of $\approx$ 10 nm radius.

To a first approximation a 10 nm Ø particle would contain $\approx 10^3$ nucleons. The total number of nucleons required would be $\approx 10^{52}$, or $\approx 10^{-5}\,M_\odot$. If that many existed, the accumulation of 10^{49} protons would give a charge of only about 0.1 V. To accumulate a voltage of 1 GeV would require $\approx 10^{39}$ dust particles, with 10^{42} nucleons, $\approx 10^{-15}\,M_\odot$. That requires $\approx 10^5$ dust particles per star charged by a proton. That is not considered an excessive figure.

The accumulation of that charge requires that electrons take longer to reach the galactic core than do protons. It has been previously shown that electrons travelling at $\approx$ 99% of the speed of protons would be sufficient to maintain a significant charge build-up. Observations show that the core of the Milky Way galaxy contains significant dust. Except for a small window, dust obscures much of the Milky Way's centre.

The hypothesis that dust particles are present in the numbers required appears to be supported by observation. It is not necessary for all of the particles to capture all the protons. Nor is it necessary for the proton accumulation to maintain a low voltage. Figure 10.9 shows that, as long as the accumulated voltage remains below $\approx 10^{10}$ eV, incident cosmic protons will make their way into the central core.

The important feature is that GeV cosmic rays entering the galactic core have a high probability of being captured, either directly by stars or dust particles, indirectly by the star's heliospheres and the galaxy's magnetic fields, or the combined charge generated potential slowing them down. At the indicated cosmic ray flux of $\approx 1/\text{cm}^2$ sec, it could take less than 1 year for proton accumulation to reach $1{:}10^{18}$ of the 10^{67} nucleons in the core. Each star has sufficient protons associated with it to repel other protons. When there is more than 1 proton for each 10^{18} nucleons in the galactic core, there are sufficient protons surrounding each star for them to overcome gravity and repel each other. That also means that, if there are other bodies with an equivalent number of negative charges surrounding them, there will be attraction between them.

The above figures suggest that there is a capacity for cosmic ray protons to accumulate in galactic centres in sufficient numbers that they can alter the effect of gravitational attraction. Varying according to the inverse square law, their attractive effect would be difficult to distinguish from a gravitational effect due increased mass.

The positive charge accumulation will be stronger in the galactic centre than elsewhere because the greater density of stars increases the ability to capture incident protons. That does not restrict positive charge accumulation to the centre, only that its effect would be stronger there. A similar effect could occur elsewhere in the galaxy, particularly in the spiral arms. It ceases when it is overwhelmed by the greater accumulation of electrons, or the charge leaves the vicinity of the accumulated mass.

If they have sufficient charge accumulation to repel other stars that have a positive charge accumulation, they have sufficient charge to attract stars that have negative charge accumulation around them. That is discussed in section 10.6.

10.5.iia. Possible indications of that effect

If galaxies were dominated by gravity, the galactic centres should also be dominated by gravity. "Typical" galaxies have a planar structure that is typically about 0.3 kpc (1,000 LYs) thick with a centre that bulges out to about 3 – 5 kpc (10,000 to 15,000 LYs) diameter. The planar nature of a galaxy's disk can be explained by gravitational attraction causing the stars to oscillate through the galactic plane. "Typically", galactic centres have a mass of about 20% of that of the galaxy in a diameter that is about 10% of the galaxy's diameter.

That gives galaxy cores a surface mass density of about 5 to 10 times that of the rest of the galaxy. The centre of the core is also the region where the stars are moving the slowest, see figure 10.3. From that, it would be expected that the galactic centre would be small and compact, as shown in figure 10.12 A. Instead, galaxies bulge out in the core region, as shown in figure 10.12 B.

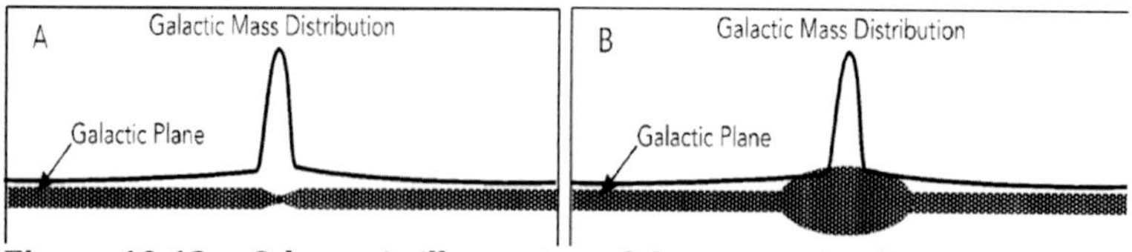

Figure 10.12: *Schematic illustration of the expected collapse of a galaxy's core, A, under its own mass. B shows the reality of bulging galactic centres.*

Examples of galaxies showing bulging cores are given in figure 10.13. The side on view of NGC 4565 clearly illustrates the bulging central core. The Sombrero galaxy shows the bulge extending all the way to the galaxy's perimeter. NGC 1300 shows an elliptical bulge with spiral arms extending out from the edge of the ellipse. Messier 81 shows the more "typical" appearance with the bright central bulge.

They show quite clearly that galaxies bulge out in their central region. If they were dominated by gravity, the additional mass would cause the slowly moving central stars to collapse in the galactic core.

It is possible that stars orbiting at an angle to the galactic plane could cause the central bulge. Over billions of years, their random paths must cross the orbits of stars moving in the galactic plane and at other angles. With gravity being the only force controlling their orbits, stars would be attracted to each other and collisions between stars would occur. Their slow velocities give them plenty of time close to each other for attraction to cause collisions.

Such collisions would form larger masses at a galaxy's centre. In turn that larger mass would attract more stars towards it. Such a mechanism is unlikely to leave most galaxies with a central bulge that is

Figure 10.13: *Photographs of galaxies with bulging centres.*

much thicker than the galactic plane. Their uniformity suggests a more stable mechanism is responsible for the central bulge in galaxies. That gravitational attraction does occur.

The centres of many galaxies have large masses ranging between $\approx 10^6$ to 10^9 $M_\odot$. They are usually referred to as "black holes". Chapter 8 showed they were just massive objects. Their mass is only a small fraction of the total number of stars in a galactic core, typically 0.1% to 1%. The only way the remainder of the stars in a galaxy's central core can bulge out is the presence of a repulsive force in its centre.

The above showed there is the possibility of cosmic protons being preferentially captured in galactic centres where the mass has greatest density. As those cosmic protons are captured, they exist in such large numbers that stars with an associated positive electric charge can cause other stars with an associated positive electric charge to be repelled from each other. 10^{49} protons in a volume of 1.5 kpc equates to ≈ 1 proton per 10 km^3.

Each star needs $\approx 10^{39}$ protons in its cloud to affect other stars that have similar charge accumulation. To a first approximation, 1 proton per 10 km^3 requires an average star separation of less than 10^{13} km, 1 LY, to achieve that accumulation.

Star separations of significantly less than 1 light year will result in gravity dominating. Masses will collapse into each other and accumulate at the centre of rotation. That explains the large masses in the centres of galaxies.

The accumulated positive charge is sufficient for neighbouring stars to exert a repulsive force on each other. That effect would not be cumulative with increasing distance because the increased charge associated with greater distances is counteracted by the increased gravity due to more mass. The effect would be to push the stars in the central region away from each other, creating an outward bulge in the galactic centre. That bulge would extend until the positive charges surrounding outer stars were either neutralized by incoming cosmic electrons or travelled out of the central core. When that occurred, gravity would take over and prevent the stars from moving further out.

The charge causing that bulge is maintained because cosmic ray protons are slowed down significantly less by a galaxy's magnetic field than are the much lighter mass electrons. As shown above, all it needs is for cosmic electrons to take slightly longer to reach the galactic centre to a significant positive charge to accumulate around the stars.

The reality is that the electrons will take significantly longer to reach the galaxy core. The positive charge repulsion will push stars away from each other. That significantly increases the distance available for the stars to orbit about their common centre of mass.

The local repulsion limits the possibility of cross orbiting stars colliding. The positive charge accumulation causes the central region of galaxies to bulge out. The incoming cosmic electrons neutralize the positive charge associated with the outer stars, limiting the size of the bulge as gravity takes over and keeps them from moving further away.

In a side effect, outer protons and charged dust particles would move away from the positive repulsion of the galaxy's centre. They will meet incoming cosmic electrons, be neutralized, come under direct gravitational attraction, and move down the outside of the galaxy's core. That would generate a stream of particles that flow down the outside of the galactic core.

It should also be noted that the positive charge accumulation within a star's region of planetary influence would not be sufficient to have any significant effect on planetary orbits. Apart from the local charge ejecting incoming cosmic protons, local stars, planets, and moons will not experience any electric field because all the objects in its immediate vicinity would have a similar potential.

10.5.iii. Cosmic Electrons on Individual Stars

For the same velocity, cosmic electrons will have their trajectories altered almost 2,000 times greater than those of cosmic protons. As shown in section 10.5.ii, cosmic protons that make their way to a galaxy's centre faster than cosmic electrons will produce a positive charge accumulation at the centre. There they produce the effects mentioned above. That leaves an excess of electrons throughout the rest of the galaxy.

Stars at the outer reaches of galaxies will be bombarded by cosmic protons, as are those in a galaxy's centre. The availability of excess electrons has two effects. Protons impinging on a star would be rapidly neutralized by impinging electrons. Electrons impinging upon the star would build up a negative charge accumulation on the star, until such time as the electric charge on the star was sufficient to repel an electron against its gravitational attraction. Using the mass of an electron, m_e, instead of m_P in equations 10.4 and 10.5 shows the much lighter electrons would be expelled from a star after its voltage accumulation was less than 1 V. That precludes stars from capturing electrons on their surface.

Stars have magnetic fields with strengths similar to, or somewhat stronger than, that associated with Earth's magnetic field (Beck 2007). Physically they are many orders of magnitude larger than planetary magnetic fields. Earth's magnetic field is sufficient to both deflect and capture protons and other charged particles in the van Allen radiation belts.

It is suggested that similar belts surrounding stars in the outer reaches of galaxies would be strong enough to capture sufficient excess electrons to give a negative charge accumulation around them. For a star with 10^{57} nucleons, the same mass as the sun, that would require in the vicinity of $10^{57}/10^{18}$ electrons or 10^{39} electrons ($\approx 2 \times 10^{10}$ C) to get electric attraction between a star and the centre of the galaxy that exceeded its gravitational attraction. The capacitance of a sphere of radius $\approx 10^{10}$ km is $\approx 10^{3}$ F.

Using equation 10.4, that would give an electric potential of $\approx 2 \times 10^{7}$ volts if the star's heliopause was a defined surface. An electric field gradient of ≈ 1 V/ 10^{3} km is not too high to rapidly eject electrons from the heliosphere. Add to that the self-capacitance of gas and dust particles and you can get significant negative charge accumulation in the heliospheres around outer galactic stars without the build-up of significant voltages.

Various estimates suggest the luminous matter associated with galaxies represents $\approx 10\%$ of a galaxy's total mass. The remaining 90% of non-luminous matter is found in the gas and dust clouds that surround the galaxy and permeate through it. As such it is suggested there is no shortage of dust particles to capture the necessary number of electrons and their self-capacitance will lower the voltage the captured electrons would otherwise produce around the star. Additionally, the stars heliosphere is immersed in the galaxy's magnetic field, which itself can also capture electrons.

Charges captured in the galactic magnetic field will influence the charge in the heliosphere and through that the star that generated it.

Negative charge accumulation around outer stars is inevitable for a galaxy to retain overall charge neutrality.

10.5.iv. Cosmic Electrons and a Galaxy's Magnetic Field

Electrons captured in a star's magnetic field that forms its heliosphere and surrounds, will be attracted to the positive charge. Its movement will influence the star's magnetic field. To determine this effect, it is necessary to evaluate the local environment associated with the outer stars.

The first feature to note is the distance between the stars. To determine this, we consider there are $\approx 10^{11}$ stars in a disc ≈ 0.3 kpc thick within a radius of ≈ 12 kpc. That corresponds to a disk $\approx 3 \times 10^2$ kpc^3, or $\approx 3 \times 10^{40}$ km^3. With an approximately uniform distribution of stars throughout the galaxy, the outer 1/3rd of the galaxy (8 – 12 kpc) contains about half of the stars in the galaxy and would collect about half the cosmic electrons associated with the galaxy. The outer third of a disk of inner radius ≈ 8 kpc, has a surface of $\approx 10^{43}$ cm^2. Being bombarded by electrons at the rate of 1/cm^2 sec, gives an electron arrival rate of $\approx 10^{43}$ e^-/sec. As mentioned above and determined from figure 10.9, the energy of the cosmic electrons is lower than the energy of the cosmic protons. Their much lower mass makes them easier targets to be captured by the magnetic fields of the stars and galaxy than are the cosmic protons.

The larger number of stars in the outer regions of the galaxy requires a larger positive charge accumulation in the inner core and inner spiral arms for attraction to occur. As shown in section 3.2, that means the delay time between the arrival of cosmic protons and electrons needs to increase from a few months to a few years. When the time taken for cosmic rays to travel from generation to galactic core is typically thousands of years, that increase is insignificant.

The $\approx 5 \times 10^{10}$ stars in the outer 1/3rd of the galaxy contain $\approx 5 \times 10^{67}$ nucleons. For electric attraction to match gravitational attraction, they require $\approx 5 \times 10^{49}$ excess electrons (1 : 10^{18}). With an arrival rate of $\approx 10^{43}$/sec, it would take about 5×10^6 sec, ≈ 2 months, to accumulate that charge excess. Even though the electron flux is generated mainly in the galactic plane, their numbers are so great that sufficient electrons would accumulate well within 1 year. As most galaxies have existed for over 10^{16} sec, there is ample opportunity for significant negative charge to accumulate in the outer reaches of the galaxy.

A radius of 12 kpc (40,000 LYs) and thickness of 0.3 kpc (1,000 LYs), equates to ≈ 3.7×10^{20} m Ø x 10^{16} m deep, giving the outer 1/3rd of the disk's radius a volume of ≈ 3×10^{56} m^3. If the electrons are uniformly spread throughout the outer galaxy plane, they have a density of < 1 $e^-/10^7$ m^3. Observations of galactic magnetic fields show they extend either side of the galaxy's plane of rotation by about one quarter its diameter (see for example Tüllmann et al., 2000; Beck 2007). This extends the thickness of the plane in which the electrons are confined. There is a large mass of gas and dust particles at the outer edges of galaxies and significant numbers throughout them. They would capture electrons. Their self-capacitance means there is no reason for any excessive high voltage build-up to occur.

It is often suggested that the magnetic field associated with stars and galaxies is generated by the dynamo motion of charged particles within the object (Brandenburg and Subramanian 2005; Beck et al. 2015). That establishes that the magnetic field is associated with the object. Where goes the magnetic field, so goes the object.

The overall situation is one in which each star has a magnetic field, its heliosphere. It is generated by the motion of charged particles associated with it. Galaxies also have a charge distribution that rotates, generating their magnetic field. Each star's magnetic field is embedded in the galaxy's magnetic field. As the galaxy rotates, its magnetic field also rotates. The stars' magnetic fields are embedded in the galaxy's magnetic field. Both have charged particles associated with them. In the outer reaches of a galaxy those particles will have excess electrons.

They will be attracted to the positive charge in the galaxy's core. Being at least partially confined to the star's magnetic field, part of that attraction will be transferred to that field. In turn, because a star's magnetic field is generated by it, that attraction is transferred to it. In that manner, the electric attraction between the positive charge at a galaxy's core and the negative charge in an outer galactic star is transferred to the star itself. Electrons would also be restrained by a galaxy's magnetic field. The attraction of that accumulated negative charge to the positive core would influence a star's heliosphere and through that, the star itself.

The charge difference is maintained by the longer time it takes for cosmic electrons to reach a galaxy's central core compared to the arrival time of cosmic protons. That produces a positive charge accumulation around the galaxy's central region and possibly in some of its spiral arms. By definition, that leaves a residual negative charge throughout the outer galactic regions. This is self-maintaining as long as cosmic rays are generated. It will cause the charged objects to be attracted towards each other. That attraction will exist as long as the cosmic ray flux continues. The greater the cosmic ray flux, the stronger that effect will be.

The accumulated negative charge in a galaxy's outer regions is held in the stellar and galactic magnetic fields. Those embedded electrons are attracted to the central positive core. That attraction is passed on to the stars through the magnetic fields.

10.5.iv.a. Possible indications of that effect

There will be a continuous passage of electrons from outer galactic regions towards the centre. There would also be protons moving away from the galactic core. There is nothing to prevent those particles from being attracted towards and neutralizing each other during their travels. When that occurs, the excess energy will be emitted as photons. That would occur randomly and would be seen by observers as coming from all directions.

The sun's position about one third the way from the galactic centre to its extremities, suggests that electrons have not been exposed to much of the high positive charge concentration at the galactic centre. This suggests the photons would have energies in the keV, to MeV, range. A background of X-rays emanating in all directions around Earth has been detected, see for example Fabians and Barcons (1992).

10.5.v. Their Cumulative Effect

The above creates the situation in which stars with a magnetic field are embedded in the galaxy's magnetic field. The galaxy's magnetic field causes protons to preferentially agglomerate in and around the galaxy's central core. That leaves electrons accumulating in the galaxy's outer reaches. As they move through the magnetic fields of the outer galactic region and their stars, the electrons' curved/spiral paths cause them to exert a force on the magnetic field as they are attracted towards the positive galactic core. That force is passed onto the stars that generate the magnetic fields.

Although it may not be much per electron, at over 10^{39} electrons per star, greater than 1 electron for every 10^{18} nucleons per star, that is sufficient for electric attraction to match or exceed gravitational attraction when the central region also has a similar ratio of protons to nucleons in each star. At an excess charge of $1{:}10^{18}$, the effect at the outer reaches of the galaxy would be that of providing a force equal to the gravitational force. That additional force will cause those outer stars to rotate faster than they would under the effect of gravity alone. A greater accumulation of cosmic rays could see that force significantly increase. Those order of magnitude calculations suggest there is plenty of opportunity for charge accumulations larger than $1{:}10^{18}$. The self-capacitance of gas and dust particles within a galaxy, limits the potential generated by that charge. It is a dynamic process that is driven by cosmic rays emitted by supernovae and the galaxy's magnetic field.

There are some important features to note about the above calculations and the process. They are orders of magnitude only and will vary between galaxies of different sizes and shapes. They serve to illustrate a mechanism by which the outer reaches of galaxies can rotate faster than would be possible under gravitational attraction alone. There is considerable scope for refining those calculations. An important feature is that the observed cosmic ray flux is large compared to the minimum required for this effect to occur.

The magnitude of the process will vary with the size, shape, and other properties of each galaxy. As such it would be expected that different galaxies would display different rotation rates. Different galaxy shapes and other properties would result in the charge distribution being different for different galaxies. In turn that would result in different regions of a galaxy rotating at different speeds. They are shown in figure 10.3, with the variations in radial velocity at different distances.

That effect occurs without the need for any additional properties such as the proposed undetected additional mass to account for the observed rotation being much faster than that due to the galaxy's mass distribution under the inverse square law of attraction.

10.5.v.a. Possible indications of that effect

In this model, a galaxy's magnetic field partially separates the cosmic ray particles. It concentrates protons in the galactic core, leaving an excess of electrons in the galactic outer reaches. Ultimately there will be equilibrium between cosmic ray flux, positive charge accumulation in the galactic core and negative charge accumulation in the galactic extremities. Their re-combination prevents further charge accumulation. The smaller electron mass suggests the most probable way of neutralization is the electrons in the outer regions of the galaxy making their way into the inner galactic regions. If that were the mechanism, it would produce two effects.

The first, mentioned in section 10.5.iv a, is the continual emission of high-energy (keV to MeV) photons when the incoming electrons meet outgoing or locally generated solar protons. In addition to that, charge neutralization is much more likely to occur by the lighter electrons making their way to the galactic core, than protons making their way to the outer galactic regions.

It has been pointed out (Tüllmann et al. 2000; Beck 2007) that the "lines of force" of the galaxy's magnetic field associated with the galactic plane, are somewhat parallel to that plane. Away from that plane, the "lines of force" tend to be less parallel and often closer to perpendicular. That would make it easier for electrons to reach the galactic core by travelling along that plane. Electrons would have two sources of energy, their own emission energy and the energy gained by attraction to the positively charged core.

With the central core having a positive voltage above 1 GeV, it should be seen in other effects. With the core positively charged to a high voltage, it attracts electrons. They make their way into the central core, and in other regions where mass concentration could capture protons. When electrons were captured by the high voltage protons, they give off gamma rays. Even though the 1 GeV protons would make it to the centre they would either impinge on the soft surfaces of the stars or be slowed down by the accumulated high voltage. Those protons would not emit gamma rays. If by chance a proton did hit a hard target, it would emit a series of pions as it gets rid of its excess energy. It would not emit GeV gamma rays. That was shown in chapter 4. That leaves attracted electrons as the major source of those GeV gamma rays.

The calculations in section 10.v.ii showed there was considerable scope for energies significantly higher than 1 GeV to be reached by the incoming positive cosmic rays. The experimental results of figure 10.9 shows that the flux of higher energy protons diminished approximately linearly with their energy increase above 1 GeV. That leaves open the possibility that protons in excess of 100 GeV could make their way into the galactic core and surrounds, even if the voltage accumulation was sufficiently high to exclude further 1 GeV photons from entering.

It is suggested those effects are detected, as shown in figure 10.14. It is a compilation of a decade of data collection by the large area telescope on NASA's Fermi gamma-ray space telescope.

It includes gamma rays of over 1 GeV. The brighter colour indicates higher gamma ray energies. The lobes either side of the Milky Way plane are called Fermi bubbles.

The brightest colour, white, is in the galactic centre. That is the region that captures the highest energy protons. The yellow band spreads out along the galactic plane, corresponding to positive charge accumulation in the dense regions of lesser energy of the spiral arms. The red regions surrounding them are where some positive charge has accumulated on the outer reaches of those regions.

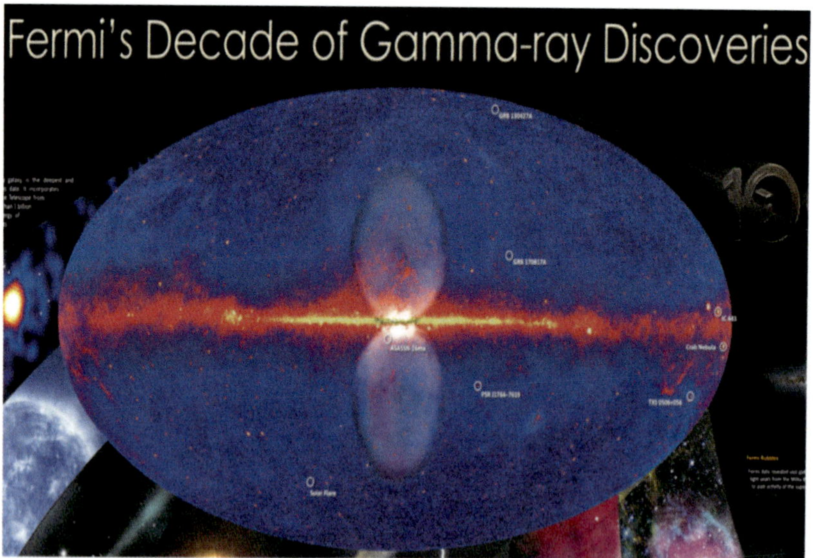

Figure 10.14: https://fermi.gsfc.nasa.gov/Fermi_Decade_Poster.pdf

One of the features associated with a positive charge accumulation in the centre of the galaxy is that they are not held in a fixed manner. They exist in a positive charge field with a voltage in excess of 1 GeV. The only thing constraining their movement out of the galactic core is the incoming protons along the galactic core.

Positively charged particles will move out because gravitational attraction is less than their mutual positive repulsion. Both diminish with inverse square of the distance. A galaxy has no preferential direction. Positively charged material will move out equally on either side of the galactic plane. They will keep going until neutralized by electrons. Their greater distance from the galactic plane and the more perpendicular (to that plane) magnetic field combine to mean it takes longer for electrons to meet and neutralize those protons. When that occurs, gamma rays will be emitted.

It is suggested that is the origin of these lobes, known as Fermi Bubbles. They are vast regions of gamma ray activity extending tens of thousands of light years from the Milky Way's plane.

Those features shown in figure 10.14, along with others mentioned, are consistent with a positive charge accumulation in the central galactic core being neutralized by electrons attracted from the outer galactic regions.

10.6. Discussion

As Beck (2007) pointed out, "*Magnetic fields are a major agent in the interstellar medium (ISM) of ... galaxies. ... Magnetic fields also control the density and distribution of cosmic rays in the ISM*". This suggests, as others have noted, that a galaxy's magnetic field plays a significant role in its structure. The above has used the magnetic fields to show how they could separate cosmic rays of different charge. That would cause positive charge accumulation around the galaxy's central core and leave an excess of electrons at the outer galactic regions. Such a situation would cause cosmic rays to give rise to some additional effects. These include:

i. The generation of X-rays in the solar corona that could indicate a temperature of several million deg K.

ii. The random emission of X-rays in space.

iii. The centre of a galaxy would bulge out instead of being attracted inwards by gravity.

iv. A stream of gas and dust particles would flow out from the galaxy's central bulge approximately perpendicular to the galactic plane.

v. The positive charge accumulation in the galactic core would attract electrons towards it. When they impinged upon another particle, they would emit high energy bremsstrahlung radiation. That radiation would be strongest and have the highest energy in the galactic plane in general and towards the galactic core in particular. It is believed that effect was detected by NASA's Fermi LAT gamma ray telescope, see figure 10.14. It shows up as the bright bands across the centre of the image.

vi. Positive charges in the galactic core could escape from it in a direction perpendicular to the galactic plane. When neutralized by electrons in the galactic plane above the rotation plane of the stars, it would generate gamma rays in that region. The Fermi lobes, also shown in figure 10.16, are the central structures perpendicular to the galactic plane and originating from the centre.

The above presentation considered the possibility that a charge imbalance between a positive core and negative outer regions of the galaxy could result in the faster than expected rotation of galaxies. While that possibility is still real, studies by the Pavlovich / Sipols team and Rourke suggest such a mechanism may not contribute much towards the rotation of some galaxies.

There are so many galaxies of different sizes and shapes that a "one theory fits all" is not feasible.

That effect may still apply in some galaxies in addition to the effects described by those authors. Figure 10.15 shows the Sombrero galaxy and galaxy UGC 12591, image courtesy of APOD. Both have at least two features in common. The distribution of luminous matter is well beyond the extent of a small bulge in the central core. Their outer rotational speeds of over 500 km/sec are approximately double of that of the Milky Way and almost double the speed of the fastest rotations shown in figure 10.3.

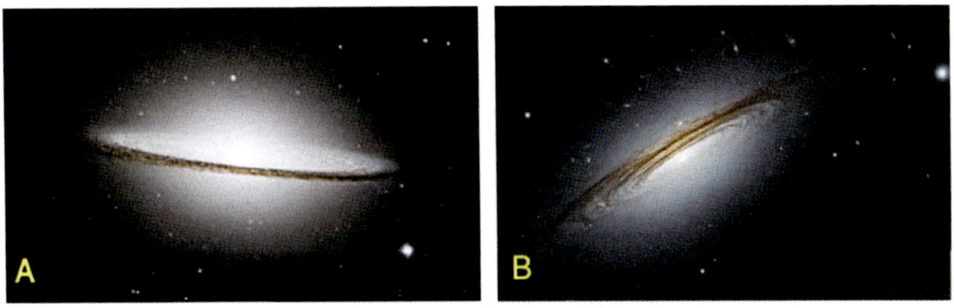

Figure 10.15: *Sombrero galaxy, A, and UGC 12591, B.*

There is no physical way by which additional dark matter, necessary for those galaxies to rotate so fast, could cause the luminous stars to be located so far from the galactic plane. Similarly, the mechanisms proposed by Pavlovich et al. (2014), Sipols and Pavlovich (2020) and Rourke (2020) do not allow for a central bulge. That leaves the possibility of a repulsive force in the centre surrounded by an attractive force at the perimeters.

It is suggested the figure 10.15 galaxies had excessive supernovae activity. A significant positive charge accumulated in the central core region, pushing stars out from the centre. It also causes a large number of protons and positive nuclei out into the surrounding space and beyond. The much lower electron mass allowed them to be captured in the outer regions of the galaxy.

That leaves a significant charge imbalance that has at least two effects. It causes the outer regions of galaxies to rotate significantly faster than would otherwise be expected from the observations in figure 10.3. It also means that the positive repulsion of stars from the galactic core could never get past the outer reaches of the galactic plane, where much of the negative charge would be located. Both of those are observed in those two galaxies.

Two recently discovered galaxies, NGC 1052-DF2 and NGC 1052-DF4 rotate sufficiently slowly that they do not need dark matter to explain their rotation. They suggest that so-called "dark matter" is not necessary for the formation of galaxies.

There are billions of galaxies in the observable universe. While they can be classified into types such as elliptical, lenticular, spiral and barbed spiral, no two galaxies within each classification appear the same.

A one theory fits all is not realistic. When it comes to galaxy rotation, it is possible that some galaxies could have additional dark matter associated with them, even if it was only an excess of helium. Non-stellar helium is difficult to detect because of its high ionization energy. Agglomerated around a galaxy or cluster of galaxies it could provide a minor change in the refractive index, sufficient to change the path of visible or near visible photons and make it appear as if there was more mass present.

It is highly probable that the Pavlovich/Sipols mechanism applies in many of them. The Rourke mechanism is a variation of their mechanism. If it applied in the Milky Way, they make searches for dark matter a most unrewarding occupation. Sipols and Pavlovich (2020) made the following comments:

"Our findings make non-baryonic dark matter unnecessary in the context of galactic rotation."

"Perhaps dark matter is like Confucius' black cat—hard to find in a dark room, especially if there is no cat".

Then there is the charge imbalance mechanism presented herein. It has the advantages of being able to explain other phenomena as well. Those phenomena are listed in i to vi above. It could also be a combination of the above. The star-to-star attraction proposed by Pavlovich and Sipols, combined with the charge imbalance effect described in this chapter provide a much better possibility of matching observations than does the hypothesis on non-detected dark matter. As chapter 9 pointed out, it is only baryonic matter that generates gravity.

It would seem vastly more productive and less costly for theoreticians to investigate those possibilities, instead of getting experimentalists to chase *"Confucius' black cat"*. Only when those and other possible suggestions were fully explored, and the results independently reviewed, should additional money be expended on what has already been an over 50-year fruitless search for something that has no reason to exist.

10.7. Summary

There are so many galaxies of different shapes and sizes that it is not considered possible to have a one theory fits all. Even hypothesized dark matter could not explain the rapid rotations of the Sombrero and UGC 12591 galaxies as well as their huge distribution of stars well outside their planes of rotation. The theories forwarded by the Pavlovich/Sipols team, supported by Rourke, suggests applying inverse square law to the attraction of each individual segment of stars on all other segments of stars gives a reasonable match of galactic surface mass density and rotation with observations of the same properties for the same galaxy.

This presentation suggested that a galaxy's magnetic field causes an uneven distribution of cosmic rays, collecting more protons to the galaxy's centre and leaving more electrons at its extremities. It showed how that could occur in such numbers that the distribution could produce an electric attraction significantly greater than the gravitational attraction. Electric and gravitational attraction both vary with the inverse square of distance. The two effects would be difficult to distinguish. It would result in galaxies rotating faster in the same manner as if there was significantly more mass associated with the galaxy.

The calculations presented show that there are more than enough cosmic rays for this to occur. Experimental evidence of other effects that would occur if this was the mechanism of rapid galaxy rotation. It is suggested this mechanism is also responsible for other effects. These include:

i. The outward bulge of the galactic core,
ii. The apparent high temperature of the solar corona,
iii. The many GeV gamma ray photons emitted from the galactic core and plane.
iv. A contribution to the Fermi lobes detected perpendicular to the galactic plane and emanating from the galactic core.

An important feature of this presentation is that there are explanations of how the galaxies rotate without the need to introduce any undetected mass particles. If continued, the search for dark matter candidates will remain unsuccessful.

Chapters 1 to 10 have taken us through from how space supports photons; how photons are responsible for the properties of matter; how those properties allow matter to form into nuclei, and atoms; how they generate gravity and on up to the structure and some properties of galaxies. Along the way, it described the mechanics involved in all those stages. The final step is to put the galaxies together into the universe of which we are a part and describe the mechanics of how it all operates.

10.8. List of References

Beck R, (2007); "Galactic magnetic fields", Scholarpedia, **2**(8) p 2411

Beck R and Wielebinski R, (2013); "Magnetic Fields in Galaxies, Planets", Stars and Stellar Systems, **5**, Springer, Dordrecht, pp 641 – 724

Beck R, et al. (2015); "Structure, Dynamical Impact and Origin of Magnetic Fields in Nearby Galaxies in the SKA Era", PoS(AASKA14)094 (http://adsabs.harvard.edu/abs/2015arXiv150100385B)

Brandenburg A and Subramanian K, (2005): "Astrophysical Magnetic Fields and Nonlinear Dynamo Theory", Physics Reports, **417**, p. 1 – 209

Delahaye T, (2010a); "Propagation of Galactic Cosmic Rays and Dark Matter indirect Detection". PhD Thesis, Mathematical Physics [math-ph] Université de Savoie, Universitá degli studi di Torino https://tel.archives-ouvertes.fr/tel-00530422v2, Ch 3 [HAL Open Science]

Delahaye T, (2010b); Ibid; p 225

Fabians AC and Barcons X, (1992); "The Origins of the X-ray Background", Annual Review Astronomy and Astrophysics, **30**, pp 429 – 456

Kecskeméty K, Zeldivich MA, et al., (2005); "Quiet time fluxes and radial gradients of low-energy protons in the inner and outer heliosphere," 29th International Cosmic Ray Conference, Pune, **2**, pp 57 – 60

Kepler J, (1609); "Astronomia Nova (New Astronomy)", Rudolph II, Prague: https://archive.org/details/Astronomianovaa00Kepl/page/n5/mode/2up?view=theater

Kepler J, (1619); "Harmonices Mundi (Harmony of Worlds)", Linz:- https://en.wikipedia.org/wiki/Harmonices_Mundi

Madore BF, (2017); "Level 5, A knowledgebase for Extragalactic Astronomy and Cosmology", NASA, https://ned.ipac.caltech.edu/level5/index.html

Newton I, (1687); "Philosophiae Naturalis Principia Mathematica", London

Newton I, (1729) Mathematical Principles of Natural Philosophy **1** ed. A Motte (London: Motte B) Propositions1 – 32, 45

Pavlovich K, Pavlovich A and Sipols A, (2014); "Newtonian explanation of galaxy rotation curves based on distribution of baryonic matter" https://arxiv.org/pdf/1406.2401

Reucroft S, (2015); "Galactic Charge", https://arxiv.org/pdf/1409.3096.pdf

Rourke C, (2020); "Inertial drag, dark matter and astrology" http://anpa.onl/media/anpa-41-web-conference/

Rubin VC, (1965); "Radial Velocity of Distant OB Stars in the Anticenter Region of the Galaxy", Astrophysical Journal, **142**, pp 934 – 942

Rubin VC, (2000); "One hundred years of rotating galaxies", Publications of the Astronomical Society of the Pacific, **112:** 747 – 750 https://ned.ipac.caltech.edu/level5/Sept04/Rubin/Rubin_contents.html

Sipols A, and Pavlovich A, (2020); "Dark Matter Dogma: A study of 214 Galaxies", Galaxies, ***8***(2), 36

Slipher VM, (1914); "The detection of nebular rotation", Lowell Observatory Bulletin, **2**, pp 66 – 66

Sofue Y, (2014); "Rotation and Mass in the Milky Way and Spiral Galaxies", Publ. Astron. Soc. Japan (2014) pp 1–34 https://ned.ipac.caltech.edu/level5/Sept16/Sofue/paper.pdf

Sofue Y, et al., (1999); "Central Rotation Curves of Spiral Galaxies", Astrophysical Journal, **523**, pp 136 – 146

Sofue Y and Rubin V, (2001); "Rotation Curves of Spiral Galaxies", http://www.ioa.s.u-tokyo.ac.jp/~sofue/htdocs/araa2001/SofueRubin2001.pdf

Tüllmann R, Dettmar RJ, Soida M, Urbanik U and Rossa J, (2000); "The Thermal and Non-thermal Gaseous Halo of NGC5775", Astronomy & Astrophysics, **364**, p. L36 – L41

Uprety Y, et al., (2016); "Solar Wind Charge Exchange Contribution to the ROSAT All Sky Survey Maps", Astrophysical Journal, **829**, pp 83 – 91: doi:10.3847/0004-637X/829/2/83

Watkins LL et al., (2010); "The masses of the Milky Way and Andromeda galaxies", *Monthly Notices of the Royal Astronomical Society,* **406**, pp 264–278: https://doi.org/10.1111/j.1365-2966.2010.16708.x

11. Mechanics of the Universe

Abstract:

The influence of gravity on the large-scale structure of the universe, based on the gravitational effects obtained in chapters 8 and 9, is discussed. It presents a view of the universe as infinite in space and time because gravity gets weaker than inverse square law as mass increases. It presents an alternative to the Doppler effect for the observed redshift. It shows that the observed SNe1a magnitudes are as expected for an infinite steady state universe. The universe is not even expanding, let alone expanding at an increasing rate.

11.1. Introduction

When early 20th century astronomers observed the universe with telescopes, they noted the nebulae were other "island universes", not unlike the Milky Way. When they analysed the light from those distant galaxies using spectrometers, they noted it was redshifted (Slipher, 1915, 1921). Hubble (1929, 1936), noted that the more distant the galaxies, the greater was their redshift. He suggested it was due to galaxies receding from us, with the recession rate linearly increasing with distance. That suggestion was a relief to Einstein. He introduced a cosmological constant, Λ, into his equations to prevent the universe from collapsing under the gravity of its own mass.

Zwicky (1929) suggested the observed redshift may be due to photons losing energy as they travelled through space. It was later called Zwicky's "tired light" theory.

Newtonian gravity indicates that an infinite universe would collapse under its own mass if gravity were inverse square. He also pointed out that gravity of a large body acted as if all the mass of that body was concentrated at its centre of mass (Newton, 1687, 1729). Laplace and Poisson extended that to a uniform distribution of masses also acting as if the mass were concentrated at its centre (Lindsay and Margeneau, 1957).

Figure 11.1 illustrates a two-dimensional slice through a three-dimensional array of galaxies within a sphere of radius *r*. Each dot represents a galaxy. It shows a distribution of galaxies around an observer at O at the centre of mass of a radius *r*. An object P on the sphere's surface, i.e., at distances *r* from O, the centre of mass in figure 11.1, will experience a gravitational attraction from the bodies as if the mass of the other bodies was concentrated at the centre of mass O. An object at the centre of mass, O, is surrounded by a large number of galaxies with an approximately uniform density. It will not experience any gravity that is not self-generated.

Under Newton's inverse square law of gravity, the gravitational attraction, g_N, exerted by mass, M, on an object at distance, r, from its centre of mass is given by

$$g_N = \frac{GM}{r^2} \qquad (11.1$$

where G is Newton's universal gravitational constant. Using the Laplace and Poisson variant, a body at point P on the surface of a sphere of radius r and density, ρ, experiences a gravitational attraction back to O, given by

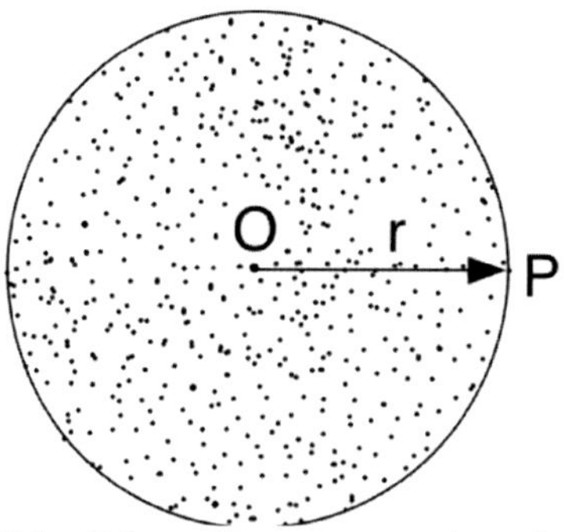

***Figure 11.1:** Schematic representation of a two-dimensional slice through the diameter of a three-dimensional sphere of galaxies.*

$$g_N = \frac{4\pi G r^3 \rho}{3r^2} = \frac{4\pi G r \rho}{3} \qquad (11.2$$

The numerator increases as r^3, while the denominator increases as r^2. As $r \rightarrow \infty$, $g_N \rightarrow \infty$. As distances increase in an infinite universe, the gravitational attraction becomes stronger. That is why an infinite universe would collapse under gravity.

Both Newtonian calculations and Einstein's conclusion, were that a universe would collapse under gravity. That made it difficult to suggest that a steady state universe could exist. Lemaître (1931, 1936) was the first to suggest that it commenced with what has become known as the Big Bang and was expanding. Some theorists, for example Hoyle (1950) maintained faith in a steady state universe.

Penzias and Wilson's (1965) discovery of the cosmic microwave background radiation was used by Dicke et al. (1965), Peebles and Ratra (2003) and others, as the residual radiation from a Big Bang. That left little option but to accept the expanding universe theory. Fred Hoyle's lasting contribution to it was that he jokingly referred to the universe commencing from a Big Bang. His description remained.

Attempts were made to quantify the rate of expansion. Hubble's early work was done using the intensity of the light from a galaxy as an indication of its distance. It was not reliable because galaxies were of varying sizes and hence intensities. Through the careful study of cephid variable stars, Henrietta Leavitt (1908) determined that their periodicity was related to their brightness (Pickering 1912). Measuring the periodicity of cephid variables in neighbouring galaxies indicated their absolute brightness. Measuring their observed brightness allowed the distance to the galaxy to be determined. That technique was good for local galaxies. It showed there was a linear relationship between redshift and distance.

Type 1a supernovae, SNe1a, were observed to have characteristic emission spectra and brightness (Colgate, 1979; Tammann and Leibundgut, 1990). They were much brighter than cephid variables and could be detected over much larger distances. Careful observation showed that they extended the linear relationship between redshift and distance to over 1,000 Mpc.

The results showed the observed redshift was consistent with the universe expanding at the rate of ≈ 70 (± 4) km/sec per Mpc (Freedman, 2021). Freedman used another way of determining distances to galaxies. It was known that, as stars aged, they expanded and became red giants. At the end of their fusion life, they collapsed. By measuring the maximum brightness of red stars in distant galaxies, she determined the distance to those galaxies. Her results showed that Hubble's constant had a value of 69.8 (± 0.6) m/sec per Mpc. That value is used in this study.

11.2. Background

The above has briefly described how distances to galaxies were determined. Measuring the redshift of a distant SNe1a allows its distance to be determined from Hubble's constant. Applying the inverse square law of distance to their known intensity allows their expected intensity when measured at Earth to be determined. Cosmic distances and intensities vary over a large range. The intensity of stars was changed to magnitude by using the scale that an intensity decrease of 100 equates to a magnitude increase of +5. The intensity of the star Vega was set at magnitude 0. Relative magnitudes are the values measured at Earth. The magnitudes of SNe1a at different redshifts were measured. The results plotted in figure 11.2 use the relationship that SNe1a at redshift z = 0.1 have a magnitude of 19.1. Their redshifts are plotted on a log scale.

The dots with error bars were the measured points, extracted from Reiss et al. (1998), Perlmutter et al. (1999) and Perlmutter (2003). Line B is a straight line drawn as an approximate average through all points. Line A is a critical density line, extracted from Perlmutter (2003) "without vacuum energy".

Perlmutter et al. indicated that line A was the critical value for the expansion of the universe under the accepted Big Bang theory. If the universe was expanding at the critical rate, SNe1a magnitudes for a given redshift would be detected on line A. If gravity is slowing down the galactic expansion, the SNe1a magnitude versus redshift points will fall below line A because they would be closer to us.

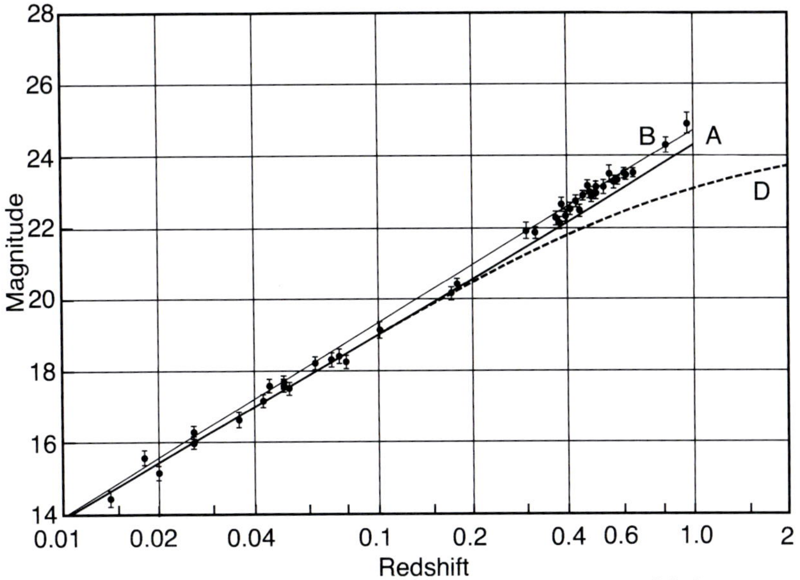

Figure 11.2: *Plot of the observed SNe1a magnitudes versus redshift.*

When astronomers measured the high redshift SNe1a in distant galaxies, $z > 0.2$, the magnitudes were above line A for their redshift. After thorough checking of the mathematics and the measurements, the conclusion reached was that the points were above the critical expansion rate line. That meant the galaxies in which the SNe1a were located, were further away than expected. A greater distance could only be due to those distant galaxies accelerating faster. The more rapid acceleration started a few billion years after the Big Bang. That was the origin of dark energy.

There can be no doubt that the experimentally measured points were accurate. Line A was calculated from mathematical considerations. The following determines the expected SNe1a magnitudes at different redshifts from physical principles.

11.3. Doppler Calculations

In an expanding universe, the greater the distance, the greater the speed of expansion. As the speed increases, time slows down and the distance increases. That slowing time means that photons take longer to be emitted. They will be seen by observers at rest, namely us, as being redshifted. The relationship between that redshift and distance is given by the relativistic Doppler equation (Clarke 1997; Davis and Lineweaver, 2003).

$$r_D = r_M \cdot \left[\frac{(1+z)^2-1}{(1+z)^2+1}\right] \text{ (LYs)} \qquad (11.3$$

where r_D is the distance calculated from the Doppler redshift and r_M is the maximum distance at which the receding galaxies have a velocity < c, with respect to observers. r_M = 13.99 x 10^9 LYs is based upon Freedman's (2021) determination, from red giant intensities, of Hubble's constant, H_0 = 69.8 km sec^{-1}/Mpc.

Equation 11.3 enables the distances associated with a Doppler redshift of any z to be calculated. Table 11.1 shows the results of such a calculation for distances up to z = 2. The distances calculated from equation 11.3 can be converted to magnitudes by using the inverse square law of intensity. The starting point of this calculation is that the magnitude of an SNe1a at redshift 0.1 is 19.1.

The equation for the magnitude of SNe1a at different redshifts becomes Mag_z = 19.1 + Δmag, with

$$\Delta mag = log_{2.512}\left(\frac{d_z}{1.329}\right)^2 \text{ or } 2.512^{\Delta mag} = \left(\frac{d_z}{1.329}\right)^2 \qquad (11.4$$

Δmag is the change in magnitude of a SNe1a at redshift z, compared to the magnitude at z = 0.1. d_z is the distance, in 10^9 LYs, to SNe1a at redshift z. The 1.329 term is the distance, in 10^9 LYs, at which luminous objects have a redshift of 0.1 due to time dilation of their receding velocity. The results are shown in table 11.1 below and plotted as curve D in figure 11.2. Curve D follows the critical density line A, at low redshifts. Above z ≈ 0.2, it drops lower than the critical density line.

Curve D in figure 11.2 depends only upon the redshift due to the special relativity corrections of distance caused by the speed of the galaxies in the expanding universe. It makes no allowance for the reduction in magnitude due to the reduction in photon flux of the receding galaxy; the reduction in magnitude during the measurement of SNe1a's intensity with redshift; or the expansion of space suggested by some cosmologists (Davis and Lineweaver, 2003).

Redshift z	$\frac{(1+z)^2-1}{(1+z)^2+1}$	Distance d_z x 10^9 LYs	Δmag	Mag_z
0.1	0.095	1.329	0	19.1
0.2	0.180	2.533	1.4	20.5
0.4	0.324	4.537	2.66	21.76
0.6	0.438	6.130	3.33	22.43
0.8	0.528	7.391	3.75	22.86
1.0	0.600	8.394	4.0	23.12
1.5	0.724	10.13	4.42	23.52
2.0	0.800	11.19	4.62	23.72

Table 11.1: *Determination of the expected SNe1a magnitudes versus redshift.*

11.4. Correcting Magnitudes for Receding Galaxies

An object moving at velocity v away from an observer, reduces the flux of the photons emitted towards the observer by 1 – v/c. When v << c, Δ*Flux = 1 – z at non-relativistic velocities. At relativistic velocities,

$$\Delta^*\text{Flux}= \frac{(1+z)^2-1}{(1+z)^2+1} \qquad (11.5$$

That change in flux produces a change in magnitude given by

$$2.512^{\,\Delta^*mag} = \left[1 - \frac{(1+z)^2-1}{(1+z)^2+1}\right]^{-1} \qquad (11.6$$

The results are given in table 11.2. Because all magnitudes are calculated from z = 0.1, the Δ*mag = 0.11 at z = 0.1 was subtracted from all the calculated Δ*mag. The results are plotted as curve E in figure 11.3.

Curve E is still below the critical line A and well below the measured magnitudes of the high redshift points. The next factor to be considered is the change to magnitude introduced during its measurement. However, it is close to the critical density line. That should not be surprising because they are calculations of the same effect.

Redshift z	$\frac{(1+z)^2 - 1}{(1+z)^2 + 1}$	Δ*mag
0.1	0.095	0.11
0.2	0.180	0.22
0.4	0.324	0.43
0.6	0.438	0.62
0.8	0.528	0.82
1.0	0.600	0.98
1.5	0.724	1.4
2.0	0.800	1.75

Table 11.2 *Calculated results for the change in magnitude due to the flux reduction associated with the motion of the SNe1a moving away from the observer.*

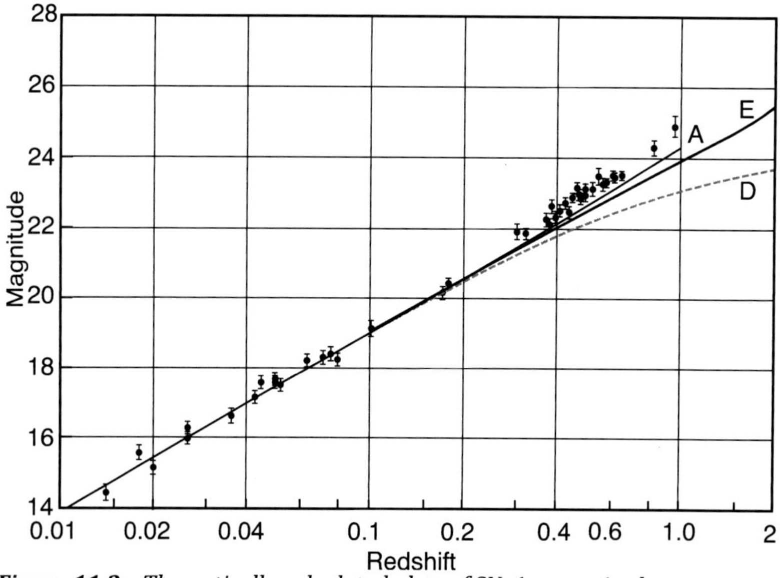

Figure 11.3: *Theoretically calculated plots of SNe1a magnitudes versus Doppler redshift superimposed on some measured values.*

11.5. Correcting Magnitudes for Redshift

A change in redshift results in different parts of the emission spectrum passing through the visible detection field, as illustrated in figure 11.4. It shows the changes to the emission spectrum of black body radiation for a temperature of 5600˚K. They are caused by it being stretched at different redshifts. The 5600˚K temperature is typical to slightly higher than the "average" temperature of galaxy NGC 3941, which is considered a "typical" galaxy.

Figure 11.4 shows that increasing redshift stretches the spectrum and moves more of it away from the visible portion of the spectrum. That takes visible wavelengths out of the measurement range that contributes to the observed magnitude of stellar objects. It also lowers the photon flux per wavelength, because the number of photons emitted remains the same. At z = 1, they are spread over twice the distance and will therefore have half the flux per wavelength. That also reduces the irradiance or power per wavelength. Spectral irradiance is the number of photons, or flux per wavelength, multiplied by the energy of that wavelength.

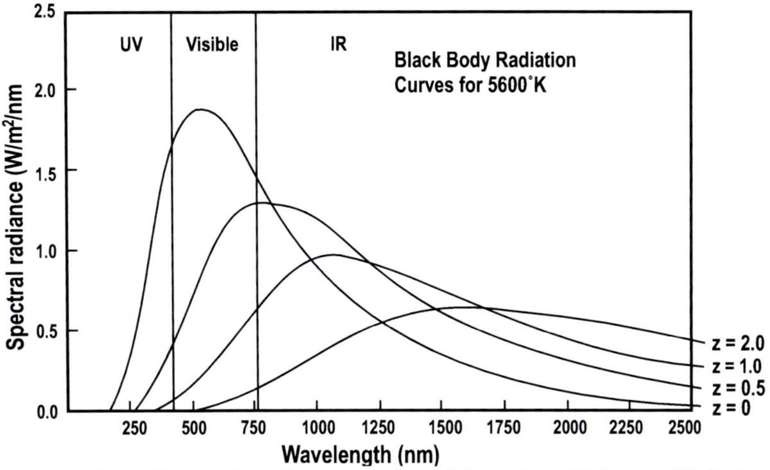

Figure 11.4: *Illustration of the variation of detection efficiency of light from a black body spectrum of temperature 5600˚K that has been red shifted as indicated.*

SNe1a emission spectra, of which figure 11.5 is considered "typical", have unique properties that are not easily confused with other cosmological spectra. It was drawn from a number of references, of which Hook et al. (2007) and Parrent (2014) can be considered typical. They have very little ultra-violet (UV) emission. Their intensity rises rapidly to ≈ 400 nm. They have absorption bands characteristic of particular elements. Their intensity diminishes relatively slowly with increasing wavelength.

The shape and irradiance of the curve varies a little between different SNe1a events and significantly with time. They typically take a few days to rise to peak intensity and a few weeks after peak intensity to decay to ≈ 20% of their peak.

The dashed curve represents the energy per photon. A 300 nm ultraviolet (UV) photon has three times the energy of a 900 nm infra-red (IR) photon. The curve tails off somewhat following that line as wavelength increases. It indicates that the photon flux, the number of photons per wavelength, is approximately constant. Figure 11.5 indicates the photon flux is a little higher towards the blue end in this spectrum. There is an abrupt beginning, with very little UV radiation.

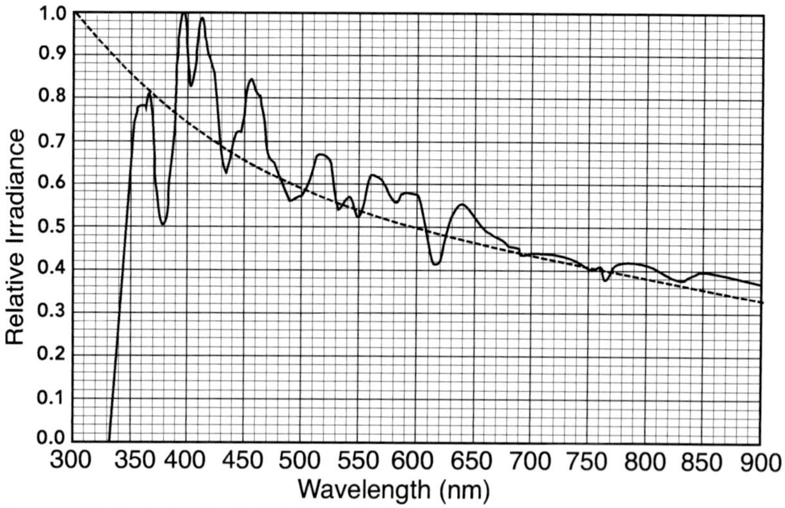

Figure 11.5: *A "typical" SNe1a emission spectrum.*

Magnitude measurements are based on the spectrum associated with their maximum intensity. The light is passed through a series of wavelength transmission bandpass filters, as shown in figure 11.6. It was copied from Bessel (1990). They cover the ultra-violet, U, blue, B, visible, V, red, R and infra-red, I. Only light between 300 nm and 900 nm is used to determine an objects magnitude.

Figure 11.7 shows the spectrum in figure 11.5 redshifted through z = 0.2, 0.4, 0.6, 0.8, 1.0 and 1.5, as well as the original spectrum.

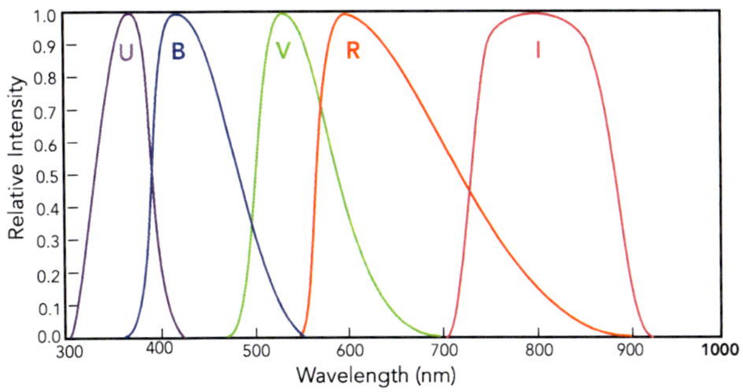

Figure 11.6: *The wavelength responses of the bandpass filters used to determine the magnitudes of stellar objects.*

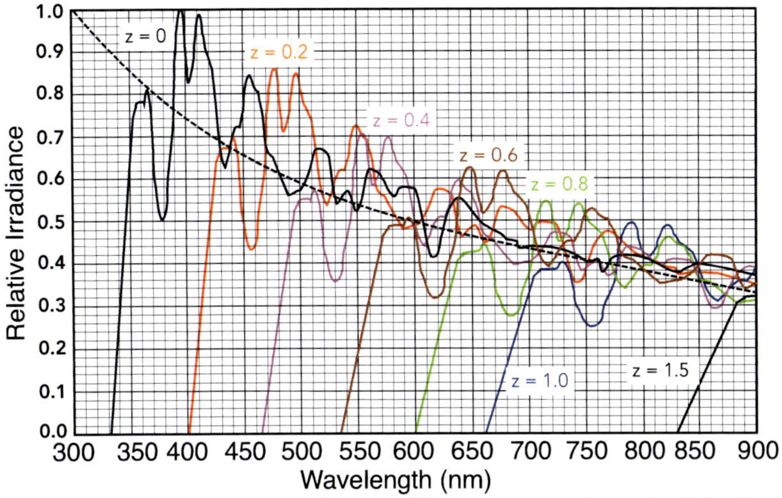

Figure 11.7: *The relative irradiance of the spectrum in figure 11.5 that has had its wavelength stretched by z for the indicated redshift values.*

As the wavelength was stretched by z, the irradiance was reduced by 1/(1+z). The total intensity is the sum of all the intensities under the appropriate curve, for each 10 nm segment between 340 nm and 900 mm.

The ratio of the redshifted intensities to the original intensity was converted to magnitude using the equation

$$2.512^{\Delta^{*}mag} = (\text{Intensity Change})^{-1} \qquad (11.4a$$

The results are given in table 11.3 and plotted in figure 11.8. Table 11.3 and figure 11.8 show there will be an additional increase in magnitude, Δ*, as the spectrum is redshifted through the wavelength range.

The relatively smooth nature of curve B in figure 11.8 indicates it can be used to determine the magnitude change induced into the SNe1a spectra at different redshifts. The figure 11.8 plot is relatively insensitive to minor changes in the shape of an SNe1a spectrum, a large UV tail excluded. The log dependence of magnitude changes means magnitudes are even less sensitive to change than are the intensity changes.

Figure 11.8 and its associated calculations show SNe1a that have been redshifted by more than about 1.7 will have moved through the bandpass filters and will no longer register at any magnitude. That only changes if there is a significant UV tail.

Redshift	Count (divisions) Relative Intensity	Count ratio Intensity Change	Δ* Magnitude
0	1510	1.00	0
0.2	1220	0.81	0.23
0.4	950	0.63	0.51
0.6	720	0.48	0.80
0.8	550	0.36	1.12
1.0	395	0.26	1.46
1.5	65	0.043	3.41

***Table 11.3:** Tabulated results showing the change in magnitude as SNe1a spectra are redshifted, considering the reduction in energy with increasing wavelength.*

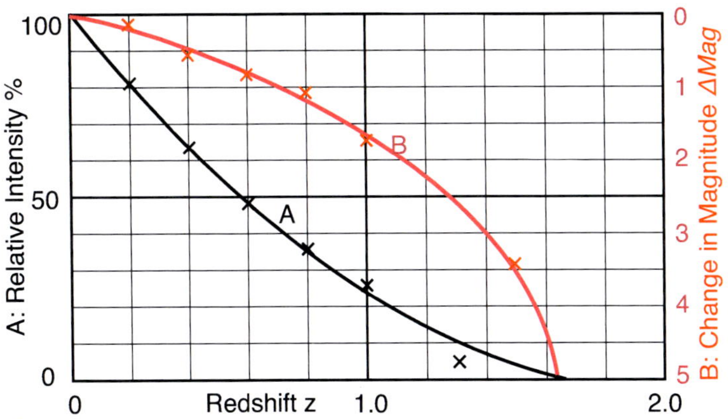

Figure 11.8: *The changes in intensity and magnitude versus redshift, due to the emitted photons being redshifted between emission and observation.*

Suggestions have been made that the UV tail is longer than shown, but the shorter wavelength photons are scattered by debris in the explosion (Brown et al., 2010).

The points in figure 11.8 do not form a smooth plot through which a smooth curve can be fitted. That is due to the irregular nature of the SNe1a spectrum shown in figure 11.5. As different absorption bands move past the 900 mm end, abrupt changes occur. The intensity does not change smoothly. That was overcome by using a smooth curve as a best fit. Those errors are significantly less than the error bars associated the measured magnitudes of the high redshift SNe1a.

The magnitude increases at different redshifts, determined from curve B of figure 11.8, are now added to curve E to give the expected measured magnitude of high redshift SNe1a. The result is curve F in figure 11.9. It shows that, if the universe is expanding, the magnitudes of the high redshift SNe1a are as expected.

The fit of curve F with the observed SNe1a magnitudes at high redshift gives no reason to contemplate any accelerated expansion. There is nothing in curve F that could in any way be construed as the universe expanding more rapidly than predicted from physical considerations. With that there is no need to consider the possibility of "dark energy", considered to have a value of about 17 times the energy equivalence of all the baryonic matter astronomers have detected.

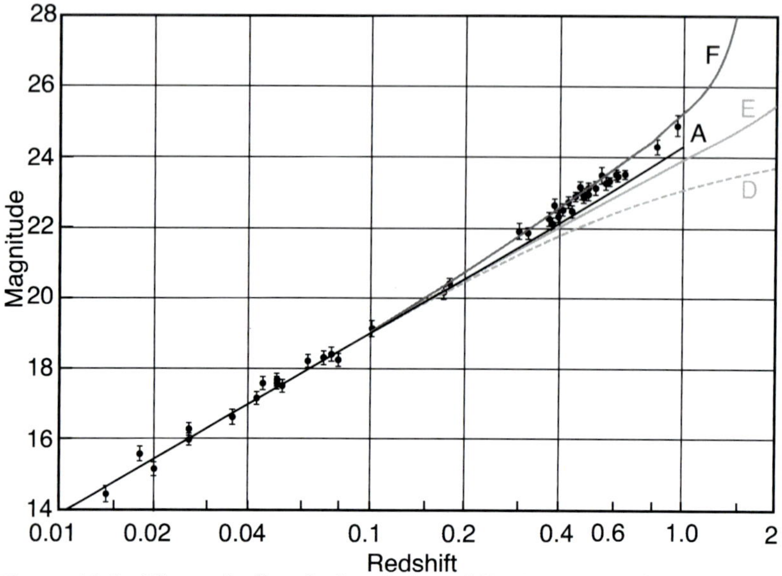

Figure 11.9: *Theoretically calculated plots of SNe1a magnitudes versus redshift superimposed on some measured values.*

In figure 11.9, curve E is the expected SNe1a magnitudes at different redshifts, determined from the relativistic redshift equation. It is close to curve A and within experimental error. Both are calculations of the same effect. Is it possible that the redshift corrections for magnitude, shown in figure 11.8, were not added to the theoreticians' calculations of curve A? That would explain why the theoreticians could find no error in their calculations and the astronomers could find no error in their measurements. That is left for theoreticians to answer!

11.6. Gravity and Space-Time Inside Matter

The visible universe has a very large mass. It is made up of a vast number of galaxies. Wherever astronomers look, galaxies are all around us. As such anywhere in the observable universe must be considered as space inside matter. A study of the universe is a study of space inside matter.

No study of gravity can be undertaken without mentioning the important work of Newton (1687, 1729) and Einstein (1905, 1911, 1916, 1917, 1952a, 1952b). The work done by both was accurate within the knowledge available to them. Chapter 8 points out how modern cosmologists' mis-interpreted their work. From Newton's work, they either do not know of, or have ignored, the simple physics of his Proposition 45 of volume 1.

A planet's orbit around the sun will precess in its direction of travel if gravity is weaker than inverse square. With Einstein's work they have shown they did not understand his derivation of his gravitational field equations. Their expansion of his field equations was founded on their flawed interpretation of Einstein's (1916, 1952a) work.

The cosmologists' mis-interpretations were pointed out in chapter 8. Combined with their flawed mathematics, they deduced that gravity was stronger than inverse square. In rectifying their mis-interpretations, the following shows a totally different universe.

The above calculations were based on the background of the accepted expanding universe theory. Curve F in figure 11.9, which matches the observed high redshift magnitude points, applies for any density of the universe. Gravity was only mentioned in its role of supposedly causing any universe to collapse under its mass. It is now time to have an independent look at the role gravity within the universe.

Figure 11.1 and equations 11.1 and 11.2 showed that Newton's gravity predicts an infinite universe will collapse. It also indicates that Newton's calculations apply to the universe at large when the Laplace and Poisson extensions are considered. Chapter 8 and Robinson (2021) showed there were four gravity attraction equations:-

$$g_Z = \frac{GM}{[r(1+z)]^2} = \frac{GM}{r^2 e^{\alpha}/r} \quad \text{Redshift} \qquad (11.7$$

$$g_E = \frac{GM}{\left(1+\frac{\alpha}{r}\right) r^2} \quad \text{Einstein} \qquad (11.8$$

$$g_N = \frac{GM}{r^2} \quad \text{Newton} \qquad (11.9$$

$$g_S = \frac{GM}{\left(1-\frac{\alpha}{r}\right) r^2} \quad \text{Schwarzschild} \qquad (11.10$$

Equations 11.7, 11.8 and 11.10 were derived in chapter 8. The terms g_Z, g_E and g_S are the gravitational attraction under the redshift metric, Einstein's metric, and the Schwarzschild metric respectively. Equation 11.9 is the standard Newtonian inverse square law expression. They are listed in order of increasing gravitational attraction.

Figure 11.1 and equations 11.2 showed that, under Newtonian gravity, mass, given by $M = \frac{4\pi r^3 \rho}{3}$ increases faster than r^2 diminishes its strength. Under Newtonian gravity, any universe that is not a single rotating galaxy, will collapse.

Equation 11.7 can be rewritten as

$$g_{rz} = \frac{4\pi G r^3 \rho}{3r^2 e^{\alpha/r}} = \frac{4\pi G r \rho}{3e^{\alpha/r}} \qquad (11.7a$$

where g_{rz} is gravitational attraction at distance $r = \sqrt[3]{\frac{3M}{4\pi\rho}}$, under the redshift metric. Substituting for $\alpha = \frac{2GM}{c^2}$ gives

$$g_{rz} = \frac{4\pi G r \rho}{3e^{\left(\frac{2G.4\pi r^3 \rho}{3rc^2}\right)}} = \frac{4\pi G r \rho}{3e^{\left(\frac{8\pi G r^2 \rho}{3c^2}\right)}} \qquad (11.11$$

Equation 11.11 shows that the numerator, upper line, increases as r, while the denominator increases proportional to e^{r^2}. No matter what the constants of proportionality, as $r \to \infty$, $g_r \to 0$. An infinite steady state universe will not collapse under gravity.

An important feature of equation 11.7 is that there is no attraction between distant galaxies. This is independent of the universe's density. In any region of the universe where there is a larger than average density, galaxies will be attracted towards it. "Great attractors" are possible.

Einstein's equation 11.8 can be rewritten as

$$g_S = \frac{2G.4\pi r^3 \rho}{3r^2\left(1+\frac{8\pi G r^3 \rho}{3rc^2}\right)} = \frac{8G\pi r \rho}{3\left(1+\frac{8\pi G r^2 \rho}{3c^2}\right)} \qquad (11.12$$

It shows that, as r increases, the numerator increases with r, while the denominator increases proportional to r^2 when $\frac{8\pi Gr\rho}{3c^2} >> 1$. Again, no matter what the values of the constants of proportionality are, as $r \rightarrow \infty$, $g_E \rightarrow 0$.

The Schwarzschild equation 11.10 can be rewritten as

$$g_S = \frac{2G.4\pi r^3 \rho}{3r^2\left(1 - \frac{8\pi Gr^3\rho}{3rc^2}\right)} = \frac{8G\pi r\rho}{3\left(1 - \frac{8\pi Gr^2\rho}{3c^2}\right)} \quad (11.13$$

It shows that, as r increases, the numerator increases with r, while the denominator decreases proportional to r^2 until $\frac{8\pi Gr^2\rho}{3c^2} = 1$. When that occurs, $g_S = \infty$. Newtonian gravity requires an infinite universe before gravitational attraction is infinite. Because the Schwarzschild metric predicts gravity is stronger than inverse square, gravitational attraction becomes infinite at a finite distance. That distance is given by

$$r = \sqrt{\frac{3c^2}{8\pi G\rho}} \quad (11.14$$

The accepted value for the density of baryonic matter in the universe is $\approx 1 \text{ x } 10^{-26}$ kgm^{-3}. Using c = 3 x 10^8 msec^{-1} and $G = 6.67$ x 10^{-11} Nm^{2}kg^{-2} in equation 14 gives

$$r = \frac{2x9x10^{16}}{8\pi x6.67x10^{-11}x10^{-26}} = 1.25\text{x}10^{26} \text{ m} \quad (11.14a$$

With approximately 10^{16} m/LY, that gives the radius at which gravitational attraction would be infinite is $\approx 10^{10}$ LYs. That shows it is not possible for a universe of density $\approx 10^{-26}$ kgm^{-3} to expand beyond 10^{10} LYs. The relativistic Doppler redshift formula, equation 11.3, gives a redshift of z $\approx$ 1.5 at 10^{10} LYs. Astronomers have images galaxies and other structures, such as supernovae, at redshifts beyond z = 1.5. The universe clearly exists beyond then. It is apparent through equation

11.14a, that Schwarzschild gravity, equation 11.10, plays no part in the universe. That includes expanding baryonic matter.

We now look at the redshift metric, knowing the Einstein metric will produce a similar result. That needs to be done in conjunction with Newton's Shell theorem. With it, Newton calculated that an object anywhere inside a hollow sphere of uniform density would not experience any gravitational attraction from that sphere if it was of uniform thickness and density. An illustration of Newton's Shell theorem, as modified by Laplace and Poisson, is shown in figure 11.10. A uniform distribution of galaxies in a hollow sphere will not exert any gravitational attraction on any mass inside it.

Observations from both telescopes and of the cosmic microwave background radiation, show that the large-scale structure of the universe is approximately uniform. The distances involved are measured in the hundreds to thousands of megaparsecs (Mpc), i.e., billions of light years (LYs).

There are obviously significant variations over distances of millions of LYs, 10^6 LYs being a "typical" distance between galaxies. That is averaged out over billions of light years. It makes Newton's shell theorem applicable to the study of space inside matter.

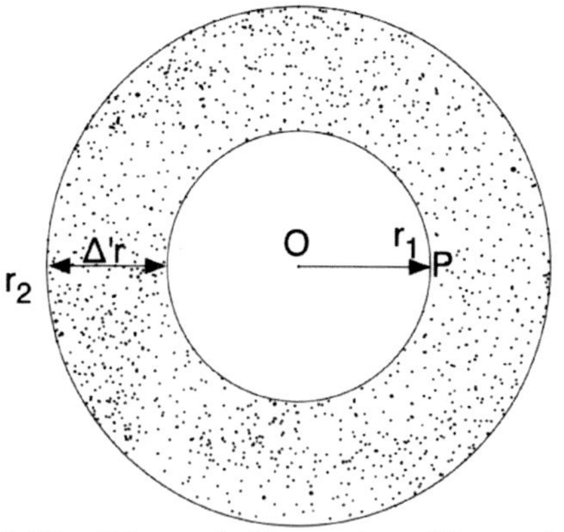

Figure 11.10: *Schematic illustration of Newton's Shell theorem, showing a two-dimensional slice through a hollow sphere of thickness Δr containing many galaxies.*

Einstein (1905) pointed out that photons had mass, $m_p = \frac{h\upsilon}{c^2}$ and were therefore subject to gravitational attraction. In his shell theorem, Newton pointed out that mass inside a spherically symmetric shell of uniform density, does not experience any gravitational pull from that shell, irrespective of its position inside it. Laplace and Poisson also demonstrated that the same principle applied for a shell composed of a uniform distribution of independent masses. A body inside a hollow sphere of radius r, see figure 11.10, will not experience any gravitational attraction from a shell of uniform density, ρ, and thickness, Δr, surrounding it. There is no limit to Δr.

Combining figures 11.1 and 11.10 means that a photon travelling away from an origin, O, anywhere within the universe will experience a gravitational attraction back towards its origin, which attraction will vary according to equation 11.11 as it travels through the universe. It will not experience any gravitational attraction from matter towards which it is travelling. Einstein (1911) and chapter 8 pointed out that photons subjected to a gravitational attraction will lose kinetic energy and hence frequency.

The mass within a sphere of radius r is given by $M_r = \frac{4\pi r^3 \rho}{3}$. Newton's shell theorem allows the galaxies in distance Δr beyond r to be ignored. That applies for all values of Δr. A photon travelling from O to P in figure 11.1, will experience a gravitational attraction back to O as it travels distance r to P. That attraction will cause it to lose kinetic energy. The change in kinetic energy, ΔKE, of a photon emitted at O and travelling distance r to P is $m_p g r$, where m_p is the photon's mass and g is the gravitational attraction at r from the mass M_r inside radius r.

The kinetic energy of the photon at O is $h\upsilon_0$, where h is Planck's constant and υ_0 is it frequency at O, when it was emitted. Under the gravitational attraction of mass M_r after it has travelled distance r, its frequency υ_r at distance r is given by

$$h\upsilon_r = h\upsilon_0 - h\upsilon_0 . gr \qquad \text{(11.15a}$$

Substituting for g under the redshift theory of gravity, namely equations 11.7 and 11.11, gives

$$h\upsilon_r = h\upsilon_0 - \frac{4\pi Gr\rho h\upsilon_0}{3e^{\left(\frac{4\pi Gr^2\rho}{3c^2}\right)}} \qquad (11.15b$$

which simplifies to

$$\upsilon_r - \upsilon_0 = \frac{\Delta\upsilon}{\upsilon} = \frac{4\pi Gr\rho.\Delta r}{3e^{\left(\frac{4\pi Gr^2\rho}{3c^2}\right)}} \qquad (11.15c$$

Equations 11.15 indicate a change in frequency as photons travel through the universe. A change in frequency indicates a change in wavelength, giving a redshift, z. Using $z = e^{\frac{\alpha}{2r}} - 1$ from equation 8.14 and substituting $\alpha = \frac{2GM}{c^2} = \frac{4\pi Gr^3\rho}{3c^2}$ into equation 11.15c, gives

$$z_r = e^{\left(\frac{4\pi Gr^3\rho}{3c^2 r}\right)} - 1 = e^{\left(\frac{4\pi Gr^2\rho}{3c^2}\right)} - 1 \qquad (11.16$$

where z_r is the redshift experienced by photons after they have travelled distance r through space. Equation 11.16 indicates that photons will be gravitationally redshifted as they move through space. It can be solved for r in terms of z. Setting $r = r_{GZ}$ to indicate the distance determined from gravitational redshift gives

$$r_{GZ} = \frac{c}{2}\sqrt{\frac{3ln(1+z).(1+z)}{\pi G\rho}} \qquad (11.17$$

Equation 11.17 can now be used to determine distances at different redshifts and densities. Using G = 6.674 x 10^{-11} $m^3kg^{-1}sec^{-1}$ gives

$$r_{GZ} = 1.794x10^{14}.\sqrt{\frac{ln(1+z).(1+z)}{\rho}}\ \text{m} \qquad (11.17a$$

Setting $\rho = 10^{-26}$ kg/m^3 and adjusting into parsecs gives

$$r_{GZ} = 5.81\text{x}10^{10}.\sqrt{\frac{ln(1+z).(1+z)}{\rho}}\ \text{parsec} \qquad (11.18$$

Adjusting equation 18 into light years (LYs) gives

$$r_{GZ} = 19.0\text{x}10^{10}.\sqrt{\frac{ln(1+z).(1+z)}{\rho}}\ \text{LYs} \qquad (11.18a$$

Equations 11.18 can be used to determine the distance an object of redshift z is away from the observer if gravity is the only source of redshift. Figure 11.11 shows the redshift versus distance results plotted for different densities in kg/m^3. Curve D shows the plot of redshift versus distance results from the relativistic Doppler effect equation associated with an expanding universe. It used the relativistic Doppler equation 11.3.

Equation 11.3 was derived from the special relativity corrections associated with time dilation at redshift z. It gives the distances at which the speed of the objects would generate the observed redshift due to its time dilation.

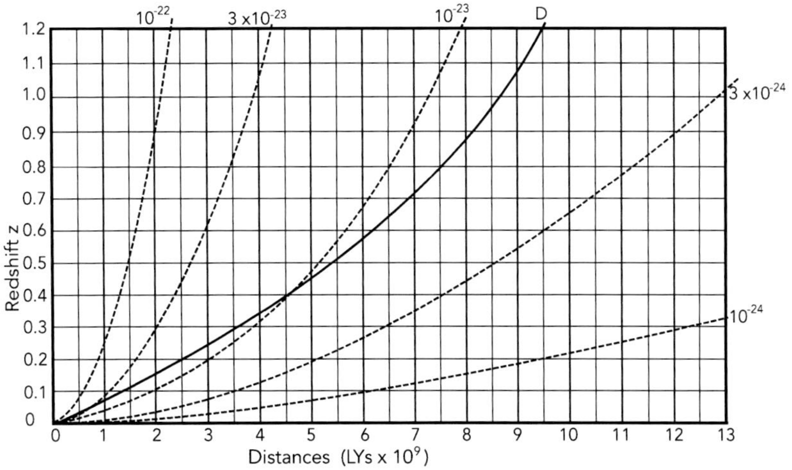

Figure 11.11: *A plot of redshift versus distance for r_G and r_D for an average density ρ between 10^{-24} and 10^{-22} kg m^{-3}.*

The maximum radius r_M = 13.99 x 10^9 LYs is based upon Freedman's (2021) determination, from red giant intensities, of Hubble's constant, = 69.8 km sec^{-1}/Mpc.

There are three important features about figure 11.11. The Doppler redshift versus distance relationship has a different curvature from the gravitational redshift versus distance relationship. The densities required to match the Doppler redshift curve are ≈ 10^{-23} kg/m^3. That is far higher than the generally accepted density of ≈ 1 x 10^{-26} kg/m^3. For gravity to match the observed Doppler versus redshift curve, it requires a continuous change in average universe density all around us.

The accepted universe density of ≈ 1 x 10^{-26} kg/m^3 was calculated from the expanding universe theory. If the density was any higher, the universe would have collapsed back onto itself a long time ago. If it was any lower, it would have moved faster over those 13.99 billion years. The visible universe would consist of the Milky Way and maybe a couple of local galaxies much further away.

Equations 11.10 and 11.13, as well as 11.14 and 14a, show that, under the Schwarzschild metric where gravity is stronger than inverse square, the universe could not expand beyond z ≈ 1.5 before gravitational attraction became infinite. An expanding universe under the Schwarzschild metric would have collapsed a long time ago! That clearly indicates that other explanations for the mechanics of the universe are required.

Observations show the SNe1a versus distance plots follow the Doppler redshift curve. The high mass density required to approximate Doppler redshift distance curve combines with the different slope curves, to rule gravity out as the sole source of redshift. That does not stop gravity from playing the important dual role of both keeping the universe together and stopping it from collapsing under the gravity associated with its mass.

Equations 11.7 and 11.8 both indicate that gravity → 0 as distance → ∞. As such, they offer the possibility that an infinite steady state universe would not collapse under its own mass. This will be explored using equation 11.11. Gravity associated with the Einstein metric, equation 11.12, is slightly stronger than that associated with the redshift metric, equation 11.11. The same principle applies for the Einstein's metric, equations 11.8 and 11.12.

11.7. A Steady State Universe Does Not Collapse Under Gravity

Equation 11.11 shows that, under the redshift metric, as distance increases, gravity decreases proportional to $\frac{r}{e^{r^2}}$. No matter what the density of space, increasing *r* will see a decrease in gravitational field attraction. Eventually it will have negligible effect. An infinite universe is schematically illustrated in figure 11.12. Each of the circles in figure 11.12 represents the observable limit, say 10^{10} LYs. Any other value will lead to the same result.

Consider the situation of an observer at X in the middle of circle A. X represents us in the middle of our observable universe, A. Each of the other circles represent a similarly sized visible universe. The number $g = 5 \times 10^{-12}$ represent the approximate gravity attraction, in msec^{-2}, at the edge of visible universe A due to the mass in A when it has a density of 1×10^{-26} kg m^{-3}. Similarly, the gravitational attraction at the edge of sphere 1, due to the mass within its radius of 2×10^{10} LYs is $g \approx 10^{-13}$ in msec^{-2}.

Continuing, a sphere of radius 3×10^{10} LYs will generate a gravitational attraction of $\approx 10^{-18}$ msec^{-2} at its surface They serve to illustrate that the gravitational attraction is diminishing exponentially.

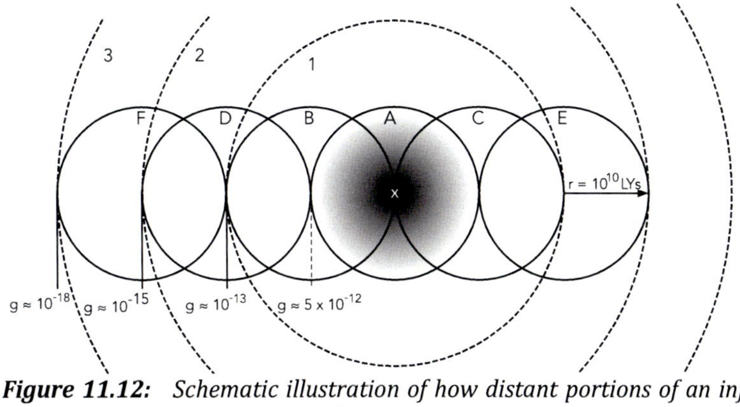

Figure 11.12: *Schematic illustration of how distant portions of an infinite steady state universe supports adjacent masses, preventing an infinite steady state universe from collapsing in on itself.*

By the time the sphere around X is 10^{11} LYs, its gravitational attraction on any mass at that distance is less than 10^{-50} msec^{-2}. At some stage, the gravitational attraction at the surface, from all the mass inside a large sphere containing a large number of galaxies inside it, will be indistinguishable from zero (0).

Visible universes B and C do not collapse into visible universe A because they are supported by visible universes D and E. Those visible universes, each of finite extent, are distributed in all three dimensions around us at X in the middle of our visible universe A. By the time centre of those visible universes are separated by 10^{11} LYs, they experience no gravitational attraction on each other. Those distant, limited size visible universes, act as pillars that stop all the limited universes between them collapsing.

That process is repeated in all three dimensions. Each of those limited universes at 10^{11} LYs separation are held by other limited universes around them. If the universe is infinite in space, time, and mass, it will not collapse under the influence of its own mass. The result is independent of the universe's average density. A different density may change the redshift distances. It does not change the trend. Einstein (1917, 1952b) considered only a finite universe. A gravitational attraction as low as 10^{-50} msec^{-2} at a distance of 10^{11} LYs would still result in a finite universe collapsing. Only an infinite universe prevents its collapse.

With an infinite steady state universe being stable, it is necessary to evaluate redshifts under a steady state universe. If they can match curve B of figure 2, there is no need to consider the other sources of redshift under the Doppler effect. Figure 11.11 rules out gravitational redshift within those redshifts and distances.

11.8. Redshift Explanations Under a Steady State Universe

Figure 11.11 shows that gravity needs space to have a locally varying density in the vicinity of 10^{-23} kg m^{-3} to have any possibility that it alone could explain those observed redshifts. It is not considered a plausible explanation.

Zwicky (1929) proposed that light got "tired" as it travelled through space and lost energy. That would increase its wavelength and give it a redshift. He gave no reason for that to occur.

Marmet (2018) evaluated some sixty proposals forwarded to explain the observed redshift of photons from distant astronomical objects. Marmet indicated that some of the proposed mechanisms gave a description of some aspects of the observed redshift effects. None gave an adequate description of all observations.

Many proposals failed because they introduced arbitrary terms. Others failed because they don't match the time dilation detected with high redshift quasars. Such attempts may have had theoretical value. They were physically meaningless without a reason for the universe to be stable against gravitational collapse. Such a reason was given above.

Along the way, Arp (1988) proposed that quasars with a high redshift could have a mass induced gravitational redshift and be relatively close. Chapter 8 pointed out the circumstances under which a massive object could redshift photons leaving them. Massive high redshift quasars would generate a redshift from their source. They could still be close. Gravitational redshift would produce the time delay through frequency decrease. That would overcome concerns about time delay, at least from quasars.

Shtyrkov (1993) suggested viscosity as a cause of redshift because photons would be slowed. Liberati and Maccione (2014) suggested the possibility of space having properties of dissipative hydrodynamics, one of which is viscosity. Viscosity reduces photon momenta, causing them to lose energy. Chapter 9 showed that a change in the speed of light requires a change in the electric permittivity of space. Photons losing momentum without losing speed will be redshifted.

For this study, the electric permittivity of intergalactic space remains ε_0. The speed of light remains constant. Figure 11.11 shows that a density of $\approx 10^{-26}$ kg/m^3, will make no detectable contribution to observed redshift up to 13 billion light years away. Other reasons for redshift must be sought.

There is nothing in the above that rules out the possibility of space having viscosity. It will be used in this study. When the study shows there is a match between viscosity predictions and observation, a possible origin of the viscosity will be presented.

Particles travelling distance, *r*, through a viscous medium are slowed down according to their momentum, i.e.,

$$R_v \propto \frac{d(mv)}{dr} \qquad (11.19$$

where R_v is the resistance to the passage of matter, mass m, and speed, v, through space. Viscosity slows down particles without scattering them. It is suggested the redshift measured by Hubble and others was due to the viscosity effect giving photons their redshift,

$$\Delta z = K\Delta D \qquad (11.20$$

where K is a constant and ΔD is the distance the photon has travelled. In non-relativistic Doppler redshift, $z = \frac{v}{c}$, where v, the velocity of the receding galaxy, is given by $v = H_0 D$. With Hubble's constant measuring a redshift due to viscosity, that becomes

$$z = \frac{H_0 D}{c} \qquad (11.21$$

Reported values of H_0 vary between ≈ 66 $km sec^{-1} Mpc^{-1}$ and 76 km $sec^{-1} Mpc^{-1}$ (Freedman, 2021). This study uses her determination, from red giant intensities, that Hubble's constant = 69.8 $km sec^{-1}/Mpc$, or 21.43 km $sec^{-1}/10^6$ LYs. It gives the distance at which galaxies would recede at the speed of light. 14.0 x 10^9 LYs.

From equation 11.21, dividing H_0 by c ≈ 2.998 x 10^5 $km sec^1$, gives K = 2.33 x 10^{-4}/Mpc, or 7.15 x $10^{-5}/10^6$ LYs. A photon travelling 10^6 LYs will lose 7.15 x 10^{-3} % of the momentum it had at its start. The wavelength, λ_0, of a photon at emission, will be $1.0000715\lambda_0$ after it has travelled 10^6 LYs. This effect is compounded. For any other distance D x 10^6 LYs, its wavelength will be $(1.0000715)^D$. Under this viscosity producing redshift mechanism, the redshift for any distance D, in units of 10^6 LYs, is

$$z = \left(1 + \frac{H_0}{c}\right)^D - 1 = (1.0000715)^D - 1 \qquad (11.22$$

The results from equation 11.22, viscosity, and equation 11.3, Doppler, are shown in table 11.4. The Doppler Dist* column is the

Doppler distance corresponding to the redshift caused by the viscosity effect at the distance indicated in the first column.

The Δ†Mag column shows the difference between the magnitudes of objects at the viscosity redshift column. The value is added to the magnitude of curve D in figure 11.9 to produce the expected viscosity magnitude equivalent to the Doppler curve E. It is shown in figure 11.15, which is displayed after other calculations are plotted to show the complete effect.

Figure 11.13 is a plot of viscosity, V, and Doppler, D, redshifts versus distance for $z < 1$. It shows the viscosity redshift approximately mimics the Doppler redshift values for redshift up to $z = 0.3$. The two curves were derived from the same constant, $H_0 = 69.8$ km/sec/Mpc (21.43 km/sec/10^6 LYs).

Not surprisingly, the two are almost identical for low redshift. As the redshift increases to $z = 1$, the viscosity curve indicates a greater distance than the Doppler curve.

Distance X 10^9 LYs	Viscosity Redshift	Doppler Dist* X 10^9 LYs	Δ†Mag
1	0.074	0.997	0.006
2	0.154	1.990	0.018
3	0.239	2.953	0.035
5	0.430	4.801	0.090
7.5	0.710	6.860	0.190
10	1.044	8.586	0.340
12.5	1.444	9.978	0.480
15	1.923	10.60	0.760
20	3.178	12.22	1.070
25	4.974	13.23	1.38

Table 11.4: *Calculated results for the change in magnitude due to the flux reduction associated with the motion of the SNe1a.*

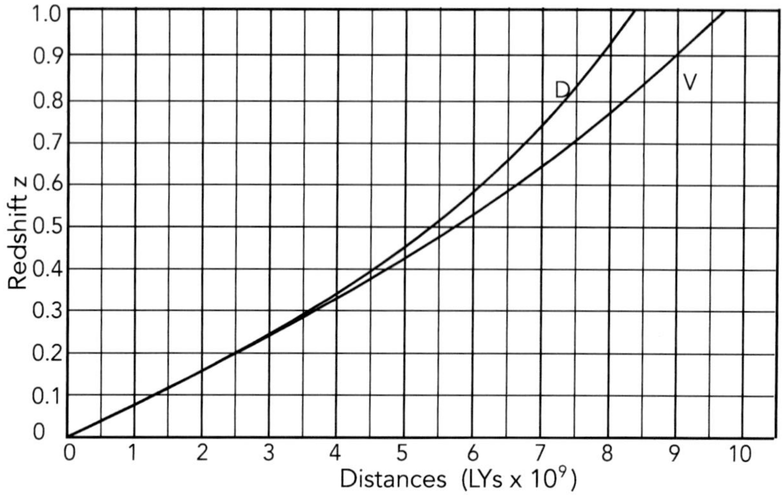

Figure 11.13 *A plot of the expected redshifts determined from the Doppler redshift (D) and viscosity redshift (V) curves for the accepted values of Hubble's constant H_0 = 69.8 km/sec/Mpc (21.4 km/sec/10^6 LYs).*

The viscosity redshift results, plotted over the results in figure 11.9 and shown in greater detail, are given in figure 11.14. Curve A and the associated points and scale are extracted from Perlmutter (2003). Curve D is the Doppler magnitude calculated from the distance versus redshift curve from equations 11.3 and 11.4. Curve E is the additional contribution to magnitude that comes from the flux change because of the receding speed of the distant SNe1a. Its values were calculated using equations 11.5 and 11.6. They are shown in table 11.3.

Curve F in figure 11.14 shows the total contribution to magnitude from the receding SNe1a at the redshifts measured, associated with an expanding universe. It shows that, if the universe is expanding, the SNe1a magnitudes for high redshift are in good agreement with the expected magnitudes. Based on that physical description of their origin, there is no need to consider any possibility of a universe expanding more rapidly than expected.

Curve V in figure 11.14 shows the magnitude expected at different redshifts, from the viscosity redshift versus distance relationship given in equation 11.22. Table 11.4 shows the expected magnitude difference values, ΔtMag, between curves D and V.

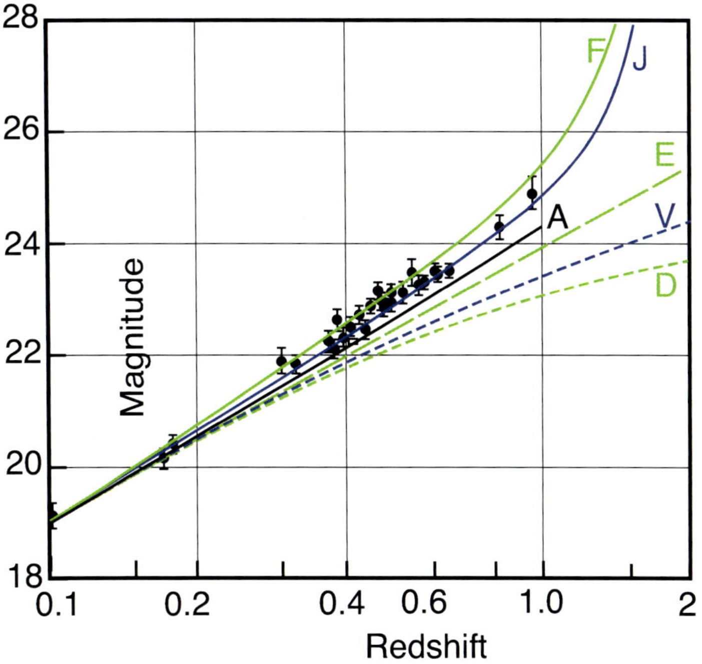

Figure 11.14: *Plots of the expected magnitudes determined from the Doppler redshift (D, G and F) and viscosity redshift (V and J) curves. They are superimposed on the grid, measured SNe1a magnitudes at high redshift and the critical value, copied from Perlmutter (2003).*

Adding the Δ†Mag values to curve D in figure 11.9 and 11.13 gives curve V in figure 11.14. Curve J was obtained by adding the magnitude changes associated with the magnitude measurement technique, obtained from figure 8, to curve V.

Curve J in figure 11.14 shows the magnitude expected if the redshift is caused by the viscosity effect in which the presence of "something" in the universe causes the photons travelling through the universe to lose momentum. Because photons are constant speed particles, their momentum loss is associated with a change in frequency, not velocity. That loss of frequency shows up as the photons being redshifted.

The final curves, F and J, both show that the expected magnitudes of high redshift SNe1a will be above the critical line presented by Perlmutter. Both show good agreement with the measured SNe1a magnitudes at high redshift. Curves D, E, F, V and J in figure 11.14 are independent of the density of the universe.

Curves F and J are dependent upon the shape of the SNe1a spectra used in figure 11.5. Excluding a blue peak around 400 to 460 nm, its irradiance approximately followed a line of equal photon flux per wavelength, absorption bands excluded.

Figure 11.15 shows a hypothetical SNe1a spectrum, in which the emission irradiance at 900 nm is ≈ 30% of the irradiance shown in figure 11.5.

With such a spectrum, a hypothetical curve J′, a revised version of curve J, would be lower than curve J by ≈ ⅓rd of the distance between curves J and A in figure 11.14. Alternatively, if the blue flux is lower than the green flux, i.e., the peaks in the 400 – 460 nm range, are lower than the dotted line in figure 11.5, while the other parts of the spectra are unchanged, hypothetical curve J″ will move higher than curve J.

The above shows the importance of careful evaluation of each SNe1a spectrum for its redshift through the magnitude determination equipment. Variations in SNe1a spectra are not significantly different from that in figure 11.5 that they would significantly alter curve J in figure 11.14.

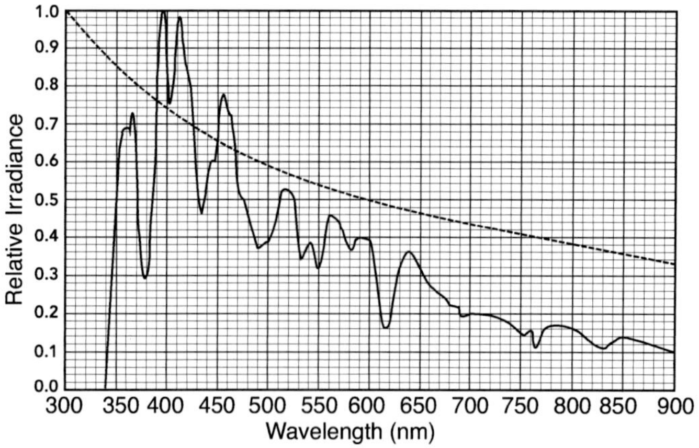

Figure 11.15: *Hypothetical SNe1a spectrum in which the red flux was significantly lower than the blue flux.*

That the experimentalists have measured the detected points is a good indication they have carefully evaluated the data from each observation to get the plotted results.

Curves F and J both show a rapid increase in magnitude beyond z = 1 as the spectra move towards the end of the detection range. Magnitudes above 26 are difficult to detect in ground based stations. SNe1a have been detected at redshifts beyond z = 1.7. No magnitudes were recorded. Under the Big Bang and expanding universe model, that could suggest the universe was expanding even more rapidly at $z > 1$ than at $z < 1$. That is not a good foundation upon which to predict that the universe has a missing energy equivalent of about 17 times the total mass of the baryonic matter, M_B, detected in the visible universe.

Curve J of figure 11.14 shows that the observed SNe1a magnitudes at high redshift are what would be expected under a steady state universe in which the redshift was caused by a viscosity effect that reduced photon frequency without scattering them.

Figure 11.16 shows the extension of figure 11.14 for redshift up to z = 10 and distances up to 25 x 10^{10} LYs. It shows that up to about z = 0.8 at a distance of ≈ 8 x 10^9 LYs, the two curves are approximately the same on this scale.

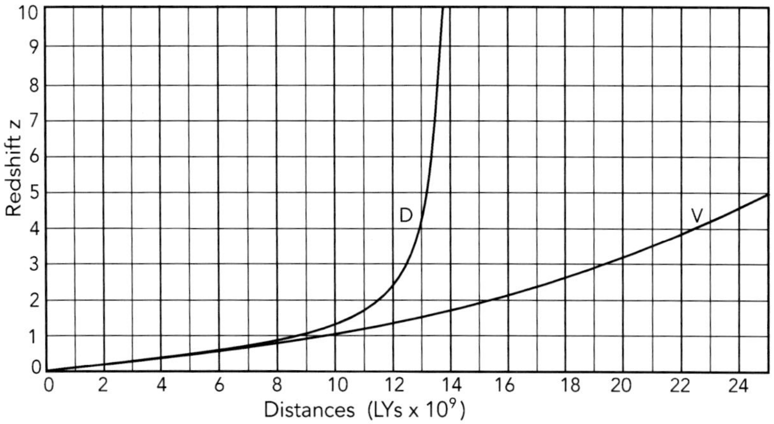

***Figure 11.16:** A plot of distances determined from the Doppler redshift (D) and viscosity redshift (V) versus distance.*

Beyond that, the Doppler redshift curve, D, rises rapidly, reaching 10 at about 13.6 x 10^9 LYs. The viscosity redshift curve, V, continues its slowly increasing rise, reaching z = 5 at ≈ 25 x 10^9 LYs. Not shown is that it reaches z = 10 at over 30 x 10^{10} LYs.

Figure 11.16 shows that, between redshift 1 and 5, the Doppler distance is less than 4 x 10^9 LYs. Under this viscosity effect, the distance is ≈ 15 x 10^9 LYs. That extra distance gives a much larger opportunity for galaxies to exist. With redshifts from 1 to 10, the Doppler distance increase is ≈ 5 x 10^{10} LYs, while the viscosity distance increase is ≈ 24 x 10^9 LYs (to ≈ 34 x 10^{10} LYs). In any high redshift field of view, the viscosity redshift mechanism will show far more galaxies than are available to be detected under the Doppler redshift mechanism.

That is shown in figure 11.17, an extract from the Hubble Ultra-Deep Field image. It shows vastly more galaxies than would be expected under the Doppler redshift associated with an expanding universe and curve D.

***Figure 11.17:** Hubble Ultra-Deep Field image (NASA)*

It is not just the distances. Curve D of figure 11.16 suggests there should be a lot of galaxies of varying redshift otherwise about the same size. Figure 11.17 shows the galaxies just get progressively smaller and dimmer. The Hubble Ultra-Deep Field image, is a very strong indication that the viscosity redshift curve, V, gives a far better representation of the universe than does the Doppler redshift curve, D.

Figure 11.11 shows that gravity will eventually increase the redshift at increasing distance. Adding that to curve V in figure 11.16 depends upon the universe's density.

11.9. Empty Space is not so Empty

The above has shown that the concept of photons losing momentum as they travel through sufficient space gives a good match between the theoretically predicted magnitudes of high redshift SNe1a and their observed values. It also gives a much more viable explanation of the Hubble ultra-deep field images produced by NASA. Curve D in figure 11.16 could produce the range of redshifted galaxies shown in figure 11.17. The range of galaxy sizes and intensities can only be produced by a great imaging depth. Curve V indicates that depth could be obtained when the galaxies are viewed over great distances. That is very strong evidence that curve V in figure 11.17 gives a much better representation than curve D.

For this to be a valid, model, it has some hurdles to overcome. The greatest of those is "How can empty space generate an effect similar to viscosity?" Empty space, often referred to as flat Minkowsky space–time, is best visualized as being halfway between neighbouring galaxies. That is typically ≈ 500,000 LYs from any galaxy. That is not a total void. There are residual, albeit extremely weak, gravitational and magnetic fields. There is also cosmic microwave background radiation and other photons passing through space. A summary of them is given in table 11.5.

The universe having a density of ≈ 10^{-26} kg m^{-3} gives it an average density, including galaxies, of all baryonic matter as a few protons/m^3. With most of that associated with galaxies, the density of baryonic matter away from them is negligible.

Waveband	Energy density of radiation (eV m^{-3})	Number density of photons (m^{-3})
Radio (300 MHz)	10^{-2}	$\approx 10^{-4}$
Cosmic Microwave Background	2.6×10^5	4×10^8
Infrared (140 – 1000 μm)	4×10^3	3×10^5
UV-optical near IR (0.16 – 3.5μm)	$\approx 10^4$	$\approx 10^4$
X-ray ($\approx$ 10keV)	20	3×10^{-3}
γ-ray ($\approx$ 1 MeV)	10	$\approx 10^{-5}$
γ-ray (> 1 MeV)	0.5	$\approx 3 \times 10^{-8}$

Table 11.5: ***Summary of the background energy and number densities of the background radiation of space, reproduced from data supplied from Hauser and Dwek (2001).***

Baryonic matter is not the only matter in space. Chapter 10 showed there was no need for dark matter. Earlier sections of this chapter showed there was no need for dark energy to explain the higher than expected SNe1a magnitudes for their observed redshifts. There are other forms of matter that is found throughout the universe. Figure 11.18 shows a plot of neutrino fluxes at different energies from different sources. The numbers were calculated from the Big Bang theory (Ice Cube, 2021). They will not apply under a steady state model.

A universe infinite in time and space would generate a lot of solar neutrinos. If nothing else, they would be slowed by gravity, eventually coming to rest. Cosmological neutrinos are those that have essentially come to near rest speeds. In a steady state universe, they would exist in an equilibrium between generation from stellar fusion and other causes, and recycling back to nucleons when captured by massive objects. Their exact density is not known.

Figure 11.19 shows another interpretation of cosmological neutrino content of the universe, suggesting their mass is 25% of the total baryonic matter. Baryons have a rest mass of $\approx$ 1 GeV. Neutrinos have a rest mass of < 10^{-4} eV. There are over 10^{12} cosmic neutrinos for each proton and neutron.

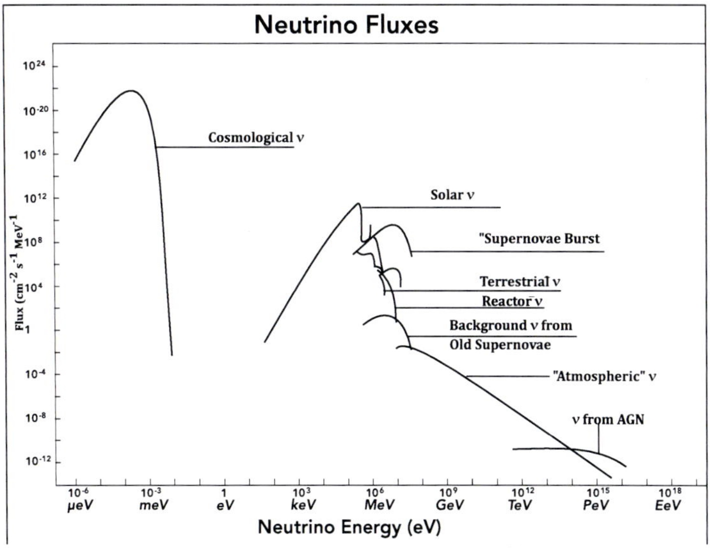

Figure 11.18: *A plot of the neutrino fluxes at different energies, from different sources.*

By numbers they are by far the universe's most numerous particles. Like cosmic microwave background radiation, they are also spread approximately uniformly throughout space, including intergalactic space. Their mass is so small that a neutrino absorbing a cosmic microwave background photon would have its speed go from rest to ≈ 0.8 c. For all practical purposes, neutrinos can be regarded as travelling at the speed of light.

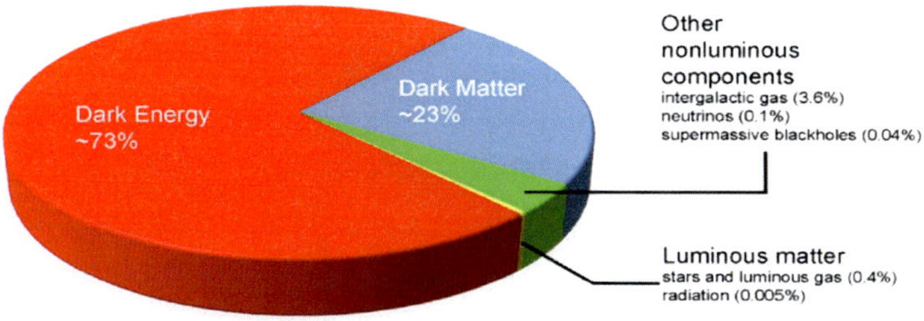

Figure 11.19: *Ratio of luminous and non-luminous matter to dark matter and dark energy required to match experimental observations under standard models. (Source: https://www.osti.gov/servlets/purl/1410794)*

A universe with an average density of 10^{-26} kg/m^3 has ≈ 6 baryons/m^3. At over 10^{12} neutrinos per baryon, there is an average of ≈ 10^{13} neutrinos per m^3, about 10^4/mm^3. Neutrinos have dimensions in millimetres. There is considerable overlap with each neutrino, its electromagnetic field and spin ½$\hbar$ occupying significantly more space than 1 mm.

A photon travelling 10^6 LYs, travels ≈ 10^{19} km or 10^{25} mm. Each of those millimetres has ≈ 10^4 neutrinos associated with it, through which a photon has to pass. This presentation suggests that a photon travelling a million light years through space passes through ≈ 10^{29} neutrinos. Over that distance it loses momentum of 0.00715% per. That is all it takes to explain the observed redshift. It does not appear to be an excessive loss rate.

As shown in chapter 9, the mechanics of gravity generation is through the high frequency nucleon oscillations. Because they have mass, neutrinos will respond to gravity, the same as do photons. No matter what mass they contribute to the universe, they will not generate additional gravity. That excludes the possibility of neutrinos being used to make up the missing mass "required" to explain galaxy rotation. Chapter 10 suggested there were suitable alternative explanations of galaxy rotation.

It could be hypothesized cosmological neutrinos provided the basis of the aethereal fluid, aether, proposed by Huygens (1690) as the means of transmission of light. They could provide the mechanism by which electric permittivity and magnetic permeability store and released energy as photons moved through space. That is for others to investigate.

This chapter suggests their presence provides space with an impedance property similar to viscosity. Flat Minkowski space-time, which is free of the influence of baryonic matter, has a contribution from cosmological neutrinos. Like all slightly viscous media, it would slow particles down according to their momentum without altering their trajectory.

It is suggested that gives a plausible reason for photons to lose momentum, and hence frequency. They will be redshifted as they travel through space. Viscosity causes particles to lose momentum. It does not change their direction. Photons will not be scattered.

The neutrino induced viscosity changes the momentum of photons. Their speed is determined by the electric permittivity of space. For all practical purposes, that is ε_0, its value in flat Minkowski space–time. In that manner it is not possible to distinguish between the redshift caused by viscosity and that caused by the Doppler effect, from observation alone.

11.10. Some Observations About Redshift

The above has demonstrated that an infinite steady state universe gives a good description of some of the universe's observed properties that were previously deemed to be due to be attributable only to an expanding universe. However, there are still other aspects of the observed universe that no previous steady state models have been able to explain.

Marmet (2018) pointed out there were over sixty attempts to explain the observed astronomical redshifts. None of them was entirely successful. Many "tired light" theories had been forwarded in the past. It was believed many failed the Tolman (1930, 1934) test.

Tolman derived a test to determine the effect the redshift associated with an expanding universe would have on the surface brightness of known intensity light sources. He derived an expression that the surface brightness in an expanding universe would decrease by a factor of $(1+z)^4$. Some cosmologists believed that separated an expanding universe from a steady state universe where, according to Tolman, the surface brightness would only diminish by (1 + z).

Figure 11.14 shows that equation 11.22 has a distance associated with its redshift, which distance is very similar to the relativistic Doppler redshift distance, equation 11.3, up to z ≈ 1. That gives it a factor of (1 + z), which continues beyond z = 1. It also has the (1 + z) reduction associated with the energy loss.

Section 11.3 showed there was also an ≈ (1 + z) reduction associated with the detection equipment. It is approximate because it depends upon the equipment, the spectrum, and the extent of redshift. Section 11.3 also showed that, for SNe1a, a redshift of over ≈ 1.7 in magnitude determinations reduced the intensity to zero. Under this physical redshift model, spectral intensities would diminish by $(1 + z)^n$. Tolman test results have shown the average intensity reduction varies as

$(1 + z)^n$, where $2.6 < n < 3.4$. That is within the range predicted by the physics shown in this model.

The Tolman test has been used to validate the expansion model over other redshift causes in a steady state universe. It does not separate a Doppler redshift from this viscosity redshift.

There are other possible ways of verifying this viscosity of space hypothesis. One is to determine the magnitudes of SNe1a at $z > 1$. They should plot close to curve J in figure 11.15.

It should be noted that magnitudes of greater than 26 are difficult to quantify through Earth's atmosphere (Reiss et al., 2018). SNe1a with redshifts greater than 1.8 have been detected. No combined magnitude and redshift measurements have been found for redshifts at or beyond $z \approx 1.8$.

Wright (2008) summarized results that showed that the time for the rise and decay of SNe1a increased linearly with redshift. That is what they expected from an expanding universe.

Equation 11.19 shows that viscosity causes particles to lose energy according to their momentum, i.e., mass times speed. Photon speed is controlled by the electric permittivity of space. Photons losing momentum lose energy and are redshifted. A blue photon, energy $\approx$ 4 eV, $\lambda \approx 400$ nm. It has twice the momentum of a red photon, energy 2eV, $\lambda \approx 800$ nm.

At a redshift of 1, blue photons have their energy reduced to 2 eV and wavelength increased to $\approx$ 800 nm. Red photons, which have half the momentum of a blue photons, have their 2 eV proportionally reduced by 50% of the blue photons, to 1.5 eV. Their wavelength is increased to 1,200 nm.

It takes twice as long for 2 eV photons to be redshifted to half their original wavelength than it does for 4 eV photons. It takes red photons twice as long to lose half their wavelength as it does blue photons. They are why the time width of $z = 1$ SNe1a is observed to be twice as long as SNe1a at $z < \approx 0.1$. That is what Goldhaber et al. (2001) and the Supernova Cosmology Project reported.

Wright took the opportunity to mention what he considered were three major problems faced by "tired light" proposals.

1) *No known interactions that could degrade a photon's momentum without blurring the image of distant objects.*

The possibility of viscosity had not been considered.

2) *Tired light models do not predict the observed time dilation of high redshift SNe1a.*

The above has clearly dispelled that idea. It predicts what Goldhaber et al. reported.

3) *Tired light model cannot produce a blackbody spectrum for the Cosmic Microwave Background without some incredible coincidences.*

The author seems unaware of the work of Assis and Nevis (1995). They listed several papers that predicted the background temperature of space at ≈ 3°K. That included Eddington predicting 3.2°K in 1926 and Regner predicting 2.8°K in 1933. Contrast those results for a steady state universe against Gamow's predictions of 7°K in 1953 and 50°K in 1961.

Another objection to an infinite steady state universe was forwarded by Olber. If the universe was infinite in time and space, it would provide sufficient time for photons from the infinite number of stars to reach Earth. The night sky would be as bright as the day sky, if not the surface of the sun. That holds true when the photons are not redshifted. Any cause of redshift will move photons emitted from distant stars out of the visible range and into infrared and microwave frequencies. There are insufficient galaxies within $z < 2$ to produce significant illumination into the night sky. Olber's paradox is not a problem for this redshift mechanism.

For objects of the same shape, viscosity causes them to lose momentum linearly according to their momentum. A particle with half the initial momentum of another loses momentum at half the rate. As shown earlier, red light, 2eV photons, lose momentum at half the rate of blue light, 4eV photons. Figure 11.5 shows that blue light is the dominant SNe1a frequency. We apply the inverse Hubble's constant, 1.0000715/MLY, to the blue photons.

This will slightly increase the intensities of the red end of the spectra at different redshifts in figure 11.7. That increase is slight because, by $z \approx 0.5$, the red end has been moved through the spectral range that contributes to their magnitude. If the effect was measurable, it would slightly reduce the magnitudes of curve F in figure 11.9 and curves F and J in figure 11.14 The natural variations in SNe1a spectra and their associated errors, make it difficult to positively detect the effect.

The speed of light is determined by the electric permittivity of free space. All photons emitted from the source will arrive at the same time. This viscosity effect reduces their momentum by lowering their energy, and hence frequency and momentum. It does not reduce their speed.

11.11 The Effect on Particles

The momentum of a photon, $E/c = hv/c = h/\lambda$, decreases as wavelength increases. For photons,
$1 + z = (\lambda_D/\lambda_0)$, where λ_D is the distance the photon has travelled (10^6 LYs). Expressed in terms of their momentum, $(1 + z) = (p_0/p_D)$, where p_0 and p_D are their momenta at origin O and distance D respectively. The momentum of a photon travelling through the same medium is given by $p_D = \frac{p_0}{(1+z)}$, giving

$$z = \frac{p_0}{p_D} - 1 = \quad \frac{H_0}{c} \qquad (11.23$$

The momentum of particles and photons are equivalent.

Inserting Equations 11.21 into 11.23 gives

$$\frac{p_D}{p_0} \approx \left[\left\{\frac{H_0}{c}.fn(E)\right\}^D\right]^{-1} \qquad (11.24$$

where $fn(E)$ is a function that relates particle energy to Hubble's constant H_0. The previous study of photons related 4eV photons to H_0. We can use 4 eV/ as a reference to H_0, giving

$$fn(E) = E/4 \qquad (11.25$$

where E is the energy of the particle that is emitted. From relativistic considerations, we have $E = mc^2$ and $p = \frac{E}{c} = mc$. From classical considerations, we have $KE = ½mv^2$ and $p = mv$. Classical and relativistic mechanics both suggest that the relationship between momentum is one of $p \approx \sqrt{E}$. The actual relationship is more complex than that and needs to be calculated for each situation.

To a first approximation, this gives the general equation for the change in momentum caused by the viscosity of space as

$$\frac{p}{p_0} = \left[\frac{H_0}{c} \cdot \sqrt{\frac{E}{4}}\right]^{-D} \qquad 11.26$$

where E is the total energy of the particle, both in eV. For example, a 1Gev proton has a mass ≈ 938 MeV/c² and motion energy of 1GeV. To a first approximation, its total energy is 2 GeV. To a good first approximation, a 1 GeV proton travelling 10,000 LYs through space will have a residual momentum of

$$\frac{p}{p_0} \approx \left[\left\{0.0000713. \sqrt{\frac{2x10^9}{4}}\right\}\right]^{-D} = (7.13 \text{ x } 10^{-5} \text{ x } 2 \text{ x } 10^4)^{-0.01} = 0.995\, p_0.$$

It would have lost about 0.5% of its original momentum.

Table 11.4 shows the approximate residual momentum 1 GeV cosmic ray protons will have different after travelling the indicated distances. It suggests 1 GeV cosmic protons will lose very little energy travelling within a galaxy. Even travelling 100,000 LYs across a galaxy, it will only lose a few percent. Travelling a typical inter-galactic distance of 1 million light years will cause it to lose about half its momentum. By the time they have travelled past a few galaxies, they have almost come to rest. Their small residual momentum means they have lost most of their speed, not their mass.

Distance (LYs)	Residual Momentum
10^4	0.995
10^5	0.95
10^6	0.7
10^7	3 x 10^{-3}

Table 11.4: *The residual momentum of 1GeV cosmic ray protons after they had travelled the indicated distances.*

The momentum of photons, $p_p = E/c$ Applying the momentum correction to equation 11.22 gives

$$z_\lambda \approx \sqrt{\frac{E_\lambda}{4}}\left[\left(1+\frac{H_0}{c}\right)^D - 1\right] \qquad (11.18$$

where E_λ is the energy, in eV, of the photon of wavelength λ. Equation 11.18 suggests blue photons will experience a slightly greater redshift than red photons travelling the same distance. They will still travel at the same speed. With the majority of the SNe1a spectra being at the blue end of the spectrum, that effect would slightly raise the magnitude of curve V in figure 11.11.

Table 11.5 shows what happens when 30 kV X-rays travel different distances through space. It shows their redshift changes very little as they travel between neighbouring galaxies. After travelling 10^{10} LYs, they would be soft X-rays with an energy of ≈ 270 eV.

Between them equations 11.15 to 11.18 can be used to determine the trends of photons and particles emitted from stellar objects, be they galaxies, supernovae, quasars, or others. A feature of equation 11.17, as shown in table 11.4, is that 1 GeV cosmic protons experience almost no resistance as they travel through a galaxy. They could reach neighbouring galaxies at 10^6 LYs, albeit with somewhat reduced energy. When they had travelled about 10^7 LYs, they would not have sufficient energy to travel much further as energetic cosmic rays. There are insufficient galaxies within 10^7 LYs to make a significant external contribution the cosmic ray flux within a galaxy. That is as observed. Most cosmic rays originate within the galaxy in which they are detected.

Distance (LYs)	Redshift z
10^6	5×10^{-3}
10^8	0.5
10^9	5
10^{10}	250

Table 11.5: *The approximate redshift of 30 kV X-rays after they had travelled the indicated distances.*

It should be noted that cosmological neutrinos, like cosmic microwave radiation, have wavelengths measures in tens of mm. They are very difficult to absorb, passing through matter as if it were not there. For that reason, their effect on individual particles, i.e., nuclei, electrons and photons can be ascertained. No attempt has been made to calculate the viscosity effect on the motion of assemblies of particles like stars.

It should also be noted that the arms of spiral always lag behind their rotation direction. Something is retarding the motion of those arms. It is not electric permittivity and its associated gravity. Spiral galaxies lagging arms is the best proof that something in space is dragging matter in an effect similar to viscosity. Table 11.4 already shows what is already known. Not many cosmic rays pass between galaxies. Table 11.5 shows a similar, although significantly reduced effect on gamma rays. The viscosity effect has its smallest impact on visible photons. Figure 11.16 and the associated equation 11.22 shows that it permits visibility beyond 35 billion light years.

When Zwicky introduced the concept that came to be called "tired light", he gave no reason for it and suggested its effect was related to the exponential of the distance travelled. This differs in that its effect depends upon the momentum of each particle. When a blue photon, E = 4 eV, has been redshifted to z = 1, it will be a red photon. It will be further redshifted the same as a red photon emitted at the blue photon's z = 1 distance.

Linear photons lose momentum and are redshifted until they reach an energy in which they are in equilibrium with the cosmological neutrinos and microwave background radiation. Rotating photons lose momentum and are slowed down until they come to rest with local mass.

An important feature in the above is not so much the actual equations. It is the prediction that something is slowing objects down as they move through intergalactic space. It is known that most of the cosmic rays detected around Earth originate from inside our galaxy. If they were not slowed down in intergalactic space, the huge number of other galaxies and their vast ages would result in most of them coming from outside our galaxy.

One current explanation for cosmic rays to slow down and not travel far in intergalactic space is the effect of cosmic microwave background radiation. It was suggested the ever present photons slow them down. The electromagnetic effect of cosmic neutrinos is much larger. Table 11.4 shows that the neutrino viscosity effect that provides the redshift for visible photons effectively limits intergalactic cosmic rays to the distance of ≈ 10^7 LYs. Only a few high energy cosmic rays would be detected in nearby galaxies. Even they would be at much lower energy than when emitted. That is what is observed!

11.12. Is this Mechanism Universal?

Equation 11.26 and its variations were used to calculate the redshift of photons and the energy loss of particles. These have some relationship to reality. If this mechanism is universal, it should be seen in other aspects of the universe. Heavenly bodies known to move in our vicinity include the planets. Mercury is the fastest moving planet. It has also been well observed.

Le Verrier (1859) and Newcombe (1882) observed that Mercury had an anomalous orbital precession of ≈ 43 arc sec/century (as/c). From his general relativity theory, Einstein derived an equation that suggested Mercury's anomalous orbital precession should be ≈ 43 as/c. It was later refined to 42.98 as/c. In equation 8.21, this study indicated it should be 42.99 as/c, based on the constants used. Mercury's orbital properties are well known. It is the fastest travelling planet, with an orbital speed of ≈ 47.36 km/sec. If the viscosity drag effect exists and can be observed, it will be seen in Mercury's orbit.

It is suggested the effect works on individual particles. It will be the same for each individual particle and will not be cumulative with any number of particles. The great majority of individual particle mass is contained in the nucleons. To a first approximation they are given a mass of 939 MeV. At their slow speed, their relativistic energy is approximately the same as their mass. Equation 11.26, indicates that, to a first approximation, the change in momentum given by:

$$\frac{p}{p_0} \approx \left[\frac{H_0}{c} \cdot \sqrt{\frac{E}{4}}\right]^{-D} \qquad (11.26$$

where p is the final momentum, p_0 is the initial momentum, H_0 is Hubble's constant, c is the speed of light and E is the energy, in eV, of the moving particle. The energy of a slowly moving nucleon is approximately the energy of the nucleon, namely 939 MeV, or 9.39 x 10^8 eV. H_0/c = 0.0000713. For a slowly moving proton, E/4 ≈ 2.35 x 10^8.

Mercury's average orbital speed is 47.36 km/sec. It takes 88 days to make one orbit, travelling a distance of ≈ 3.601 x 10^8 km. In 100 Earth years, it makes 415 orbits and travels 1.494 x 10^{11} km. Over that time period, Earth travels Mercury distance multiplied by (Earth speed/Mercury speed), or 1.494 x 29.78/47.36 = 0.94 x 10^{11}. Referenced from Earth, Mercury has only travelled 0.554 x 10^{11} km.

There are ≈ 9.46 x 10^{12} km/light year. In 100 years, Mercury travels 5.86 x 10^{-3} LYs relative to Earth.

That gives D = 5.86 x 10^{-9}. Putting these into equation 11.26 gives

$$p \approx p_0 \left[0.0000714.\sqrt{2.35x10^8}\right]^{-0.00000000586}$$

$$\approx [7.14\,x\,10^{-5}x1.53x10^4]^{-0.00000000586}$$

$$\approx 10^{-0.00000000586}$$

$$\approx 1 - 1.4 \text{ x } 10^{-8} \qquad (11.27$$

Equation 11.27 indicates that the momentum of a nucleon will be reduced by a factor of 1.4 x 10^{-8}, over its starting momentum. Because of its low relativistic speed, all of that momentum change will be due to a loss of speed. Its speed reduction of 1.4 x 10^{-8}, is what it experiences at the end of 100 years. The deceleration it experiences is uniform. The average speed reduction will be half or that, namely 7 x 10^{-9}.

As seen from Earth, the total distance travelled is 5.54 x 10^{10} km. With no change in time co-ordinates, that equates to a reduction of 7 x 10^{-9} in the distance Mercury travels, or ≈ 388 km. Mercury orbit's semi major axis is 57.91 x 10^6 km, and its eccentricity is 0.2056. That gives it an effective radius of 55.46 x 10^6 km, and circumference of 3.485 x 10^8 km. There are 1.296 x 10^6 arc sec in a 360° circumference. Each arc sec in Mercury's orbit represents a distance of 269 km. Mercury's orbit reduction of 388 km represents 1.44 arc sec. Under this study, that reduces Mercury's anomalous perihelion precession prediction from 42.99 as/c to 41.55 as/c.

The latest available information put Mercury's anomalous perihelion precession as 41.8 (±0.65)as/c (Park et al., 2017). The predicted result of 41.55 is well within the measured error range. That suggests the mechanism applies to all matter and is applicable to each individual particle, be it photon or nucleon, through the same equation. Earth is subject to the same effect. It is our reference frame. As small as it is, the effect is inbuilt into calculations of our orbital travel.

There is one other observation that suggests its effect is universal. The arms of spiral galaxies drag behind the galaxy's rotation direction, see figure 11.20. The galaxy rotates with the spiral arms lagging behind. All spiral arm galaxies show the same effect.

It could be argued that gravity is pulling them in. Their theory of galaxy rotation requires over six times as much dark matter than detected baryonic matter. Gravity can be dismissed. The neutrino viscous drag effect will contribute to, if not cause stars to follow each other at the outer ends of the spiral arms. That is seen in many galaxies.

An alternative to neutrinos for the observed viscosity effect is cosmic microwave background photons. They will not penetrate through stars or solid objects like Mercury. That leaves cosmological neutrinos as the only viable explanation. They penetrate everywhere, pervading everything, including planets and stars. The effect of equation 11.26 and its variants is universal.

Figure 11.20 *Image of a spiral galaxy (Wikipedia)*

11.13. Entropy in an Infinite Universe

One of the great criticisms of an infinite universe comes from entropy and the laws of thermodynamics. Entropy is a measure of the state of disorder of a system. Thermodynamics second law is that entropy always increases with time.

That is far from a correct statement. Gravity reverses entropy. Cosmologists are aware that stars and planets condensed out of giant gas and dust clouds. Our solar system is in a much more ordered state than the cloud of gas and dust particles from which it condensed. It is at that stage thermodynamic scholars qualify their second law. The second law of thermodynamics works well in its applicable realm. Gravity is outside its realm.

Another comment has been that an infinite universe would have fused all hydrogen into heavier elements. There would be no more fuel for stars. It is here that gravity again comes to the rescue.

Einstein mentioned that mass distorts space–time to generate gravity. Chapter 9 showed the origins of gravity and why it is only attractive. It also showed how it gave rise to space–time distortion. Equation 9.14 gave Newton's inverse square law as a first approximation and Einstein's gravity as a second approximation.

The exact solution showed that when an object such as a neutron star becomes sufficiently massive, space–time distortion becomes so great that gravity becomes weaker. The cross-over between stronger and weaker gravity is at half the Schwarzschild radius.

The massive object at the centre of galaxies attracts all mass, including all the heavy atoms, along with photons and neutrinos. Gravity compresses everything into neutrons. Contrary to the standard model, neutrons don't collapse to a point particle. They just get increasingly compressed.

With sufficient mass, the strongest gravity is outside the mass. This is shown in figures 8.6 and 11.22. Gravity outside the mass attracts neutrons towards it. This lowers the force on neutrons in the centre. The lower force allows neutrons decay back to hydrogen. The hydrogen pressure builds up. It eventually forces a break in the neutron shell around it. The gas then escapes at high pressure. It is mostly hydrogen, although it contains a few lighter elements like magnesium (Whiting, 2006). It also releases large numbers of photons.

Once the hydrogen breaks through, it moves at high speeds and is accelerated by the gravity attracting it. As it moves past the massive object's gravity begins to slow it down. It travels due to its high pressure. The mass of the body in which that process occurs is very large. It has a huge reserve of mass and continues to emit hydrogen, a few light elements and photons for a very long time.

They are the quasars that are observed. They are so massive that they also have a gravitational redshift in their own right. Halton Arp was right in his interpretation of the quasar next to galaxy NGC 4319. Quasars recycle everything. An infinite universe has all the time required to recycle everything.

That is the manner by which the universe recycles everything. Electron neutrinos, which are produced when hydrogen fuses to helium by generating neutrons from protons, can be absorbed when gravity compresses protons and electrons back to neutrons. Photons, including cosmic microwave background radiation, are absorbed into the particles, increasing their energy.

Cosmic microwave background radiation is not the residual nucleon/anti-nucleon annihilation in the early universe. It is the background temperature of space, as calculated by Eddington to be ≈ 3˚K in 1926.

11.14. Euclidean versus Curved Space-Time

If the universe was Euclidean in nature, it would be the same in all directions, for all speeds and at all positions. The first variations to that came from Michaelson and Morley. They found that the speed of light was always constant, irrespective of the velocity of the observer.

From that Lorenz and Lorentz determined that lengths and time changed with a particle's speed. Einstein extended that further by pointing out that mass must also increase with increasing speed. From that he derived the equivalence between mass and energy through $E = mc^2$.

Those determinations were all based on mathematical calculations that many scientists found too difficult to follow. There are still those who try to suggest that they are in error.

Chapter 2 showed the physical principles of Einstein's special relativity corrections. All matter is composed of rotating photons. As Einstein pointed out, photons convey inertia between emitter and receiver, giving them the property of mass. A photon rotating twice within its wavelength gives it angular momentum.

Photons, whether linear or rotating can only ever travel at the speed of light set by the electric permittivity of the local space. When a particle moves, its rotating photon spirals through space. That spiralling rotation obeys Pythagoras' theorem and gives rise to the special relativity corrections of length, mass and time. As it travels faster, the spiral's radius gets smaller. That adds a fourth correction, with lateral diameter reducing with increasing speed.

All the special relativity corrections are due to the rotating photon structure of matter. The nothingness of space still remains Euclidean. The photonic nature of mass means it is measured differently for different relative speeds of the observer and the observed. Thus, its reference as a theory of relativity.

No absolute rest has led to some confusion about which party is at rest and which is moving. That is solved by cosmic rays. GeV energy protons arrive at Earth from all directions. They are clearly moving. Earth is at rest. Moving satellite special relativity corrections show they are moving around Earth.

Equation 1.1 showed that the speed of light depends upon the electric permittivity of space. The higher is the electric permittivity of the medium through which a photon travels, the slower is its speed. Chapter 9 showed how the high frequency of the rotating photons that are the nucleons that make up mass change the electric permittivity of matter free space. The closer to a massive object, the greater is the electric permittivity of space.

The absolute value of electric permittivity determines a photon's speed. Changes in electric permittivity deflect photons. It is that deflection that gives mass their property of gravity. The radial differential of electric permittivity of space gives Newton's inverse square law of gravity as a first approximation and Einstein's gravity as a second approximation.

The exact solution predicts the donut shape detected at the centre of galaxies. It also predicts the recycling of fused heavy nuclei back to hydrogen, thus enabling the steady state universe to continue forever.

Are space and time curved? In the absence of mass, the universe would be void and space and time would be constant. Mass doesn't distort Euclidean space. It changes the electric permittivity of Euclidean space. In turn those change photon trajectories. Those trajectory changes cause gravity. Regions of higher electric permittivity slow photons, including their speed of rotation in particles. Slower photons mean slower time.

The only way we have of observing distant events is through photons. Using a definition of Euclidean space as the trajectory of photons, mass distorts space-time. It does it through a change in the electric permittivity of space, not by curving Euclidean space.

In the same manner, travelling at high speed doesn't shorten Euclidean space. A distance of 10 light years remains 10 light years, even if high speed travellers travelled that distance in 5 years, as measured by an accurate Earth time clock on the travelling craft. The travellers would not see the spiralling nature of the rotating photons that are the matter of which they were composed. All measurements made of local mass, length and time within their craft would be the same as measured on Earth prior to departure.

Einstein's calculations have not removed logic from physics. His corrections have very logical explanations. Both his special and general relativity corrections are due to the rotating photon nature of all matter particles. Applying those logical corrections means the relativity theories are easy to understand using common sense.

At the other end of the universe is the seemly irreconcilable phenomenon known as quantum foam. In it, space itself seems to have properties akin to fluctuations in space-time at the smallest scale. Space itself is nothing, the absence of any material substance. How can nothing possibly fluctuate? Again, the answer lies in the nature of individual particles, particularly nucleons.

Their electric field oscillates at approximately 10^{23} Hz. Equation 9.16 suggests there will be a distortion in the electric permittivity of space in the local vicinity of each nucleon. There will also be a distortion in the magnetic permeability. Although empty space is nothing, its properties of electric permittivity and magnetic permeability will fluctuate at a frequency of $\approx 10^{23}$ Hz. Those fluctuations could have properties similar to those attributed to quantum foam.

Although very minor, the local quantum gravity effect per nucleon will be larger than the large scale gravitational effect per nucleon. In that manner, quantum foam is a logical effect that can be explained by the same common sense that explains Einstein's relativity theories. As well as uniting Einstein's relativity effects with quantum foam, this model shows they are both quite logical effects that are easily understood using classical concepts introduced by Newton and Maxwell.

Space is Euclidean. Its properties of electric permittivity and magnetic permeability are distorted by the rotating photon nature of mass particles. Remove mass from any region of space and it reverts to being Euclidean in nature.

On the large scale, that is seen by the sun bending light when photons move close to its surface. As Earth moves away from an alignment between a distant star and the sun's surface, normal Euclidean trajectories are observed. In the same manner, quantum foam effects are only detected near particles. Away from them, quantum foam effects are not detected.

The sole exception is the presence of cosmological neutrinos. They cannot be removed. They limit the unrestricted transmission of photons through space by causing them to lose momentum. It is that which causes photons to experience a redshift after they travel through millions of light years.

11.15. Discussion

This topic is literally as vast as the universe. It is believed the above has given sufficient information to separate an expanding universe from a steady state universe. The concept of an expanding universe commencing from a Big Bang first arose from Hubble's (1929) determination that galaxies at greater distances had their light redshifted by a greater amount when observed on Earth. His first conclusion was that the redshift was produced by a Doppler effect of galaxies moving away in an expanding universe. Zwicky (1929) suggested the redshift was due to photons losing energy as they travelled through space. He gave no reason for the energy loss. Lamaitre (1931) was the first to suggest the universe commenced from *"l'atome primitive', "the primeval atom"*. Thus began the Big Bang theory, as Hoyle called it.

That laid the foundations for modern cosmology. Newton's gravity theory predicted an infinite steady state universe would collapse due to the gravity of its infinite mass. Einstein predicted that the universe would also collapse. He only considered a finite universe. Others who expanded on his field equations, did not understand them. They concluded his work predicted that gravity was stronger than inverse square when they predicted the existence of black holes. In doing so, they showed a lack of knowledge of Einstein's simple mathematics and Newton's simple physics, as pointed out in chapter 8.

Einstein's field equations predict that gravity is weaker than inverse square. It is suggested Einstein knew that. It is required for a planet orbiting the sun to have its perihelion precess in its orbital direction. The detection of Mercury's anomalous orbital precession, and star S2's precession around the massive object in Sagittarius A, are two reason why black holes physically cannot exist.

Not understanding the mathematics or physics associated with Einstein's gravity, cosmologists proceeded on the basis of gravity will cause a universe to collapse. Penzias and Wilson (1965) discovered the background radiation with an energy equivalent to a temperature of ≈ 3°K. It was claimed as the residual from the Big Bang, even though Eddington (1926) calculated that the background temperature of space for an infinite steady state universe was ≈ 3.12°K. Assis and Neves (1995) reviewed other calculations of the temperature of space in a steady state model. All were closer to the observed value of ≈ 2.7°K than any of the predictions made prior to the Penzias and Wilson determination. With no alternative to gravity causing the universe to collapse under its mass, all cosmological measurements were fitted to a universe expanding from a Big Bang.

Along the way that included:-

- A period of hyperinflation, in which the universe expanded faster than the speed of light.

- Odds of the universe surviving ≈ 14 billion years of ≈ 1 : 10^{60} (Hawkins, 1998).
- That produced an assertion that our visible universe was part of a multiverse of at least 10^{60} other universes (Rees, 1997). Our universe was the only one with physical properties that supported the structure we observed. Those odds are so low as to be zero.

The list goes on to include:-

- Space and time were created with the Big Bang.
- Space is expanding faster than the galaxies, stretching the photons to give extra redshift.
- Space and galaxies aren't expanding into anything because there was no space before the Big Bang.

The above is mathematicians' "talk". Like so much else in modern cosmology, they have no physical foundation.

Cosmologists don't know what caused the Big Bang. They require parity rules to be violated to produce a slight excess of matter particles over anti-matter particles. We can add to that their findings that the universe must be expanding at an accelerating rate starting a few billion years ago. That requires an energy equivalent to about 17 times the energy equivalent of all the baryonic matter that astronomers have observed.

They and much more are the mathematician's explanation of the universe observed by astronomers. It all originates because they don't understand Einstein's simple mathematics or Newtons' simple physics.

Cosmologists did not understand that Einstein's gravity field equations predict that gravity is weaker than inverse square. Orbital precession such as Mercury's anomalous orbital precession and star S2 around the massive object in Sagittarius A, require gravity to be weaker than inverse square. The physical explanation is that when gravity is weaker than inverse square, the planet/star orbits a little further away from the attracting mass. It has to travel further under a lesser attractive force. It returns to its perihelion/aphelion positions a little later in each orbit. That is the simple physical explanation of precessing orbits. It was first pointed out by Newton in 1687.

It appears theoretical cosmologists were impressed by Einstein's field equations. They tried to expand on them to show their scientific prowess, aka mathematical skills. Equations 11.11 to 11.13 show that effect of gravity inside matter can be calculated directly for any density of the universe and any distances travelled by photons.

The question could be asked as to why cosmologists never applied Newton's shell theorem to the Schwarzschild interpretation of Einstein's gravity field equations. Had they accurately expanded the Schwarzschild metrics, they would have found that gravitational attraction would become infinite at a finite distance. That is demonstrated by the equation 11.14a solution to equation 11.10. Newton's shell theorem was an essential part of determining that an infinite steady state universe would collapse under his gravity theory. It cannot be ignored because gravity is less than inverse square.

Despite all its defects, the Big Bang and expanding universe theory remained the only acceptable cosmology theory of the origins and properties of the universe because Newton's gravity theory and Einstein's prediction both indicated a universe would collapse under the gravity of its mass. Newton's shell theorem provided the means by which his inverse square law would collapse the universe. Einstein considered only a finite universe.

Over many decades, dozens of alternative redshift theories were presented to explain the observed redshift. Even if some were successful, they would not overcome the problem of gravity causing the universe to collapse under its great mass. Equations 11.11 and 11.12 show that, under the redshift and Einstein's metrics, gravity is weaker than inverse square. Under either metric, an infinite universe will not collapse.

That still requires an explanation of the observed redshift with distance. Figure 11.11 shows that gravity cannot cause it. The observed redshift requires a much larger density and a non-uniform mass distribution. The above presentation shows that a universal viscosity effect explained the observed SNe1a magnitude versus redshift points.

The generation of such a redshift must have a physical reason. It has been known that cosmic rays are slowed down as they travel the huge distances through intergalactic space. Most of the detected cosmic rays originate within our galaxy. Only a few make it from neighbouring galaxies, which are typically over a million light years away. Space slowing down particles was well known. It was attributed to the cosmic rays being scattered by the cosmic microwave background.

It is suggested photons are influenced by the high density of cosmological neutrinos. They are vastly more numerous and at far greater energy density than cosmic microwave background radiation photons. They uniformly distribute a combination of electromagnetic field with angular momentum throughout space through which photons pass. The interaction is exceedingly week. After photons have travelled for 10^6 LYs, $\approx 10^{22}$ m, they have lost 7.13×10^{-3} % of their initial momentum. That figure was obtained from H_0, Hubble's constant divided by the speed of light. Whether that loss is due to cosmological neutrinos, or some other effect, is not as important as that an energy loss effect occurs. There does not appear to be any alternative to cosmological neutrinos.

As mentioned earlier, an infinite steady state universe collapsing under its own mass depends upon Newton's shell theorem. His shell theorem applies whether gravity is inverse square or is stronger or weaker than inverse square. It provides a much easier way of determining gravitational field strengths for assemblies of galaxies. Equations 11.7 and 11.11 show that when gravity is weaker than inverse square, it tends to zero as distance tends to infinity. An infinite steady state universe will not collapse under its mass when either of those equations applies.

Equations 11.13 and 11.14 shows that, when gravity is stronger than inverse square law, its field strength tends to infinity in a finite distance. For an average universe density of 10^{-26} kg m^{-3}, that occurs at $\approx 10^{10}$ LYs. Such a universe would have long since disappeared. That would be independent of any amount of dark energy applied to it.

That clearly hasn't happened! In the same manner, gravity stronger than inverse square would cause the perihelion and aphelion of Mercury and star S2 to regress, not precess as is observed. On both counts, gravity being stronger than inverse square can be ruled out as a workable cosmological solution to Einstein's gravitational field equations. That leaves gravity weaker than inverse square as the only viable option that allows the universe to exist.

Einstein's "relief" when Hubble proposed that the universe was expanding was because he considered only a finite universe. A finite universe will collapse under any gravity mechanism because gravity only attracts.

Figure 11.11 shows gravity is not strong enough, and does not have the correct format, to cause the observed redshift. The only alternative is something akin to viscosity. It causes photons to lose energy/momentum, while maintaining their constant speed. They are redshifted as they move through space. That is the redshift with distance originally observed by Hubble.

Over the decades, observations have been fitted to the Doppler redshift theory. It remained a favourite for several reasons. Included in them was the Tolman test and the assumption that because gravity is only attractive, the universe must collapse under its own mass.

Another was the statement by Wright (2008), that the time width of observed SNe1a spectra increased as (1 + z). As shown in section 11.9, that is an automatic consequence of any natural redshift mechanism. Einstein predicted it for gravity, and it has been observed. Time dilation is an automatic consequence of any natural photon frequency change.

The cosmic microwave background radiation is cited as another proof of the Big Bang theory. It is almost a perfect black body radiation spectrum for 2.715°K. Assis and Neves (1995) surveyed the theoretical calculations prior to the Penzias and Wilson (1965) discovery of the background radiation. They published early predictions of the temperature of space under a steady state universe. They were ≈ 5.6°K, 3.12°K and 2.8°K, by Guillaume in 1896), Eddington in 1926) and Regner in1933. Early predictions of the temperature of space from the Big Bang residual were Gammow, 50°K in 1948 and 7°K in 1953 and 6°K in 1956. Alpher and Herman calculated 5°K in 1948 and 28°K in 1950. The "Big Bang" estimates were much higher than the "steady state" estimates. "Big Bang" cosmologists have modified their mathematics to match observation ever since the discovery of the cosmic microwave radiation and its refinements through satellites such as WMAP.

In this "steady state" model, cosmic microwave radiation is the background temperature of space. The closer to a galaxy that the temperature is measured, the warmer it will be. Eddington's 1926 answer is out by ≈ 0.4°K compared to the +2000 CE measured value. That is often used as an indication he was wrong, and the universe did start from a Big Bang. Regner's 2.8°K work is conveniently ignored.

Cosmologists' treatment of the cosmic microwave background radiation is one of the better examples of fitting a theory to an observed result and searching for excuses to reject an alternative.

Olber's paradox is an often-quoted pillar of the Big Bang theory. Any effect that redshifts photons will prevent visible light from making its way unchanged across the universe. A redshift greater than 2 moves all visible light into infra-red. There are not enough stars within redshift 2 to illuminate the night sky.

In the late 1930s, Nernst proposed a model in which the universe was infinite and steady state. He proposed that "luminiferous ether" would remove energy from photons, thus red shifting them. He failed to find a solution to an ever increasing entropy (Assis and Neves, 1995).

Another pillar of the Big Bang theory was its calculation of the hydrogen to helium ratio. It predicts 74% hydrogen and 26% helium, by mass, formed from the Big Bang theory. That closely matches the ratio observed in the sun and other stars. The hydrogen to helium ratio in cosmic rays is ≈ 76% hydrogen, 23% helium and 1% everything else. With exploding stars being the source of heavy metals for the planets, the hydrogen to helium ratio they emit is not much different from that calculated in the Big Bang.

Other "support" for the Big Bang theory is that distant galaxies appear relatively smaller than their nearby counterparts. There are two physical reasons for that. Photon redshift means less visible light is detected. Images of distant galaxies have less intensity, making them appear smaller. Figure 11.16 shows that distances at $z > \approx 1$ are significantly greater under the viscosity redshift than under the Doppler redshift. Calculating size based on the Doppler redshift distance will give a much larger size than under the viscosity redshift distance.

That is often accompanied by statements that distant galaxies have only hydrogen in their detected spectra. They do not show any metal in their spectra. Figure 6.1 shows that the sun has a hydrogen abundance of $\approx 10^4$ greater than oxygen, the next most abundant element after helium. The most abundant metals, Mg, Si and Fe, have abundances of typically 10^{-6} that of hydrogen. Accurately detecting such small intensity peaks over great distances is a daunting task. A null result does not necessarily mean no metals.

This mechanism does not prevent hydrogen clouds from having a less than average abundance of heavier nuclei when thermonuclear fusion commences. New stars would not have time to fuse a detectable abundance of heavier nuclei.

Observationally, those "pillars of the Big Bang" model have equivalent explanations under this steady state model. Theoretically, the Big Bang model relies heavily upon two major predictions.

i) Newton's shell theorem and the inverse square law of gravity results in an infinite steady state universe collapsing under the gravity of its own mass.

ii) Einstein's pronouncement that a finite universe would collapse.

Between them, they did not allow for a steady state universe. There was no option but to interpret the detected redshift from distant objects as due to the redshift of an expanding universe.

Applying Newton's shell theorem to the redshift of Einstein's gravity, shows that an infinite universe will not collapse under gravity. That leaves the situation open for other interpretations of the observed redshift. This presentation interpreted Hubble's constant as due to a viscosity like effect. It gave a good match with the observed SNe1a magnitude versus redshift plot.

Under the Big Bang theory, those measurements have been explained as the accelerated expansion of the universe. That requires about 17 times as much energy as the energy equivalent of the all the baryonic mass astronomers have detected. Yet, as curve F in figures 11.9 and 11.14 show, a physical evaluation shows that the observed magnitudes of high redshift SNe1a are exactly as expected.

Curves A and F in figure 11.14, both were calculated for the same feature, the magnitude of SNe1a with redshift. Their results are within experimental error, even at z = 1. That explains why the theoreticians' calculations were believed to be correct. The astronomers' measurements were also correct. Is the prediction of dark energy due to neither party understanding the implication of magnitude increase with redshift through the measurement equipment, as shown in section 11.5?

The Big Bang model requires a stringent density of ≈ 10^{-26} kg m^{-3} for the universe to exist in its current form about 14 billion years after it was formed. The odds of survival are about 1 : 10^{60}. Calculations under this model were density independent. Any density universe would be stable against gravitational collapse.

This model does rely upon matter free space having a property similar to viscosity. Viscous fluids provide a drag that slows particle motion. The suggested medium is the electromagnetic field provided by cosmological neutrinos. In this model, neutrinos at rest are rotating plane polarized photons that have a rest mass approximately the same as the cosmic microwave background radiation, ≈ 10^{-4} eV/c^2. Their combined numbers can be obtained from table 11.5 and figure 11.18. Figures 11.18 and 11.19 indicate the cosmic neutrino flux is much higher than the cosmic microwave background radiation.

That leaves the large-scale structure of the universe controlled by its infinite size and gravity. The observed redshift of light from galaxies up to z ≈ 4, is mainly due to background cosmological neutrinos being of sufficient density to give a slight viscosity to baryonic-matter-free space. At greater redshifts, gravity will start to make a detectable contribution. The distance depends on the universe's density. The gravitational effect will occur for any density.

The above physical explanation was made without complex gravitational field equations. Much of the work on theoretical cosmology has been based around "extensions" of Einstein's field equations. It creates some major problems. Field equations are complex. That leaves open the possibility of the experts implying "*Trust us. We are the experts. We know what we are doing*." That would not be much of a problem if their field equations and interpretations were accurate.

As chapter 8 pointed out, none of those who studied Einstein's field equations understood the simple mathematics associated with his gravitational field equations. Their expansions and revisions of them have all commenced with no knowledge of the interpretational errors that were introduced from misunderstandings of his work. The result is a litany of errors that could have been avoided if at least one of them knew how to recognize the approximations Einstein admitted he used. Alternatively, it would have been thought that at least one would have read Newton's work on gravity, in which he indicated how orbital precession required gravity to be weaker than inverse square.

In this approach, we have arrived at the stage where the properties of galaxies and the mechanics under which they operate have been described in three space dimensions and time. There was no need to introduce any unknowns or undetected material such as other dimensions or dark matter to explain galactic properties. We can now put the hundreds of billions of observable galaxies together into the universe.

In addition to all of the above that favours an infinite steady state universe, there is still one unanswered question that favours the universe having a beginning. How does the universe replenish its supply of hydrogen required for the continual supply of fuel for fusion? The answer lies in figure 8.6, reproduced as figure 11.21 with Einstein's gravity added.

As mass accumulates at the centre of galaxies, it reaches a stage where gravity overpowers electromagnetism. All matter is compressed, forming a neutron star. Under this model, neutrons can be compressed, but they have no sub-structure to which they can collapse. Increased mass compresses the neutrons further.

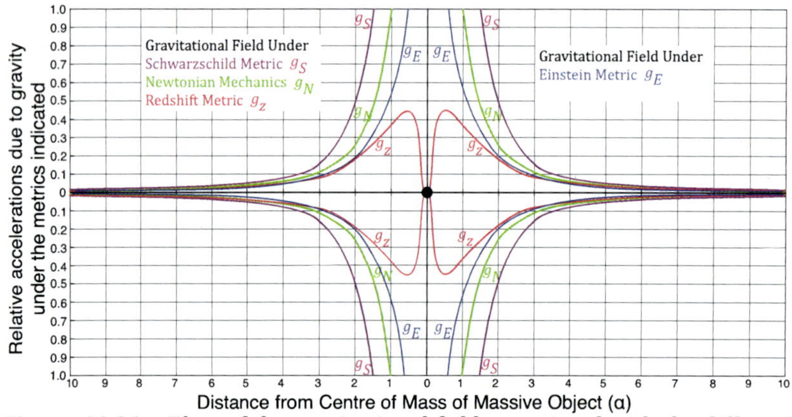

***Figure 11.21:** Plots of the gravitational fields associated with the different metrics. The redshift metric, g_z, is the only one that remains accurate for all distances.*

The neutrons form a crystalline neutron nucleus. Think of it as a massive version of figure 5.32, uranium nuclei, without the protons. When rotating, their axis will point in their rotation direction. That varies at different positions around it. They can't form a "perfect crystal" nucleus. It will have imperfections.

With enough mass, i.e., > millions of solar masses, the Schwarzschild radius exceeds its physical radius. Gravity gets weaker, as shown in curve gz of figure 11.21. Gravitational attraction is strongest at $r= 0.5\alpha$. The mass at that distance is attracting neutrons in the centre towards it. This greater pressure is at $r = 0.5\alpha$, not at the centre, as is the case of Newtonian gravity as for example in the sun or on planets.

With the greatest gravitational attraction away from the centre towards $r = 0.5\alpha$, pressure in the centre is maintained by the neutrons imperfect crystalline structure. As that mass increases, the pressure in the centre decreases. The lowered pressure will reach a stage where there is insufficient force to prevent neutrons decaying into protons and electrons, namely hydrogen gas. A gas pressure builds up inside the neutron star's imperfect structure.

Each neutron decaying to a proton and electron releases about 0.7 MeV energy. That raises the temperature, creating conditions where fusion to helium and higher Z nuclei occurs. The more energy released further raises the pressure.

As the hydrogen and helium gas pressures build up, it breaks through the neutron shell at its weakest point. The weakest point would be at or near the rotation poles. That is where the greatest neutron mis-alignment occurs. From then on, hydrogen, helium, and the occasional other light elements, are ejected at high pressure and accelerated by the strong gravity at r = 0.5α. Most of the elements that were fused as per chapter 6, will emerge primarily as hydrogen. Once the process has started, the strong gravitational attraction at r = 0.5α prevents gravity from stopping the process until its mass becomes small enough that its Schwarzschild radius is well inside its physical radius. Gravity stops the process when it again dominates electromagnetism.

Neutron stars would have different defects within their structure. Their size and extent mean the process could occur at different masses. It is suggested that process occurs in quasars. It is suggested that is the origin of quasars. The different imperfections in the massive neutron stars would result in the effect causing different gravitational redshifts. As such, their redshift is not an indication of their distance, supporting Arp's (1998) contention. It is suggested, but not shown, that this source of energy would take a long time to exhaust the supply of neutrons decaying to hydrogen. That suggests the intensity variations of quasars would not vary over short time periods.

It is also possible that process could generate conditions favourable to the rapid fusion of hydrogen to helium, an enormous hydrogen bomb. That could blow the neutron shell apart. The energy released by that process would be massive. Such a mechanism could account for the strongest gamma ray bursts.

Concerns have been raised about ever increasing fusion of hydrogen in an infinite steady state universe. Ultimately it would run out of hydrogen for further fusion. The above description suggests gravity captures the cosmic photons and neutrinos, along with the fused heavy elements, converting them all into neutrons. The process requires considerable energy, ≈ 0.7 MeV per neutron. That energy is provided by relativistic speeds to which all matter is subjected as it falls into the massive object, see Equations 8.31.

Inside the massive neutron star, they are changed into neutrons. As the neutron star gets larger, they are released primarily as high energy hydrogen that spreads out to ultimately form new stars. As the process continues, the same mechanism captures cosmic photons and neutrinos. The cycle can repeat forever.

11.16. Summary

Like everything else, the universe functions through the mechanics that control it. Chapter 8 showed how gravity controls space and time outside mass. Chapter 9 showed the origins of gravity, being the differential of the change in the electric permittivity of free space caused by the high frequency oscillations induced by the structure of the nucleons. Gravity dominates the universe.

This chapter used Newton's shell theorem to show that, when gravity was weaker than inverse square law, as it is under the redshift, equation 11.7, and Einstein, equation 11.8, metrics, gravitational attraction diminished with increasing distance. It was expressed as $g_r \rightarrow$

∞ as $r \rightarrow 0$. It applies for any universe density. Along with figure 11.12, it showed an infinite static universe was stable against collapse under its own gravity, independent of its density.

The evidence in favour of an expanding universe starting from a Big Bang depended primarily upon two factors known by 1930. The universe would collapse under the gravitational attraction of its own mass and the linear nature of photon redshift increasing with distance. They formed the foundation of modern cosmology. With an incorrect understanding of Einstein's gravitational field equations, expansions of his work, or the derivation of alternative field equations, were not soundly based in either Einstein's mathematics or in physical principles. It is suggested that cosmological derivations based upon field equations have serious errors. The greatest is the prediction that gravity can be stronger than inverse square law.

Applying Newton's shell theorem to gravity stronger than inverse square means that gravitational attraction will be infinite at a finite distance. At the estimated universe density of $\approx 10^{-26}$ kg^{-3}, that distance will be $\approx 10^{10}$ LYs. That is well within the range of visible and radio wave telescopes. No sign of said collapse has been detected. It is another reason why the accepted Schwarzschild metric is wrong, as mentioned in chapter 8.

Equations 11.11 and 11.12 have shown that an infinite steady state universe will not collapse. There is no reason for the universe to have to expand to prevent its collapse. There is nowhere for an infinite universe to expand. That required an alternative explanation for the linear nature of the observed redshift versus distance. A version of Zwicky's "tired light" suggestion was forwarded. It was based on space having a property similar to viscosity. It would cause photons to lose momentum. Being constant velocity particles, light losing momentum would lose energy, i.e., frequency, but not velocity. They would be redshifted.

Hubble's constant was used as measure of the redshift induced in photons as they travel through space. It showed that such a mechanism gave a good match with the observed magnitudes of high redshift SNe1a, without the need for the universe to either expand or expand at an accelerating rate. It was also shown that some of the other pillars of the Big Bang and expanding universe theory were explained by this mechanism. That included the Tolman test, Olber's paradox and the observed time delay associated with redshifted events.

A reason was suggested to cause photons to lose energy as they travelled vast distances through empty space. That reason was because empty space was not empty. It is occupied by cosmic microwave background radiation and cosmological neutrinos. It was suggested that, by a mechanism not described, such background would cause photons to be redshifted. That is the only hypothesis used in this model. Suggested reasons were given for other "problems" supposedly related to a steady state universe. These included how neutrinos and fused elements were recycled back to hydrogen. Both involved gravity and quasars. The mechanism was not further explained, other than it required the redshift gravity of equation 8.28.

It also answers a number of enigma's associated with the Big Bang theory. Perhaps the greatest is the $1{:}10^{60}$ odds of the universe surviving in its present form almost 14 billion years after the Big Bang (Hawkins, 1998). They are so low as to be essentially zero. It could not happen as described. That, along with other problems like the breaking of the conservation of parity; the necessity for a period of faster than light expansion soon after the Big Bang and nothing, not even empty space exists outside the expanding universe. The expanding universe creates space. They can have mathematical explanations. They have no physical validity.

None of those explain NASA's Hubble ultra-deep field image, shown in figure 11.18. It is a physical reality. It is amusing to read the mathematics introduced by some authors to explain figure 11.18 in terms of an expanding universe. Some violate known physical principles.

An infinite steady state universe does not require any changes to physical laws, such as hyperinflation soon after the Big Bang. Nor does it require an explanation about what was before the Big Bang. The Hubble ultra-deep field image is explained by equation 11.22 and displayed in curve V in figure 17.

This infinite steady state universe has no need to hypothesize about either what caused, or existed before, the Big Bang. There is no need to be concerned about the violations of the laws of physics that allowed more matter to form than anti-matter, or the hyper-inflationary period shortly after the Big Bang that caused space to expand faster than the speed of light.

Perhaps the only bright spot for the Big Bang theory was curve F in figures 11.9 an 11.15. It also gives a good fit for the measured magnitudes of high redshift SNe1a. It comes at a huge price. It requires that the universe is expanding when equations 13 and 14 show it cannot possibly expand beyond 10^{10} LYs before it reaches infinite gravitational attraction back to its origins. At least that is what is predicted from the Schwarzschild solution to Einstein's field equations. However, if the universe is expanding, there is no accelerated expansion. The observed SNe1a magnitudes at high redshift are the same as the predicted values under the Big Bang theory.

Those who wish to continue the idea of the Big Bang have problems. They are stating they like the idea of a universe with a 1 : 10^{60} probability of surviving after its origin in a Big Bang from unknown sources.

Theoretical cosmologists' predictions come from their use of gravity field equations, of which they have demonstrated a lack of understanding. They claim to understand their own equations without even knowing they have the wrong interpretation of Einstein's field equations, upon which their own work was founded.

There is just one universe existing in three space dimensions and time. It extends forever in all four dimensions. It has no boundary across which the laws of physics or the values of the physical constants change. It just is, always was and always will be, the universe. The unique nature of gravity under the redshift mechanism recycles the fused hydrogen through quasars to provide a continuous source hydrogen for fusion.

Everything within the universe operates on the basis that it follows its path of least resistance until it reaches a stable state. Figure 11.22 gives an indication of the large scale composition of that stable state.

It should be compared with the figure 11.19 representation that has emerged from theoreticians' studies. There is no dark matter that generates gravity and has no other effect. There is no dark energy blowing the universe apart at an increasing rate. The universe isn't even expanding.

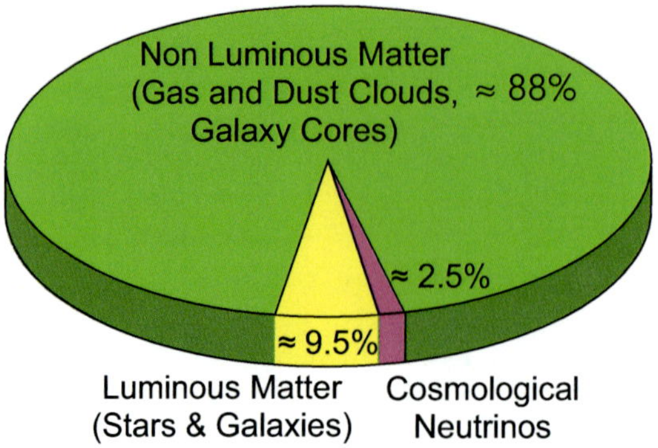

Figure 11.22: *The distribution of matter in the observed universe, as required to explain astronomers' observations of its large scale structure.*

The universe's most stable state is to exist forever in space and time, happily recycling itself. Its mysteries, if they can be called such, are the failure of "experts" to recognise the properties of photons that make it all happen and allow us to see it!

11.17. List of References

Arp HC, (1988); "Quasars, Redshift and Controversies" ISBN 0-521-36314-4

Assis AKT and Neves MCD, (1995); "The Redshift Revisited", Astrophysics and Space Science **227**, pp 13 – 24

Bessell MS, (1990); UBVRI Passbands, "Publications of the Astronomical Society of the Pacific" 102:1181-1199: https://iopscience.iop.org/article/10.1086/132749/pdf

Brown PJ, Roming PWA, Milne P et al., (2010); "The Absolute Magnitude of Type 1a Supernova in the Ultraviolet", Astrophysical Journal, **721**, pp 1608 – 1626.

Clarke S, (1997); "Redshift", University of Hertfordshire Press, Herts, UK, Ch 4.

Colgate SA, (1979); "Supernovae as Standard Candles for Cosmology", Astrophysical Journal, **232**, pp 404 – 408

Davis TM and Lineweaver CH, (2003), "Expanding Confusion: common misconceptions of cosmological horizons and the superluminal expansion of the universe": https://arxiv.org/abs/astro-ph/0310808

Dicke RH, Peebles PJ, Roll PG and Wilkinson DT, (1965); "Cosmic Black Body Radiation", Astrophysical Journal, **142**, pp 414 – 419

Eddington AS (1926); "The Internal Constitution of Stars", Cambridge Science Classics

Einstein A, (1905); "Ist die "Trägheit eines Körpers von seinem Energiegehalt abhäbgig?" Annalen der Physik, **17**, pp639 – 641

Einstein A, (1911); "Uber den Einfluss der Schwerkraft auf die Ausbreitung des Lichtes", Annalen der Physik, **35**, pp898 – 908

Einstein A, (1916); "Die Gründlage der Allgemeinen Relativätstheorie," Annalen der Physik, **42**, pp 769-822.

Einstein A, (1917): "Kosmologische Berachtuneg zur algemeinen Relativtätstheorie", Sitzungberichteder Preussuschen Akadamie der Wissenschaften, pp 142 – 152

Einstein A, (1952a); "The Foundation of the General Theory of Relativity," in Lorentz HA, Einstein A, Weyl H and Minkowski H, "The Principle of Relativity", translated by Perret W and Jeffery GB, Dover Publications, New York, NY, pp 109-164

Einstein A, (1952b); "Cosmological Considerations on the General Theory of Relativity," in Lorentz HA, Einstein A, Weyl H and Minkowski H, "The Principle of Relativity", translated by Perret W and Jeffery GB, Dover Publications, New York, NY, pp 177-188

Freedman WL, (2021); "Measurements of the Hubble Constant: Tensions in Perspective", The Astrophysical Journal, **919**, 16: https://doi.org/10.3847/1538-4357/ac0e95

Goldhaber G et al., (2001); "Timescale Stretch Parameterization of Type 1a Supernova B-band Light Curves", https://arxiv.org/abs/astro-ph/0104382

Hauser MG and Dwek E, (2001); "The Cosmic Infrared Background: Measurements and Implications", Annual Review of Astronomy and Astrophysics, **39**, pp 249-307: DOI: https://doi.org/10.1146/annurev.astro.39.1.249

Hawkins M, (1998); "Hunting Down the Universe", Abacus, London, Ch 2

Hook IM, Howell DA, Aldering G, et al., (2007); Spectra of High-Redshift Type 1a Supernovae and a Comparison with their Low-Redshift Counterparts", The Astronomical Journal, **130** (6): 2778: https://www.researchgate.net/publication/230966303_Spectra_of_Hi

gh Redshift Type Ia Supernovae and a Comparison with Their Low-Redshift Counterparts

Hoyle F (1950) *The Nature of the Universe – a series of broadcast lectures*, Basil Blackwell, Oxford 1950 *Frontiers of Astronomy*, Heinemann Education Books Limited, London, 1955. The Internet Archive. HarperCollins, ISBN 978-0060027605

Hubble EP, (1929); "A Relation Between Distance and Radial Velocity Among Extra-Galactic Nebulae", Proceedings of the National Academy of Science, **15**, pp 168 – 173

Hubble EP, (1936); "Effect of Redshift on the Distribution of Nebulae", Astrophysical Journal, **84**, pp 517 – 555

Huygens C, (1690); "Traite de la Lumiere (Treatise on Light)", https://www.gutenberg.org/ebooks/14725

Ice Cube, (2021); "IceCube MasterClass", https://masterclass.icecube.wisc.edu/en/learn/detecting-neutrinos

Lemaître G, (1931) "L'Hypothèse de l'atome primitive" in , Lemaître, The Primeval Atom – an Essay on Cosmogony, D. Van Nostrand Co, 1946.

Leavitt HS, (1908); "1777 variables in the Magellanic Clouds". Annals of Harvard College Observatory. **60**: 87 – . Bibcode:1908AnHar..60...87L

Le Verrier U, (1859); "Lettre de M. Le Verrier à M. Faye sur la théorie de Mercure et sur le mouvement du périhélie de cette planète"

Liberati S and Maccione L, (2014); "Astrophysical Constraints on Planck Scale Dissipative Phenomena", Physical Review Letters, **112**, 15130: https://doi.org/10.1103/PhysRevLett.112.151301

Lindsay RB and Margenau H, (1957); "Foundations of Physics", Dover Publishing, New York, Ch 6

Marmet L, (2018); "On the Interpretation of Redshift: A Quantitative Comparison of Redshift Mechanisms", http://www.marmet.org/cosmology/redshift/mechanisms.pdf pp 1-57

Newcomb S, (1882) US Nautical Almanac Office Astronomical paper 1 (Washington) p 473 https://www.relativitycalculator.com/pdfs/mercury_perihelion_advance/S.Newcomb.pdf

Newton I, (1687) "Philosophiae Naturalis Principia Mathematica," London,

Newton I, (1729) Mathematical Principles of Natural Philosophy **1** ed. A Motte (London: Motte B) Propositions1 – 32, 45

Park RS, Folkner WM, Konopliv AS, et al., 2017; "Precession of Mercury's Perihelion from Ranging to the Messenger Spacecraft", The Astronomical Journal, **153**, pp 121 – 127

Parrent J, etal., (2014), "A Review of Type 1a Supernova Spectra": https://arxiv.org/abs/1402.6337

Peebles PJE and Ratra B, (2003); "The cosmological constant and dark energy". Reviews of Modern Physics, **75**, pp 559–606. doi:10.1103/RevModPhys.75.559.

Perlmutter S, Aldering G, Goldhaber G, et al., (1999); "Measurements of Omega and Lambda from 42 High-Redshift Supernovae", Astrophysical Journal, **517**, pp 565 – 586

Perlmutter S, (2003); "Supernovae, Dark Energy and the Accelerating Universe", Physics Today, pp 53- 60: https://supernova.lbl.gov/PhysicsTodayArticle.pdf

Penzias AA and Wilson RW, (1965); "A measurement of excess antenna temperature at 4080 Mc/s", Astrophysical Journal, **142**, pp 419 – 421

Pickering EC, (1912); "Periods of 24 Variable Stars in the Small Magellanic Cloud", Harvard College Observatory, Circular 173

Rees M, (1997); "Before the Beginning", Helix Books, Reading MA, Ch 10

Reiss AG, Filippenko AV et al. (1998); "Observational Evidence from Supernovae for an Accelerating Universe and a Cosmological Constant", Astronomical Journal, **116**, 1009: https://iopscience.iop.org/article/10.1086/300499/meta

Reiss AG, Rodney SA et al., (2018); "Type 1a Supernova Distances as Redshift > 1.5 from the Hubble Space Telescope Multi-cycle Treasury Programs: The Early Expansion Rate", The Astrophysical Journal, 853, 126

Robinson VNE, (2021); "Physical Explanations of Einstein's Gravity", Journal of Physics Communications, **5** 035013, https://doi.org/10.1088/2399-6528/abee2f

Tolman RC, (1930); " On the Estimation of Distances in a Curved Universe with a Non-static Line Element", Proceedings of the National Academy of Science, **16**, pp 511 – 520

Tolman RC, (1934); "Relativity, Thermodynamics, and Cosmology", (Oxford: Oxford University Press)

Slipher VM, (1915); "Spectrographic Observations of Nebulae", Popular Astronomy, **23**, pp 21 – 24

Slipher VM, (1921); "Two Nebulae with Unparalleled Velocity", Popular Astronomy, **29**, p 128

Shtyrkov EI, (1993); "A New Interpretation of Cosmological Redshifts: Variable Light Velocity" in Arp HC et al., ed., Progress in New Cosmologies: Beyond the Big Bang, p 327, New York, Plenum Press.

Tammann GA and Leibundgut B, (1990); "Supernova studies IV, the global values of H_0 from supernovae 1a and the peculiar of field galaxies", Astronomy and Astrophysics, **236**, pp 9 -14

Whiting MT, et al. (2006); "Multi-object spectroscopy of the field surrounding PKS 2126−158: discovery of a z = 0.66 galaxy group "Monthly Notices of the Royal Astronomical Society, **368,** 341–350

Wright EL, (2008); Errors in tired light cosmology: http://www.astro.ucla.edu/~wright/tiredlit.htm

Zwicky F, (1929); "On the Redshift of Spectral Lines Through Interstellar Space", Proceedings of the National Academy of Sciences, **15**, pp 773 – 779

12. Describing the Universe

Abstract:

The above has briefly presented the universe from its smallest sub-atomic particle structures, through the formation of atoms, on to gravity and how everything combines to form the large-scale structure of the universe. This chapter describes how physics fits them together, compared to the way mathematicians have tried to describe them.

12.1. Introduction

The previous chapters have been a step-by-step guide through the various aspects of the universe. This chapter is about reviewing how they all fit together and work.

The first step to understanding the universe is not about the components described. Nor is it about the mechanics, be it described by Newton, Maxwell, Einstein, Feynman, quantum, and others. The basic rule of the universe is that everything in it seeks its path of least resistance to a stable state of existence. The mechanics explained in all the above is how that rule applies in the specific situations described. They are all interlocked.

The second step to understanding the universe is that it has existed forever. It doesn't do that because it is finely tuned. It does that because everything in it is inherently stable state for the circumstances in which it finds itself. Stars only explode to supernovae because they were no longer stable. If a theory requires "fine tuning" or has a low probability of something happening, it needs a good dose of common sense, often called Occam's razor.

Another fundamental aspect of the universe is that things happen for physical reasons. Mathematics doesn't make things happen. There has been an increasing reliance on mathematics to describe physical events. Mathematics is essential to our understanding of the universe. In his Principia, Newton described physical principles and then used mathematics to derive the answers to those physical situations. Maxwell derived his electromagnetic equations from physically observed results. Both their mathematics could be extended to make predictions. The ability to extend their mathematics and make accurate predictions was part of the beauty of their work.

Scientists seeking the "ways of the universe" should not tell the universe how it should operate. Telling electrons their orbits are quantized by integers of orbital angular momentum or de Broglie wavelengths is not how electrons function. It is a good start in the absence of other information. As chapter 7 showed, reasons for quantization should have been found in the quantized charge of nuclei.

Treating electrons and photons as point particles with properties attached, is a mathematician's approach to a problem. Quantum mechanics became wave mechanics. With waves extending over great lengths and many ways of manipulating them, it was "easy" to calculate results that match observation.

When describing the universe, the role of mathematics should be to describe the magnitude of a proposed physical effect. It is not to tell the universe what it needs to have to function the way they think it should. To truly appreciate the physical universe, it helps to understand how mathematicians, including theoretical physicists, describe it.

12.2. A Mathematician's Universe

In his special and general relativity theories, Einstein was one of the first people to use pure mathematics to make significant physical predictions that were later verified. They were seen as mathematics without a physical foundation. Those who understand his special relativity theory will see the physical relationship between his mathematics and the constant speed of light. His general relativity theory and gravity has caused greater concern.

The mathematical complexity of the gravitational field equations from his general relativity theory has led to some major problems. Those who call themselves experts, "extended" his calculations with no acknowledge of the approximations he introduced. Chapter 8 showed how simple the mathematics could have been if he used calculus instead of tensors and field equations for his calculations.

Despite their claims to excellence, none of those who studied his gravity field equations understood the mathematics he used. None had any idea of the physics he used. They were proud of their prediction of black holes from their "strong field" solution to Einstein's field equations. They did not have sufficient knowledge of Newton's work, first published in 1687, to realize that gravity stronger than inverse square law would lead to Mercury's anomalous orbit regressing its perihelion against its orbital direction.

Being unable to follow Einstein's mathematics, they relied upon his interpretation that gravity would cause a finite universe to collapse. They could not follow Einstein's mathematics sufficiently to realize he was predicting gravity became weaker than inverse square as the mass increased.

With Newton's theory predicting large scale gravitational collapse and Einstein only considering a finite universe, cosmology developed along the lines that the universe must be either expanding or collapsing. Hubble discovered that the redshift of light from distant galaxies increased with their distance. The redshift from distant objects was interpreted as a Doppler redshift due to galaxies moving further away. The Big Bang and expanding universe theories were firmly established.

Galaxies were seen to rotate faster than predicted from their knowledge of gravity. That was explained by the assumption that there must be dark matter within galaxies that caused the more rapid rotation. Their prediction of galaxy rotation was based upon Kepler's laws of planetary rotation. That required Newton's shell theorem. In turn, it required mass to have a uniform density. Any image of a galaxy shows they do not have a uniform density. Kepler's laws of planetary motion do not apply to galaxy rotation.

Seemingly unaware of that, mathematicians applied Kepler's laws and determined the galaxies were rotating faster than they expected. To solve the problem, they invented that fictional substance called dark matter.

On top of that, it was observed that the magnitude of high redshift, $z > 0.2$, type 1a supernovae were higher than expected under their critical density theory. That was explained by the dark energy blowing the universe apart at an ever-increasing rate.

Figure 12.1 shows the mass breakdown compiled for the standard cosmological model of the universe. Its highlights are that the observed matter, all the stars in all the galaxies, makes up 0.4% of their universe. The galaxies are surrounded by halos of cold gas and molecular dust that are about 10 times the mass of the stars in each galaxy. Together, their baryonic matter is thought to make up about 4% of the universe's expected mass. The dark matter and dark energy they require to match their mathematical calculation makes up the remaining 96% of their universe. Dark matter and energy are "mathematical fudge factors" introduced to explain why their calculations don't match observation.

Despite decades of searching, experimentalists have not found the slightest evidence of the existence of either. That doesn't include the 10^{60} parallel universes required to give our visible universe a reasonable probability of existing 14 billion years after the Big Bang. Other predictions have been made about different universes arising from quantum effects. All those cosmological predictions are based upon mathematical calculations.

Figure 12.1 is how mathematicians predict the universe. According to their own work, they know nothing about 96% of the universe they think exists. The standard mathematical model of the 4% we can see is also interesting. It is made up of protons and neutrons that form nuclei, the electrons that are attracted to them and the photons we use to observe their assemblies.

Photons are treated as point particles with wave functions mathematically attached, as per the 1926 Copenhagen convention interpretation. It is good for mathematically matching observations. It doesn't give a physical description of what is happening.

Electrons have so many apparently contradictory properties that it was easier to follow the Schrödinger and Dirac approach and treat them as waves. They are now considered to be point particles with

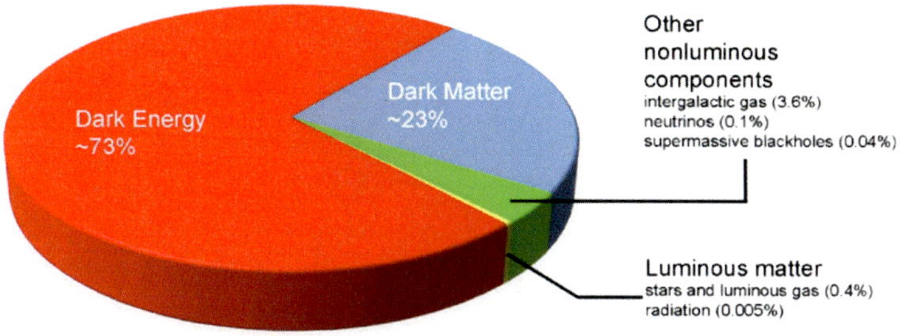

Figure 12.1: *Ratio of luminous and non-luminous matter to dark matter and dark energy required to match experimental observations under standard models. (Source: https://www.osti.gov/servlets/purl/1410794)*

properties, including waves functions, mathematically attached. Wave functions can be manipulated in so many ways, it is easy to match observation. Beyond that, standard model practitioners don't have much idea of their physical nature.

Protons and neutrons have fared somewhat better. They are considered to be composed of quarks held together by gluons. Quarks are considered to be spin half particles of relatively low mass, held together by gluons. Table 12.1 lists the 61 fundamental particles required to explain experimentally observed results. Figure 12.2 shows how they are expected to interact. Apart from their huge complexity and mathematical foundation, they have several physical problems.

Particle	Types	Generations	Anti-particles	Colours	Total
Quarks	2	3	Pair	3	36
Leptons	2	3	Pair	None	12
Gluons	1	1	Own	8	8
W	1	1	Pair	None	2
Z	1	1	Own	None	1
Photon	1	1	Own	None	1
Higgs	1	1	Own	None	1
Total					61

Table 12.1: *List of fundamental particles in the rotating photon model.*

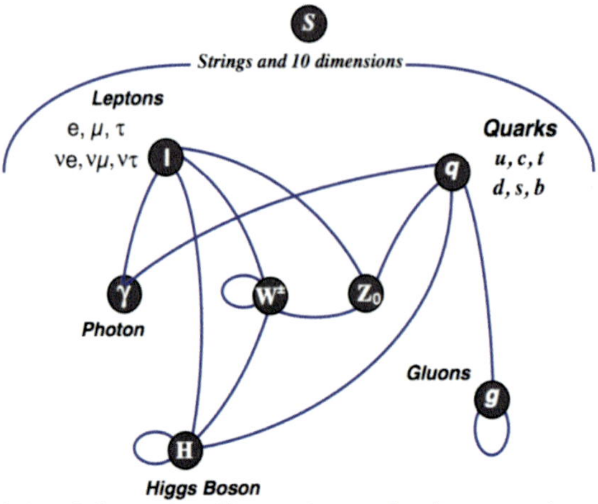

***Figure 12.2:** Schematic illustration of the reactions between the fundamental particles under the standard model.*

Quarks and gluons have never been separately isolated and identified. They are said to have fractional charge, +⅔e and – ⅓e. Fractional charges have never been detected.

They, and many other complexities are forwarded by some as the triumph of the standard models of physics. It is a mathematical triumph that its proposers use to bluff non-mathematicians into believing they know what they are doing.

Table 12.1 and figure 12.2 show a complex mathematical presentation that is unintelligible to most people. Mathematicians don't know what they represent physically. For solutions, some turn to string theory. It requires extra spatial dimensions that have never been detected and uses particles too small to ever be detected.

After a couple of decades of extensive mathematics research, an unbiassed assessment would suggest they have not made any progress.

One look at table 12.1 and figure 12.2 shows they are so complex they couldn't possibly survive for billions of years without a little "hiccup" somewhere along the line.

With a half-life of less than 10^{-20} sec, the Higgs bosons are supposed to create fields that gives mass to everything in the universe, photons excluded. Really! In 1905 Einstein determined that photons have mass. Photons having mass have been measured experimentally many times.

We shall ignore the 10^{60} parallel universes the Big Bang theory needs to match observation with reality. What else does the mathematics tell us about the observed universe. Cosmologists have developed a mathematical theory that requires 25 times the observed mass/energy balance of the observable universe to explain observations. As an exam mark, some people could regard that as less than 4%.

What of the standard model of sub-atomic particles? Quarks and gluons have never been separately identified. Among the leptons they know a lot about electrons, except the physical structure that gives them their strange properties. They have detected three types of neutrinos as well as their anti-neutrino equivalent. Only the electron and anti-electron neutrinos are stable. They know they have spin $\frac{1}{2}\hbar$ and that they have a mass upper limit of ≈ 0.1 eV. Standard model practitioners know very little else about them. They have no physical reason for their spin.

The W and Z bosons have half-lives of $< 10^{-24}$ sec, while the Higgs boson has a half-life of $\approx 10^{-22}$ sec. During their brief existence, standard model mathematics enables them to perform significant actions.

Almost two thirds of the 4% of figure 12.1, as described in table 12.1, the quarks and gluons, have never been detected. Most of the remainder, electrons, photons, and neutrinos excluded, have exceedingly short half-lives. It is a fair comment to say that the standard model, which its proponents acknowledge knows nothing of its origins, would, by their own standard, be struggling to get a mark of 1%.

The standard model is a mathematical treatise that does not tell us much about the physics of individual sub-atomic particles. It does tell us that atoms are made of a central positive nucleus composed of protons held together by neutrons and surrounded by electrons. With that knowledge physicists have laid the foundations for the modern technological world we live in.

Our engineering marvels are founded on Newton's work. Our modern, electricity-based benefits started with the work of Faraday and Maxwell. The transistor and our computer-based world were founded on the experimental work of Shockley, Bardeen, and Brattain when they developed the world's first working transistor. They were all developed before the standard model of matter.

The efforts of countless numbers of scientists, engineers, technicians, and others have expanded on those basic developments to produce our modern lifestyle.

That does not mean to say that the standard models have not had an impact upon our modern life. Their biggest impact can be summed up in their mathematical complexity. They are so complex that only those who classify themselves as the very brightest claim to be able to understand them. That has resulted in some unfortunate comments that include, but are not limited to:-

1) "*Trust us. We are the experts. We know what we are doing!*"
2) "*Zero isn't zero. That is what the mathematics tells us.*"

The first comment means experts don't have explain the logic of their reasoning or decisions. That has spread to many other fields of life, often accompanied by those very words instead of explanations. The second comment has resulted in complex mathematics being used to get any desired answer. As long as the mathematics is complex enough, it is accompanied by "*That is what the mathematics tells us, so it must be right*"! Too many mathematicians and mathematical physicists actually believe mathematics is infallible. It could be if it was based on sound physical principles, and they did not make mistakes.

They could get away with it if it wasn't for a couple of major blunders. Generally, the scientific community regards Einstein's gravitational field equations as one of the most difficult scientific topics. Its practitioners were highly respected. Until chapter 8 and Robinson (2021), they could show their brilliance. Those presentations show none of them understood Einstein's work. They made black holes a cornerstone of standard cosmology model,

Another cornerstone was the Big Bang and expanding universe theory for the formation and mechanics of the universe. In 1687, Newton predicted that, if gravity were weaker than inverse square, a planet's perihelion would precess in its direction of travel. Le Verrier's 1859 and Newcomb's (1882) studies showed Mercury's orbit had an additional precession of ≈ 43 arc sec per century, beyond Newton's prediction. Mercury's anomalous orbital precession should have alerted

20th century mathematicians to the necessity for solutions to Einstein's gravitational field equations to be weaker than inverse square.

Einstein didn't believe black holes were feasible predictions from his field equations. Again, that should have alerted mathematicians that something was wrong with their calculations when they derived a solution to his field equations, in which they proudly boasted gave gravity a stronger than inverse square law. Instead, they chose the "*Einstein was wrong, and I am right*" approach.

If it only affected black holes, it would not be so bad. As chapter 11 showed, applying gravity weaker than inverse square law to Newton's shell theorem, an infinite steady state universe is stable against gravitational collapse. There is nothing difficult about the mathematics that describes its physics. It took two simple failures to conclude the Big Bang and expanding universe theories were correct. The first was to not understand Einstein's simple mathematics. The second was not to understand Newton's simple, common sense physics.

Mathematics is essential for physics. A physical principle is forwarded to explain an observation. A mathematician or theoretical physicists calculates the magnitude of that effect. If the calculation matches observation, the principle has some validity. That is mathematics putting physics on a sound basis.

When those passing themselves across as the world's best mathematical physicists can't understand Einstein's simple mathematics, what does it say for the rest of them? Some would say: "not much"! As a minimum, their calculations should be restricted to three space dimensions and time; known and detected particles; and only changes that follow a path of least resistance to reach a stable state.

It has been a goal of theoretical physicists to unite everything into a single set of equation that explain everything in a continuity from the smallest sub-atomic particles to the large scale structure of the universe. It has been variously known as Grand Unified Theory, GUT, Unified Field Theory, UFT and Theory of Everything, TOE, among others. A relatively recent update was given by Lincoln in the reference associated with figure 12.1. Lincoln's comments suggested a mathematical approach was unlikely to yield meaningful results in the foreseeable future.

So much for what mathematics tells us about the universe. It is now time to see what physics tells us about the universe! What follows was interspersed with some mathematical interpretations that allowed readers to better understand the physical universe.

12.3. A Physicist's Universe

One of the first things that modern physics tells us is there are only five stable particles, photons, electrons, protons, neutrons, and neutrinos. All other particle decay to one or more of those five. When a particle meets its anti-particle, they annihilate each other, producing photons. It doesn't take too much to realize the most probable reason is that all individual sub-atomic particles are made from photons. A brief physical introduction to photons was given in chapter 1, along with a little supporting mathematics.

Chapter 2 proposed that all individual sub-atomic particles were composed of photons of the appropriate energy making two revolutions per wavelength. It gave a physical reason for that to occur. It also showed how that structure gave rise to Einstein's special relativity theory. $E = mc^2$ because energy, E, is the photon travelling at c in a straight line and mass, m, is the same photon travelling at c in a circle, making two revolutions per wavelength.

That structure gives all individual sub-atomic particle their angular momentum ½$\hbar$. It automatically leads to particles being subject to the special relativity corrections with increasing speed. It went on to show the properties all individual particles would have because of that rotating photon structure. They matched known particle properties. Their properties are due to their structure under the rotating photon model. They are not something that needs to be added on mathematically to explain their behaviour.

Under this model, the derivations of the special relativity corrections were straight forward. They took only a few equations and some descriptive illustrations and words. They require not much more than Pythagoras' theorem and high school mathematics. Readers can physically see how they occur. One additional correction was derived.

Einstein's mathematical derivation used dozens of such complex mathematical equations that over a century later, only advanced mathematicians can follow them. He only derived three corrections. This shows that, when the physics is understood, the mathematics is quite simple.

Chapter three showed how those properties varied for each of the stable particles, electrons, protons, (bound) neutrons, and neutrinos. It shows they are rotating electromagnetic toroids that generate their own electric and magnetic fields. They are not the little "billiard ball" structures often portrayed in standard model illustrations. Their mass is in their oscillation frequency. Electrons are straight forward, single oscillation toroids.

The more massive protons and neutrons have their core oscillation plus a series of one third, one ninth and one twenty seventh frequency harmonics. They give those nucleons their charge distribution and magnetic moment. It also shows that protons have two positive structures and one negative structure in their inner core. Neutrons have two negative structures and one positive structure in their inner core. They could be interpreted as quarks.

This model predicts that neutrinos will have a mass of $\approx 10^{-4}$ eV/c^2, as it derives their properties. The latest experimental observations suggest electron neutrinos have a mass $< \approx 0.1$ eV.

With all the particles made from the same material, it was easy to show how protons and neutrons could interchange by emitting or absorbing the appropriate electron/neutrino or their anti-particles. The physics is straight forward. It doesn't need mathematics to understand and use it. More information may be gained by some calculations.

Chapter 4 gave the briefest introduction to the origins of the "elementary" particles such as pions, muons, Ξ, Σ, Λ, etc. When charged nucleons are accelerated to high energies, their mass increases and time slows down. When they are stopped, their time frame reverts to their rest time frame. They now have extra mass they must remove. They do that by emitting linear photons, bremsstrahlung, and rotating photons, pions, and muons. They cascade through a series of quasi-stable states until they finally reach their stable nucleon rest mass. Those quasi-stable states have half lives in the region of 10^{-20} to 10^{-6} sec.

All those "elementary" particles and bosons that are detected, are not in the nucleons at rest, muons and pions excluded. They are induced into them. When the particles are accelerated, they gain mass. When stopped, that mass is released. Spin ½ particles are single rotating photons, such as a muon. Spin 1 and spin 0 particles, such as pions, are two spin ½ particles. There is no reason why, with sufficient energy, three or more spin ½ particles could not agglomerate for a short period to give higher spin "particles". Their structure and properties are irrelevant because they are not part of the nucleons at rest. No number of mathematical calculations will alter that physics.

It is in the field of nuclear physics where this model can make a great contribution. It is over 80 years since the first mathematical models of nuclei and nuclear binding were proposed. Over that time there have been about 40 versions of nuclear structure and binding theories proposed. None have been successful in describing nuclear structures, properties, and binding mechanisms across all nuclei.

The unique nature of the nucleons under this model makes it easy to see how they bind together to form nuclei. Chapter 5 shows the few simple rules to apply to determine the structure and properties of any nucleus of any A and Z.

Nuclear binding is entirely electromagnetic. It is a physical property of the rotating photon model that electric charge is generated at the circumference of the rotating photons. When negative segments of the neutron overlap positive segments of the proton, there are opposing charge segments with effectively no distance between them. That was called "division by zero" electric attraction. The nucleons orient their spin so that their magnetic fields reinforce that attraction. That is the origins of the strong nuclear force. Its requirement of direct overlap and the inverse cube of magnetic field attraction, limits its range to $\approx$ 5.5 fm.

The nucleons form in strongly bound layers. The layers are held together by a weaker residual magnetic field between the two layers. The nucleons remain as individual particles within the nuclei. That strong binding force has a strictly limited range. The calculations are limited to adding numbers. The fit with known nuclear properties is very good. It explains a few unknowns, such as the origins of nuclear binding, the structure of nuclei, the nuclear skin effect and more, making dozens of testable predictions.

When nucleons bind, they come sufficiently close that their most stable state is one in which a neutron's negative electric field overlaps a proton's positive electric field, with very little distance between them. The nucleons align their spins to give maximum internal magnetic binding. Once formed, over 200 are quite stable and will remain that way for billions of years.

Chapter 5 takes nuclear physics away from the exclusive realm of nuclear physicists and enables many ordinary people to gain a reasonable understanding of it.

The next step in understanding the universe was how the heavy nuclei that make the rocky planets were formed. That is covered in chapter 6. It shows that the two-dimensional distribution of charge within nucleons means there is no charge barrier to be overcome to fuse hydrogen and other nuclei into larger nuclei. It shows how fusion within the sun gives a close match to the observed elemental abundances in it. It shows why the important features in fusion are proximity and alignment. Energy is of secondary importance, although it is necessary to get proximity and alignment.

Understanding this presentation up to and including chapter 6 is essential for anyone who wants to make improvements in nuclear fusion. It requires some mathematics to determine the abundances. For most of the elements, the predicted relative abundances were within a factor of $\approx$ 3 of the measured abundances, over several orders of magnitude.

Once they are fused, they are either distributed for capture into rocky planets, or recycled to hydrogen in quasars. Whiting et al. showed that quasars emit mainly hydrogen and helium and only a few low Z nuclei like Mg. Quasars are not giant black holes emitting much radiation as mass falls into them. They are massive objects that have grown very large. Their gravity has weakened so much, as indicated in curves g_Z in figures 8.5 and 8.6, that it can no longer hold neutrons against their spontaneous decay to protons and electrons.

The universe will never run out of hydrogen as a fuel for future fusion. Nor will it lose energy as photons are redshifted. Their energy loss is transferred to the cosmological neutrinos. They, like photons, will eventually be captured by massive objects. There, gravity will recycle them through the mechanism discussed in section 11.9. The same fate awaits all stars and planets. It takes billions of years. Time is no problem in an infinite universe.

The next step in understanding the universe was to put electrons around the nuclei. With equal numbers of protons and electrons, the light weight negatively charged electrons are attracted to the heavy positively charged nuclei. The standard model uses quantized orbits and other quantum mechanical properties to describe how electrons form around nuclei.

None of that is necessary. Chapter 7 shows that electrons are far too large to fit into even the largest nucleus. They don't need quantized orbits to stop them falling into any nucleus. An electron's spin of $\frac{1}{2}\hbar$ means its whole mass is rotating at the speed of light in a radius hundreds of times larger than the radius of a proton.

That makes electrons very powerful gyroscopes for their mass. Like all gyroscopes they can only move in a straight line unless they have something else with which to exchange angular momentum. The best way to do that is to pair with another electron. That is why electrons always form pairs whenever possible.

The number of protons in a nucleus is limited. The first two electrons around a nucleus form a tight knit pair. They have each other and do not need another electron. They will screen out the positive charge of two protons. Each additional electron screens out the charge of one additional proton. As more electrons come in, they each have a unique binding energy based upon the total number of unscreened protons and the positions of the inner electrons.

Electron orbits aren't quantized because of some artificial law. Their unique binding energy is determined by the limited number of protons in the nucleus and the number screened out by inner electrons. After that, it is a free-for-all, almost anything goes, as the electrons move around the nucleus, screening out its positive charge. Their behaviour is limited by their de Broglie wave. It is automatically generated as it moves.

Chemical binding is caused by unpaired outer electrons seeking a partner, or group of partners, with which to share angular momentum and so move in other than straight lines. Some examples of how that causes chemical reactivity are given. Some mathematics was necessary to determine the binding energies to show how they matched measured energies. There is huge scope for expanding this to gain a greater understanding of chemical binding.

That gives all the matter in the universe. The next step is to show how gravity holds them together. The standard model follows Einstein's explanation that mass distorts space–time to generate gravity. Neither Einstein nor general relativity practitioners described the property of space–time that mass distorted to produce gravity. Chapter 8 showed that space–time distortion was the redshift of photons. That was used to calculate the deflection of photons passing massive objects, the redshift of photons from the sun's surface and the anomalous precession of Mercury's perihelion. The results obtained were the same as Einstein obtained. The physics and mathematics were much simpler to follow.

Understanding Einstein's mathematics showed he used approximations when he derived his field equations and his solution for the bending of light rays by the sun.

Black hole practitioners are proud of their ability to derive the Schwarzschild metric that predicts the existence of black holes. Einstein never believed in them. Chapter 8 shows it was not possible to use accurate mathematics from Einstein's field equations and derive the black hole solution. Black holes require what is called a strong field solution. They need gravity to be stronger than Newton's inverse square law.

In 1687, Newton pointed out that, if gravity were weaker than inverse square, a planet orbiting the sun would have its perihelion precess in its orbital direction. If gravity were stronger than inverse square, its perihelion would regress. The observation that Mercury's anomalous orbit was a precession, as was that of star S2 orbiting about the massive object in Sagittarius A, should have alerted physicists to gravity being weaker than inverse square law. That it didn't is a testimony to the dominance of mathematics over physics in the physical sciences. It has led cosmologists down a "blind alley" for decades because they didn't understand Einstein's simple mathematics.

The Event Horizon Telescope collaboration showed an image of a torus shape supposedly predicted as the image from a black hole. The same torus shape is predicted from an exact mathematical solution to the redshift that gives the space–time distortion used by Einstein, see figures 8.7 and 8.11.

Chapter 8 is another example where knowing the physics involved makes the calculations much simpler than those used by

Einstein. It also made it much easier to understand the topic and correct the mistakes made by the myriad of black hole practitioners who were unable to follow Einstein's mathematics.

Chapter 8 only pointed out that Einstein's mass distorting space–time to generate gravity was through photon redshift. It did not point out how mass distorted space–time. Neither did Einstein. Chapter 9 proposed a mechanism based on Feynman's lecture on the origins of refractive index. In it he pointed out that the high frequency oscillations of the atomic electrons within the medium, changed its refractive index, by changing the electric permittivity of the medium.

That physical principle was extended to suggest the high frequency electric field oscillations of nucleons would change the electric permittivity in the space surrounding a large number of nucleons, i.e., a massive object.

The refractive index of a medium is a function of its electric permittivity. Through its electric permittivity, the refractive index determines the speed of light in the medium. Changes in the electric permittivity deflect and change the speed of light. Chapter 9 shows that the differential of the electric permittivity of space surrounding an assembly of nucleons, with respect to distance, generates a proportionality that has both Newton's inverse square and Einstein's redshift correction in the one equation. That gives the deflection of photons. With individual sub-atomic particles composed of rotating photons, they will be similarly deflected.

It requires a sufficient mass, measured in milli grams, that the out of phase nucleon oscillations generate alterations in the electric field that approach the Planck frequency, the inverse of Planck time. Following Feynman's presentation, that frequency, $\approx 10^{40}$ Hz, is so high that it treats the significantly lower nucleon, $\approx 10^{23}$ Hz visible photon, $\approx 10^{14}$ Hz, frequencies in the same manner.

Gravity is a deflection of the trajectory of photons caused by changes in the electric permittivity of space surrounding a massive object. Linear photons pass by the massive object and are barely deflected. Rotating photons stay in the one position and are similarly barely deflected each cycle. If they are not supported, the deflection effect is cumulative. They are accelerated through space. If they are supported, the downward deflection generates a force on the supporting mantle. It is not cumulative.

The high frequency electric field oscillations of the rotating photons that are the nucleons, changes the electric permittivity of the space in and around a massive object. Using Newton's gravitational constant as the proportionality constant, that principle has all the features associated with gravity.

Knowing the origins of gravity, we can begin to describe the cosmological universe. When viewed through a suitable telescope, its most notable feature is the huge number of galaxies. They are many in numbers and varied in size and shape. Standard cosmology has made reasonable progress in understanding many of their properties. One big exception is in their rotation speed.

Galaxy rotation prevents them from collapsing in on themselves. Astronomers have observed that galaxies rotate much faster than they should under the influence of Keppler's law. (That means the gravitational force on stars at radius *r* is treated as if the mass of all the stars within radius *r*, was concentrated at the galaxy's centre.) With cosmology being dominated by mathematicians, they did what mathematicians tend to do when they don't understand the physics. Invent something that seems plausible and then tell physicists to go look for it. In this case, they came up with dark matter. It was matter that exerts gravity but has no other influence. They calculated it needed about six times as much mass as the totally baryonic matter.

That was the mathematical approach. Now we need to look at what a physical approach would reveal. Chapter 10 described two physical approaches. The Pavlovich and Sipols team approximated the results of the effect of Newtonian gravity of individual stars in a galaxy on all the other stars within the galaxy. They published that a large percentage of galaxies in their study rotated at their detected speeds for their mass distribution. Rourke's frame dragging approach showed how galaxies could develop spiral arms which rotated as observed. In those two approaches, the authors applied mathematics to known physical principle and matched observation.

Chapter 10 also presented the idea that galactic magnetic fields caused cosmic protons to be preferentially captured in the galactic core, leaving a residual negative charge in outer galactic reaches. With electric attraction also being inverse square, it would be difficult to determine the difference between the two. Sufficient mathematics was used to show it was feasible for such a mechanism to also increase the rotational speeds of galaxies. It also predicted the results such an effect would have on other internal galaxy properties. It showed that some of those predictions were observed. They included why the solar corona appears to have a temperature of millions of degrees C when the sun's surface temperature is only about 5,500°C; the origins of the GeV gamma rays from the galactic plane; the origins of random X-rays generated from "nowhere".

With so many galaxies and variations, it is not a case of one theory fits all. Some galaxies have been observed to rotate as expected under Kepler's law. As well as explaining some other galaxy effects, chapter 10 gives a strong indication of why no signs of dark matter have ever been detected in a laboratory. It does not exist! It is not needed to explain galactic rotation! That eliminates 23% of the mathematician's universe.

That gives us all the components of the universe. It is built from known particles with known properties all and within three space dimensions and time. Its greatest unknown is the mass density of cosmological neutrinos. All suggestions are that it is a significant fraction of the universe's total baryonic mass. Some suggestions are that it may exceed it. That is something physicists need to investigate further.

With all properties up to and including galaxies known, we can now describe the complete image of the universe. It is not simply a matter of putting them all together and saying the universe is the sum of all it parts. Its huge mass has a collective influence. Chapters 8, 9 and 11 showed that gravity became weaker as mass increased. Chapter 11 showed why an infinite universe would not collapse under the gravity of its own mass.

The 20th century started out with astronomers observing distant galaxies. First impressions were that the universe was steady state and infinite. Knowing only Newton's inverse square law it soon became apparent that the universe would collapse under the gravity of its own mass. Einstein considered only a finite universe. A finite universe would also collapse from its own mass under any gravity theory. It obviously had not collapsed. Other explanations were sought.

Hubble observed that the redshift from distant galaxies got larger as the distance increased. That was interpreted as a Doppler redshift associated with an expanding universe. The problem was solved. An expanding universe would not collapse. Observations showed the expansion was linear. Its expansion coefficient was called Hubble's constant, H_0. Mathematicians calculated the critical density required for the universe to be still expanding after 13.98 billion years. It was so critical that the odds of a universe existing after that long time were about 1 : 10^{60}. That prompted the suggestion that there were 10^{60} parallel universes. We were fortunate to reside in one with the correct conditions to support life. Those fanciful ideas can only be dreamed up by mathematicians with no sense of physical reality.

Astronomers measure the redshift and intensity of galaxies and distant objects. They translate their intensity to a magnitude. Using stars and events of known intensity they plotted the redshift versus magnitude of distant events. Mathematicians noted that the plotted high redshift, $z > 0.4$, SNe1a were above the critical density line. They were further away than mathematically expected. That was interpreted as the universe expanding at faster than predicted from Hubble's constant.

Mathematicians could find nothing wrong with their complex calculations. Astronomers found no error in their measurements of redshift and magnitude. Together they announced the only explanation was that the universe's expansion was speeding up. That required energy equivalent to about 16 times the measured baryonic mass of the observed universe. Two decades of physicists searching for the mathematicians' prediction have been fruitless.

Curve F of figures 11.9 and 11.14 shows a physical derivation of the expected redshift versus magnitude for SNe1a. Its straight-forward derivation was described in chapter 11. All the SNe1a magnitude observations are close to, or slightly below, that line. If the universe is expanding, its expansion is not accelerating. That alone eliminates the need for dark energy. It reduces the physical universe back to being composed of the observed baryonic matter and cosmological neutrinos.

Is the universe expanding at all? Or was Zwicky correct when he indicated light lost energy as it passed through the universe? To investigate that, it was hypothesized that cosmological neutrinos existed in such numbers and were of such electromagnetic nature that they add a slight viscosity to space. Like all viscous media, they slow down objects according to their momentum, without scattering them.

It was suggested that Hubble's constant was a measure of the viscosity effect neutrinos had on photons travelling through space. Hubble's constant of 69.8 km $sec^{-1}Mpc^{-1}$ was used as a measure of that viscosity. When that was applied to the magnitude of SNe1a and combined with the magnitude reduction due to the redshift, it gave an almost exact match with the observed SNe1a magnitudes at the different redshifts. It is shown in curve J of figure 11.15.

When that is the source of the observed redshift, curve F in figure 11.14 shows that the SNe1a magnitudes at their observed redshift was as would be expected. It does not need an expanding universe to explain their measured magnitudes versus redshift. That gives the additional distances required for the numerous galaxies seen in the Hubble Ultra Deep Field image, reproduced in figure 11.17. The speed of light is controlled by the electric permittivity, ε_0, of matter space. It does does not vary. The momentum loss shows up entirely as the observed redshift.

The effect of cosmological neutrinos on photons travelling through intergalactic space is so small that a photon travelling one million light years through intergalactic space would pass through $\approx 10^{29}$ neutrinos and have its momentum reduced by $\approx$ 0.00715%.

If figure 12.1 is correct, it shows their mass is about 25% of the total baryonic matter mass. Fortunately, it isn't. Chapter 9 shows why they make no contribution to gravity within the universe.

Chapter 11 suggested a mechanism by which quasars both recycled large nuclei back to hydrogen and helium. That would maintain a renewable source of hydrogen to allow the universe to fuel stars forever. It also suggested the process would reverse the high entropy of assorted particles to a lower entropy state of a single neutron star.

It also indicates there is no such thing as empty space. Intergalactic space is free of baryonic matter. It, like all matter, irrespective of density, is never free of cosmological neutrinos. Figure 11.23 indicates there are tens of thousands of cosmic neutrinos in every cubic millimetre. That includes that most cosmological neutrinos have radii of over 1 mm and a lateral spread of several mm.

Those who wish to question that the above describes a physicists' universe should revisit figures 11.23 and 12.1. Despite decades of searching, no signs of dark matter or dark energy have been positively detected. Modern experimentalists with modern equipment have made marvellous discoveries. Common sense points to problems with the theory!

12.4. Summary

The above has shown we live in a universe that exists in three space dimensions and time. It is infinite in all four dimensions. It has neither a beginning nor an end. It is neither expanding nor collapsing. It goes about its eternal activities with every aspect of it seeking its path of least resistance towards a stable state.

It is composed of five fundamental particles, photons, electrons, protons, neutrons, and neutrinos. The latter four are all composed of photons of the appropriate energy making two revolutions per wavelength. All other short-lived particles are nucleon frequency harmonics, or combinations thereof.

There is only one force, electromagnetism! Neutrons bind protons together to form nuclei. The rules are few, simple and universal. The strong nuclear force is a combination of the direct overlap of two dimensional segments of a proton's positive charge and a neutron's negative charge, combined with magnetic field orientation. In its simplest form, the weak nuclear force for β^- decay is a variation in which the charges don't overlap sufficiently to prevent a neutron emitting an electron.

The positive nuclei attract electrons. They don't enter the nuclei because they are simply too large to fit inside them. Even if they could fit, most nuclei do not have energy levels to accommodate them. That is standard electric attraction in action.

Electrons bind around nuclei with fixed energies. They are based on the number of protons in each nucleus, the number of inner electrons that screen out the charge of some of those protons and the arrangement of those inner electrons.

For their mass, electrons are among the most powerful gyroscopes in the universe. Like all gyroscopes, they cannot change their direction of travel unless they associate with another electron, or in a group of electrons, with which they can exchange angular momentum through photons. Beyond that, electrons are in a "free for all", modified by the de Broglie wave they generate as a result of their motion. The quantum rules they are said to obey, are simply the natural consequences of their structure and screening effect. That is the basis of all chemical binding, be it to form gases, molecules, metals, crystals, organic structures, or others.

Gravity is not a force in the conventional sense. It is a displacement of photons caused by the varying electric permittivity of space. In turn, that is produced by the high frequency oscillations of the rotating photons that are the nucleons. Through the mechanism shown in chapter 9, they respond to the changed electric permittivity in the same manner as do all photons and particles. Its origin is still electromagnetic.

Chapters 1 to 11 and the above summary are the briefest statement about the physicists' universe. It describes the physics of the universe and uses mathematics to determine the magnitude of the physical effects described.

This physicists' universe is very different from the mathematicians' universe. It provides a physical continuum between the smallest sub-atomic particles and the classical world we all know so well. That continuum extends to the properties of galaxies and the large-scale structure of the universe. Chapter 2 showed the connection between the sub-atomic world and Einstein's special relativity theory. Chapters 9 and 11 showed the continuity between the sub-atomic world, gravity, and the structure of the universe.

At the beginning of this presentation, I mentioned the universal law by which everything in the universe functions. Everything within it takes its path of least resistance to reach its most stable state. Nothing in it obeys complex laws. When complexity exists, it is because humans don't understand it.

This description of the universe has been the simplest possible to accurately cover many of its important properties. Some mathematics was inevitable to show the magnitude of the physical effects described. It is vastly simpler than the mathematics used by most standard model practitioners while it describes the universe with much greater accuracy. Vast opportunities are available for other interested parties to fill in the missing detail.

The physical universe is not something that is finely tuned by laws such as quantum mechanics or other artificial constraints. Its fundamental and only guiding law is that everything in the universe seeks its path of least resistance to a stable state. The universe is inherently stable.

Everything that experimentalists and observers have discovered relates to how objects within their study regime have followed that fundamental law. That law is eternal in time and space. Over the past century, experimentalists and observers have documented a lot of the universe's behaviour and properties. The physicists' universe is well observed. It is the mathematicians' universe that is lacking in detail.

There are two properties of "empty space" that influence photons. The speed of light depends upon the electric permittivity of the space through which the photons are travelling. It is generated by the high frequency oscillations of nucleons in massive objects. Gravity is caused by changes in the electric permittivity with distance.

For those seeking the ultimate GUT or TOE, this presentation gives just one type of particle, photons, and just one force, electromagnetism. In turn, photons are electromagnetic fields confined to making two revolutions per wavelength. Those wishing to mathematically derive a grand unified field theory of everything (GUFTOE), would do well to start with a photon field equation, such as in equation 1.8, and confine it to two revolutions per wavelength.

Those wishing to tackle it from a mathematical approach could do well to follow the work of John Williamson. A partial summary of his work is that the sum of all changes is zero (0). That is essential for the universe to be infinite in space and time. As mentioned above, the nature of gravity under this model is that it enables fused nucleons to be recycled in quasars and entropy to be reduced.

All the particle variations come from variations of photons. Interchange between particles, protons, electrons, neutrons, and neutrinos is a re-arrangement of photons. Unification between the smallest sub-atomic particles and the large scale structure of the universe, including gravity, redshift, special relativity corrections, is an automatic consequence of the rotating photon structure of all particles. Treating photons and electrons as particles gives them their structure and properties.

The electromagnetic field generated by the rotating photon nature of cosmological neutrinos adds a slight viscosity to matter free space. That changes the momentum of photons as they pass through it, without deflecting them. Because their velocity is determined by the electric permittivity of space, which remains almost unchanged, the loss of momentum produces a loss of mass, and hence frequency, of photons. In linear photons, that shows up as a redshift that is proportional to the distance the photons have travelled.

In summary, photon speed is controlled by the electric permittivity of intergalactic space. Their redshift is produced by a viscosity effect which reduces their momentum. Those two effects combine to give the universe its observed properties with a 100% probability of existing in its current form. It is not possible to distinguish the viscosity redshift from a Doppler redshift. Compare that 100% certainty to the 1 : 10^{60} odds associated with the current cosmological Big Bang model.

For those who think such odds are realistic, an opportunity is given for them to "put their money where their mouth is", in a challenge of chance. A "Big Bang" believer (BBb) is invited to place some of his/her "own money," a minimum of US$100, in a space allocated to a well-known public charity to which they have no affiliation.

The chance is determined by the toss a coin with distinctive head and tail sides. Their contribution gives the BBb ten coin tosses under correctly controlled circumstances. Before each coin toss, the BBb is invited to call which side of the coin will appear uppermost when it comes to rest. If their called side finished uppermost, they get another toss. If the BBb wins all 10 tosses, they get to make one statement, no longer than 100 words, in defence of their theory. Or they can ask one question, no longer than 100 words.

The odds of anyone picking all 10 tosses are about 1 : 1,000. They are a mere 10^{57} times longer than the 1 : 10^{60} chances under the Big Bang theory. Any takers?

12.5 Footnote

Newton was both a theoretician and experimentalist. He carried out experiments and the developed theories that both explained his observations and allowed calculations to predict related events. Faraday carried out many experiments on electricity and magnetism. Maxwell compiled the results into a set of equations. They were used by others to make significant advances.

Einstein was both a theoretician and experimentalist. As an experimentalist he observed Brownian motion as irrefutable evidence of the atomic theory. He used Planck's theoretical work to show Maxwell's predicted electromagnetic waves were quantized into discrete packets of energy now called photons. He made theoretical predictions based on the speed of light being constant, his special relativity theory. Using his derivation that photons have mass, he developed his general relativity and theory of gravity. Chapters 2, 8 and 9 showed they were natural consequences of the constant speed of light and the rotating photon structure of particles. They were the days when mathematics was used to explain existing observations and make logical deductions that predicted new effects.

Post the 1970s, mathematicians took control of physics. It originated about the time Einstein published his relativity theories. People were impressed by his ability to derive answers that matched observations from entirely mathematical arguments. His gravitational field equations were particularly complex.

Starting about 1930, theorists like Schrödinger and Dirac tried to calculate electron properties. They had no idea of their structure, but knew they behaved like waves. They used wave equations to describe them. Wave equations can be manipulated in many ways. Most reasonable mathematicians could transform them in different ways to match observation.

Their "success" resulted in wave equations being applied to all other particles. Quantum mechanics became wave mechanics. It was applied at all sub atomic and some molecular levels. Physics was transformed into a mathematician's "paradise". Leading mathematicians and mathematical physicists took delight in presenting mathematics in complex manners.

Their "triumph" in the sub-atomic world is seen in their "standard model" shown table 12.1 and figure 12.2. Of its 61 particles, 44 have never been detected and 3 have half-lives less than 10^{-20} sec. They refer to photons and electrons as point particles to which the attach properties. The muon and tau leptons have lifetimes of $\approx 10^{-6}$ and 10^{-13}

sec respectively. Beyond electron neutrinos having spin ½$\hbar$, they know very little about them. Figure 12.2 shows sub-nucleon particle interactions that are so complex they could not possibly sustain themselves over billions of years.

Their record in the cosmological universe is just as bad. Figure 12.1 shows they require 25 times as much mass/energy as astronomers have observed to explain the astronomers' observations. They make almost no mention of the universe's most numerous particles, electron neutrinos. Neither could be considered as remotely equating to a pass mark. Such is mathematicians hold on physics that theoretical papers that aren't written with extensive mathematics are frequently considered as "not rigorous enough" to published in the "mainstream" literature.

Give me "a physicist's universe" any day!

End Note

Newton's "Mathematical Principles of Natural Philosophy" formed the basics of classical mechanics. He laid the foundations for modern mechanical engineering. His theory of gravity is widely used almost everywhere. His "classical mechanics" works in what is called Euclidean space. It has three orthogonal space dimensions, usually denoted x, y and z. Time is independent of them and light travels so fast as to exhibit no time delay between a local emitter and observer. It is easy to follow and works wonderfully well for our little corners of the world.

Based largely on Faraday's experimental work, Maxwell developed his electromagnetic field equations. Between them, they laid the foundations for the electrically powered world which provides us with most of the modern benefits that were not available to our ancestors three or four generations ago.

Newtonian and Maxwellian mechanics, and all their offshoots, worked well when everything was treated as a continuum. The early 20th century saw significant changes. The atomic theory was finally accepted. That placed a limit on how small things could be before they were no longer continuous. Atoms appeared to not obey Newtons laws. The sizes of Maxwell's electromagnetic waves were limited by the dimensions of photons. Einstein played a significant role in both of those predictions.

It was also discovered that atoms were not indivisible, as originally thought. JJ Thomson was the first to "split the atom", when he showed they had electrons within them. In 1905, Einstein described the mechanics of how an electron would travel, as he introduced his special theory of relativity. It was complex mathematics that not many people can follow over a century later. He introduced the relationship of energy to mass through $E = mc^2$. His special relativity calculations applied for all matter, not just electrons. They differed from Newtonian calculation only when the speed of particles was a detectable fraction of the speed of light.

Einstein went on to develop his new theory of gravity, in which mass distorts space time to attract other masses. His extensive use of mathematics meant that none of his followers understood it over 100 years after he published it. Other specialists before him used what could be called complex mathematics. Most were easy to follow when studied carefully. Einstein set a different standard.

Prior to about 1900, Newton's classical mechanics and Maxwell's classical electromagnetism were easy for experts to follow. It all "sort of made sense". Between about 1900 and 1950, the continuum upon which classical idea were founded were replaced by new limits.

In the sub-microscopic world, the continuum gave way to discrete particles that didn't always obey classical ideas. Atoms were experimentally confirmed. They were shown to be composed of even smaller sub-atomic particles. Quantum mechanics with all its mathematical complexities was developed to explain particle behaviour. It developed into wave mechanics as theoreticians found they could manipulate waves to get an answer that matched observation. The mathematics is difficult to follow.

In the macro-universe, distances were so large that that the limited speed of light had to be considered. Einstein's special and general theories of relativity helped to resolve some of the problems. His mathematics was also difficult to follow, even for most "experts".

Einstein described his work on gravity as a new theory, in which Newton's gravity theory was a first approximation and Mercury's anomalous orbit precession was a second approximation. For his era, it was a magnificent calculation. His idea that mass distorted space–time to produce gravity was certainly visionary. But his gravitational field equations were not absolute. They were variations to Newton's gravity. It was a modification to Newton's theory. Until my work in chapter 8 was published, none of the experts knew that space–time distortion was photon redshift.

Between them, the theories of quantum mechanics and relativity established the idea that only complex mathematics could be used to understand those topics. They required experts. With the major complexities established by 1950, the stage was set for mathematics to form a significant part of future physics progress. Mathematics dominated physics.

Practitioners of Einstein's relativity theories are only too willing to tell others of their intellectual skills in understanding and working with his complex mathematics. When he derived his relativity theories, physics was not as advanced as it is now. With the advances in this book, chapter 2 derives $E = mc^2$ in one line and the special relativity corrections in a few easy-to-follow pages. It also derived a fourth correction with velocity. Pythagoras' theorem was about the most complex mathematics used. That is progress. Many topics are much easier to understand over a century later.

Practitioners of his gravitational field equations are very pleased with their ability to follow his equations and make predictions he didn't. They portray his gravity field equations as one of the most difficult topics in the world to follow. To a casual reader, that could seem to be correct. They are complex. Einstein had to enlist the help of others to complete his study using tensors and field equations.

Mathematicians expanded on their interpretation of Einstein's field equations. They believed an exact solution to those equations predicted black holes when masses were large enough. Astronomers observed massive objects at the centres of galaxies. With no understanding of Einstein's mathematics, mathematicians called them black holes. Black hole practitioners are proud to announce that there is no physics underpinning them. They call them "exact mathematical solutions to Einstein's successful gravitational field equations", so they must exist. Black holes have now entered everyday language as a fact of cosmology. With that, mathematics positioned itself as the arbiter of physics.

Cosmology is now firmly founded on the Big Bang and expanding universe theory. Using complex mathematics that authors appear not to understand, they further predict the universe is expanding at an accelerating rate. Their whole theory is full of so many unknowns they ignore anything that doesn't fit it. Equation 11.11 was derived by applying Newton's Shell theorem to the more complete explanation of Einstein's gravity. It shows that gravity gets weaker as mass increases. An infinite static universe won't collapse. There is no need to consider an expanding universe as an explanation of its non-collapse or the observed redshift from distant galaxies.

Physics has been at the cutting edge of science since Newton's work, if not earlier. Physical discoveries have taken civilization into the modern era. Think of mechanically powered transport and all things electrical and electronic. How many of the modern medical, pharmaceutical, and other health benefits would we have without them?

They and many other features have allowed modern physics to be regarded as the forefront of modern science and technology. Other branches of modern science, medicine, engineering, and technology look to physicists as their model for conducting science.

The tax paying public want to know about the origins of the universe and many things in it. Yet the "experts" who tell them, don't understand either Einstein's or Newtons gravity. They fascinate the public with ridiculous "stuff" like black holes, Big Bang and the expanding universe, dark matter making the galaxies rotate faster, dark energy blowing the universe apart at ever increasing rates. They use words like: "*That is what the mathematics tells us, so it must be right*"! Or "*This is what our modelling tells us.*" "*So, it must be right*"! When questioned they use the trusted, centuries old line: "*Trust us. We are the experts. We know what we are doing.*" If only it was true!

The top experts in the forefronts of physics have convinced the public the topics they are studying are so complex that only they can understand them. When questioned about anything, they merely refer to mathematics of similar complexity to Einstein's relativity equations, or Dirac's relativistic quantum wave equations. They are intimidating enough to make most people back down. They don't describe the physics because they don't understand it.

That in itself is not so bad. After all, people make mistakes and leading physicists are only human. The problem is the same approach has been adopted in many other fields. "Experts" in other fields are copying them. The make pronouncements about aspects of their field of expertise and accompany it with: "*That is what our computer modelling tells us*". The public have been conditioned by physicists to accept that as a satisfactory answer.

Everything within the universe operates on physical principles. Mathematics is required to determine the magnitude of the physical principle. Mathematics does not make anything happen!

Perhaps the modern computer age and the internet have changed that in some ways. Hackers can use mathematics to hack into bank accounts and transfer money out of some accounts into their own. Even that requires the physical principles through which computers operate.

Understand the physical principles. Use mathematics to determine the magnitude of the physical principle. Mathematics can then extend the understanding of the physics beyond the small macro-world in which we all live. Unfortunately, physicists have allowed mathematicians to take over their profession. Their resulting standard models show they know very little about it outside the macro-world in which we live.

The late 19th century to the mid 20th century was the era of great scientific and technological progress. It was a time when "common sense" ideas were acceptable. Many people understood it and used it. Developments beyond then have centred around improvements rather than break throughs. Mathematics has replaced the common-sense approach to physics.

If theoreticians wish to use mathematical or computer modelling to demonstrate their findings, they should give a step-by-step presentation of the physics or science involved. The mathematics or computer modelling should only be used to show how they blend the physical or scientific principles into the result. Anyone purporting to be an expert on a particular topic must be able to give those explanations. If they can't or won't, they don't understand the topic!

The greater the impact their modelling has on society, the more they should be questioned. Those "*experts*" who believe "*the science is settled*" should have a look at chapters 8, 9 and 11 to see just how wrong the "*settled science*" really is. Those who believe the "*science is settled*" in their field of expertise, should take every opportunity to explain the science. Those who won't, do so because they don't understand the science!

Those who want to see how complex topics can be explained in a simple manner should have a look at chapter 5. Nuclear physics has long been regarded as a very difficult topic. Even the "experts" have difficulty explaining it to other "experts". Chapter 5 shows the topic presented in a simple manner such that some good high school matriculants could work out the structure and properties of some nuclei.

Nuclear physics and cosmology have long been regarded as among the world's more difficult technical topics. This presentation shows how they can be presented in a simple, yet accurate manner. Those who truly know topics of lesser complexity should also be able to present them in an easy to understand manner.

A long, involved report is not a substitute for a short, direct, and technically correct answer. Einstein's gravitational field equations is an example. They have long been regarded as one of the most complex topics in in the world, understandable only by advanced mathematicians. Einstein used dozens of tensor equations to derive answers none understood. Chapter 8 described the physical principles involved and used simple calculus to get a more accurate answer that leaves no doubt about the origins and accuracy of Einstein's work. In over 100 years, none of those who considered themselves as the world's best mathematical physicists, were able to follow his work.

The complexities of other topics pale into insignificance compared to his work. Surely someone who is an "expert" in a field can give the taxpayers a better answer than "*Trust us, we are the experts. We know what we are doing. The science is settled.*"

Black holes and the Big Bang/expanding universe theories are firmly established as pillars of modern cosmology. Chapter 8 showed black holes are the result of mathematicians not understanding the simple mathematics associated with Einstein's gravitational field equations. Chapter 11 showed that the introduction of an expanding universe was because mathematicians did not understand either Einstein's or Newtons work on gravity. Chapter 12 showed the simple physical errors mathematicians made because they didn't understand the physics the gravitational work by either Einstein or Newton.

Since the 1960's theoretical cosmologists have regarded themselves as the "best of the best" because of what they claim is the complexities of Einstein's general relativity theory. If they did not understand the simplicities in Einstein's and Newton's work, where does that leave the "experts" in other fields.

The next time "experts" tell you "The science is settled", ask them to explain the science in simple terms. If they can't explain it in simple terms, they don't understand it and the science is not settled.

The same applies to mathematical and computer modelling. If "experts" tell you that is what their modelling tells them, ask them to explain the physical nature of the parameters they used and how their modelling interconnects them. If they can't do that in simple terms, they don't understand the science behind their models and their results probably of questionable value. All too often the predictions are based on a narrow range of parameters designed to give a biased answer. In many cases, the "experts" who made predictions that didn't happen, never apologise, and keep their government funded position.

About the Author

Dr Robinson is a PhD physicist with over 50 years post-doctoral experience. His main claim to fame in this publication is that he was the first person to publish a paper that gave the physical explanation of Einstein's gravity (V N E Robinson 2021 *J. Phys. Commun.* **5** 035013; https://doi.org/10.1088/2399-6528/abee2f). In the over 100 years since Einstein published his papers on general relativity, it was a topic of great interest and debate among many of the world's best mathematicians and physicists. None were able to work it out. Not only did Dr Robinson work it out, but he also explained it in easy to understand physics. He reduced the mathematics to a level where many good high school matriculants can calculate some gravity effects under Einstein's gravity theory.

It takes a certain skill level to read through Einstein's mathematics and understand it and physics well enough to write the above paper. He applied the same skills to other aspects of fundamental physics. He derived a structure for particles that was responsible for Einstein's special relativity correction, including $E = mc^2$. His physics knowledge was sufficiently broad that he showed that other particle properties were also a natural consequence of their structure under that model.

That broad knowledge enabled him to extend his study beyond the standard models of physics, into new realms, such as the origins of gravity. He applied common sense to the field of interest and then used the simplest mathematics necessary to determine the magnitude of the deduced effect. If the match was no good, he modified the physics and reapplied the mathematics until the match was reached. That way, all the mathematics he used was supported by physics. That approach is different from many others who modify the mathematics with no regard to the physics that supports it.

He suggests his common sense approach came from a country upbringing in an environment where you had to improvise for things you wanted. A mechanical aptitude developed in the days before electronics made many things easier. His PhD, completed in 1969, involved both theory and experiment. In his professional career he never learned to program computers. Whenever he wanted to solve theoretical problems, he had to rely on "old fashioned" maths. When all else failed, a good approximation could be achieved by numerical integration.

After completing his PhD, he worked for 15 years in an academic environment. During that time, he developed equipment and processes that world experts in the field said would never work or were of no commercial value. He left academia in 1984 and spent decades making a successful commercial life from those developments. Running his own high technology companies provided a different insight into life. There is no back up. Succeed and all is well. Fail and you are on your own. In industry you are only as good as the worst thing you do. You learn quickly that if something is wrong, you want to be the first person to know. He learned how to recognize the key points of technical/scientific situations and present them to others in a simple yet accurate manner.

Throughout his commercial career, he never lost sight of scientific integrity. His commercial success gave him the resources to fund his own research, including this study. He suggests his different background and approach enabled him to succeed where others had failed. The work presented forms the foundations of "The Theory of Everything". It is not a Unified Field Theory" because particle generates fields. His philosophy was that "understand the particles and the fields follow". That shows up best in his work on the origins of gravity.

The uniqueness of his capabilities shows up in his ability to recognize the reasons why Einstein never believed that black holes could be derived from his field equations. It seems that none of the experts recognized that, as Newton predicted, gravity had to be weaker than inverse square law for Mercury's perihelion to preces. He was able to recognize that and went on to show that an infinite steady state universe would not collapse. That changes the whole field of cosmology.

He worked on the basis that: "*If it needs doing, do it yourself*". He read widely on the topics of Einstein's theories of relativity, cosmology, and nuclear and particle physics, trying to understand them. A common thread appeared to be photons. This treatise is his presentation of the role of photons in the universe and how they can, with the appropriate simple mathematics, explain most of the observed universe, from the structure of the smallest sub-atomic particles all the way to the properties of an infinite stable universe.

Made in the USA
Las Vegas, NV
15 November 2026

6c270d18-a1d5-4099-9ca2-84864a185aa0R01